Novel Approaches in Landslide Monitoring and Data Analysis

Novel Approaches in Landslide Monitoring and Data Analysis

Editors

Jan Blahůt
Michel Jaboyedoff
Benni Thiebes

MDPI • Basel • Beijing • Wuhan • Barcelona • Belgrade • Manchester • Tokyo • Cluj • Tianjin

Editors

Jan Blahůt
Czech Academy of Sciences
Czech Republic

Michel Jaboyedoff
University of Lausanne
Switzerland

Benni Thiebes
German Committee for
Disaster Reduction (DKKV)
Germany

Editorial Office
MDPI
St. Alban-Anlage 66
4052 Basel, Switzerland

This is a reprint of articles from the Special Issue published online in the open access journal *Applied Sciences* (ISSN 2076-3417) (available at: https://www.mdpi.com/journal/applsci/special_issues/landslide_monitoring_data).

For citation purposes, cite each article independently as indicated on the article page online and as indicated below:

LastName, A.A.; LastName, B.B.; LastName, C.C. Article Title. *Journal Name* **Year**, *Volume Number*, Page Range.

ISBN 978-3-0365-3787-0 (Hbk)
ISBN 978-3-0365-3788-7 (PDF)

Cover image courtesy of Jan Blahůt

Contents

Preface to "Novel Approaches in Landslide Monitoring and Data Analysis"

Recent progress in landslide science is shown not only by the constant increase in the number of published papers but also by the novel methods, approaches or best practices applied. This Special Issue, which is focused on the novel methods in landslide monitoring, modelling and data analysis, brings together scientists from around the world and shows the state-of-the-art research in this fast-evolving branch of landslide science. We hope it will help to manage the significant landslide hazards around the world.

Jan Blahůt, Michel Jaboyedoff, and Benni Thiebes

Editors

applied sciences

MDPI

Editorial

"Novel Approaches in Landslide Monitoring and Data Analysis" Special Issue: Trends and Challenges

Jan Blahůt [1,*], Michel Jaboyedoff [2] and Benni Thiebes [3]

1 Institute of Rock Structure and Mechanics, The Czech Academy of Sciences, V Holešovičkách 94/41, 182 00 Prague, Czech Republic
2 Risk Analysis Group, Institute of Earth Sciences, University of Lausanne, 1015 Lausanne, Switzerland; michel.jaboyedoff@unil.ch
3 German Committee for Disaster Reduction (DKKV), Kaiser-Friedrich-Str. 13, 53113 Bonn, Germany; benni.thiebes@dkkv.org
* Correspondence: blahut@irsm.cas.cz; Tel.: +420-266-009-394

Keywords: landslide; monitoring; modelling; susceptibility; InSAR

1. Introduction

The purpose of this Special Issue is to bring together recent studies related in particular to landslide monitoring and data analysis. In engineering geology, geotechnical engineering and geomorphology, landslide monitoring using standard techniques is quite common. However, the rapid development of both hardware and software solutions, including miniaturization or remote sensing techniques, brings new possibilities for increasing monitoring accuracy, real-time or near-real-time data analysis and early warning.

2. Summary of the Special Issue Contents

The Special Issue topics can be sub-divided into three groups according to the main topic covered by the articles. The majority of the articles (seven) are focused on landslide monitoring, monitoring data analysis and surveying, while a further two papers are focused on slope stability modelling using large-scale analog models and the remaining four papers deal with landslide susceptibility and detection.

2.1. Landslide Monitoring, Monitoring Data Analysis and Surveying

Thiery et al. [1] performed airborne electromagnetic measurements for rapid surveyance of the volcanic tropical environment of La Martinique, an island in the Caribbean. They combined their findings with a physical-based model to obtain improved and integrated information about the internal structure of landslides, founding a better understanding of landslides' initiation conditions.

Gili et al. [2] have monitored the Vallcebre landslide in the Pyrenees in NE Spain since 1987. A range of classical and novel methods have been used to that end (e.g., triangulation, photogrammetry, wire extensometers, GNSS-GPS, satellite DInSAR and terrestrial GBSAR). They conclude that while some methods give higher-precision results than others, all systems play valuable roles in landslide movement interpretation, and provide meaningful monitoring results at different study stages.

Fang et al. [3] present a monitoring system installed on the Pingding landslide in Taiwan. Their system consists of a GPS array combined with inclinometers, extensometers and rainfall data. The system is emergency response-centered and provides a basis for local early-warning indices. The paper tackles the important issue of multilateral cooperation among different subjects and disciplines involved in landslide disaster management.

Qiao et al. [4] focus on early-warning methods for largely abandoned rockfill slopes. These pose a significant threat in areas of large construction works. The authors used

Citation: Blahůt, J.; Jaboyedoff, M.; Thiebes, B. "Novel Approaches in Landslide Monitoring and Data Analysis" Special Issue: Trends and Challenges. *Appl. Sci.* **2021**, *11*, 10453. https://doi.org/10.3390/app112110453

Received: 3 November 2021
Accepted: 4 November 2021
Published: 7 November 2021

ground-based InSAR to monitor slope deformations and verified the method on five landslides in the area of Huangdao, China.

Dabiri et al. [5] used object-based image analysis to map geomorphological features, and assessed the applicability of Sentinel-1 data to the fast creation of post-landslide digital elevation models. Their findings revealed that—without further post-processing—the automatically derived results need to be interpreted with care, as the automatic generation of a digital elevation model is influenced by several factors.

Blahůt et al. [6] propose a methodology for analyzing time-series monitoring data from a large, slow-moving San Andrés landslide on El Hierro, Canaries, Spain. They used precise 3D dilatometric data and compared them with possible landslide-triggering factors (e.g., seismic, rainfall) to allow for fully automatic processing, thus decreasing the subjectivity of the analysis.

Troiani et al. [7] applied different surface analysis and monitoring methods to decipher the structural controls of rock slope stability in coastal areas. They worked on the Adriatic coast of the Conero promontory in Central Italy, and their results stress the need to analyze slope stability over a long timescale, to understand the current processes.

2.2. Slope Stability Modelling Using Large-Scale Analog Models

Feng et al. [8] determined a soil-water characteristic curve for landslide seepage under varying hydrodynamic conditions. They used large-scale experiments combined with finite element modelling. Consequently, they evaluated the uncertainties in the modelling using the Bayesian approach.

Tang et al. [9] assessed the influence of an intermediate coarse layer on slope stability during heavy rainfall. They found that the unsaturated hydraulic conductivity in the coarse layer was much lower than that of a fine layer, which led the capillary barrier to work at a lower water content. They also revealed that the coarser layers may have negative effects on slope stability.

2.3. Landslide Susceptibility and Detection

Fabbri and Patera [10] searched for uncertainties associated with the prediction patterns of landslide susceptibility maps. They conclude that the properties of prediction patterns are mostly unknown, but nevertheless, are critical for interpreting and justifying prediction results.

Lai [11] performed an automated data-mining procedure to differentiate the landslide sources and runout zones of landslides triggered by Typhoon Morakot in Taiwan. The author's models revealed that the detection of landslide sources provided accurate results, while the extraction of the runout areas achieved excellent accuracies.

Li et al. [12] explored the influence of multitemporal digital elevation models on the generation of susceptibility maps in the southern Sichuan Province in China. They conclude that the susceptibility assessment level of an area with historical landslides decreases in the short-term and that the usage of multitemporal digital elevation models has a serious impact on susceptibility results.

Li et al. [13] prepared a spatial, proximity-based, geographically weighted regression susceptibility model for the Qingchuan area in China. Their results suggest that the newly developed model shows higher predictive accuracy than five other commonly used models.

3. Bibliometric Analysis of Current Trends

A simple bibliometric analysis was performed in the ISI Web of Knowledge: "Web of Science Core Collection" (1900–Present) to capture the main trends in the Special Issue-related topics. The statistics were gathered on 2 November 2021, so only the first ten months of the year 2021 were included in the search. We followed a similar approach to Jaboyedoff et al. [14] and selected keywords related to the topic of this Special Issue. A number of papers published in every year have been analyzed for the query "TOPIC" (Table 1).

Table 1. List of queries used in simple bibliometric analysis. Timespan: 1900–Present. Indices: SCI-EXPANDED, SSCI, A&HCI, CPCI-S, CPCI-SSH, BKCI-S, BKCI-SSH, ESCI, CCR-EXPANDED, IC.

Keyword	ISI Web of Science Core Collection (1900–Present) Query for "TOPIC"
Landslide	(landslide * OR rockslide * OR rockfall * OR rock-fall *)
Monitoring	(monitoring * OR surveying *) AND (landslide * OR rockslide * OR rockfall* OR rock-fall *)
Susceptibility or modelling	(susceptibility * OR modelling * OR modeling * OR assessment *) AND (landslide * OR rockslide * OR rockfall * OR rock-fall *)
InSAR	(InSAR* OR DInSAR * OR GBSAR * OR GB-InSAR *) AND (landslide * OR rockslide * OR rockfall * OR rock-fall *)

*: Search keywords.

Before 1990, only a small share of the total number of papers had been published. For the topic "landslides", only 585 papers (1.65%) were published before 1990 out of 35,440 (till 10/2021); for the "landslide + monitoring" topic, only nine papers (0.17%) were published before 1990 out of 5210 (till 10/2021); and for the topic "landslide + susceptibility, modelling or assessment", only 13 papers (0.11%) were published before 1990 out of 12,248 (till 10/2021). In the case of the "landslide + InSAR" topic, the first publication appeared in 1998, and up to the end of October this year, 1079 papers had been published. Figure 1 shows the yearly growth in the number of papers in the respective topics since 1990. The presented graphs illustrate the fast-growing exponential trend in all the analyzed topics. However, the fastest growth in recent years can be seen in the "landslide + InSAR" topic.

One of the basic variables for assessing the trends in a time-series is the growth rate [15]:

$$k_i = \frac{y_i - y_{i-1}}{y_{i-1}} \cdot 100 \tag{1}$$

where k_i is the growth rate in %, y_i is the value at the time i and y_{i-1} is the value at time $i-1$.

Positive values mean a growth in percentage terms compared to the previous year while negative values mean a decrease compared to the previous year in percentage terms. As can be seen from Figure 2, rapid growth in topics "landslides", "monitoring" and "susceptibility" in the 1990s has stabilized since 2006; studies on these topics are now increasing at a relatively stable rate of around 10–15% per year. The "InSAR" topic is a special case as the first paper appeared in 1998, which consequently resulted in very high growth rates. However, during the last five years, this trend seems to have stabilized, and the topic now seems to follow the trend of the other three topics with a yearly growth rate of between 15 and 20%.

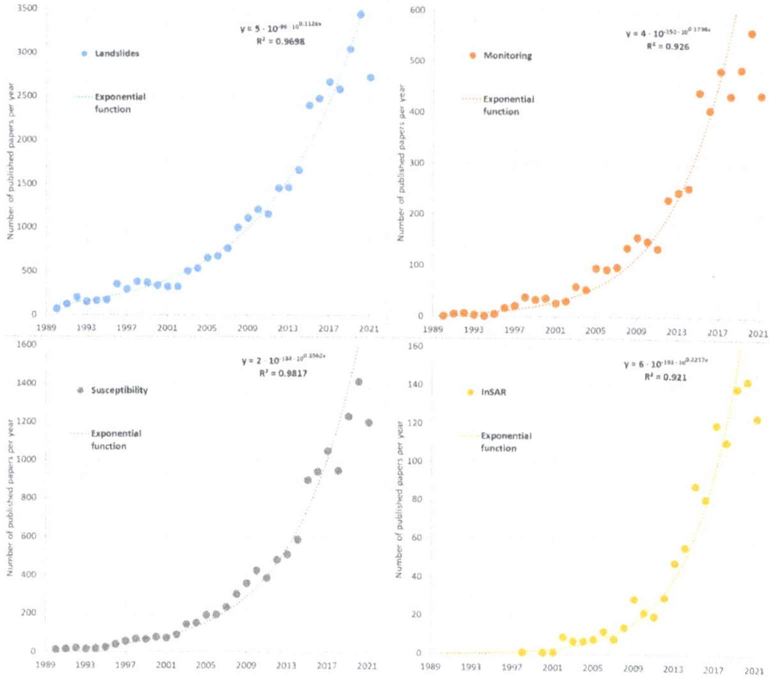

Figure 1. Number of published papers per year on the selected topics since 1990.

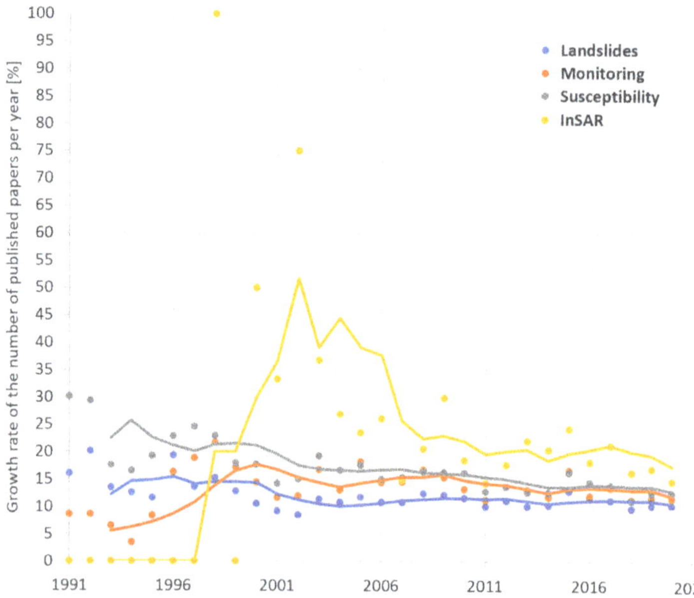

Figure 2. Growth rate of the number of published papers per year on the selected topics since 1990. Lines show the five-year moving average.

4. Future Challenges

Future trends are always hard to predict. It can be expected, however, that the current trend of yearly growth in the number of papers published on the Special Issue topics, of between 110% and 120%, will continue during the next five to ten years. This could result in more than 10,000 papers published per year on the "landslides" topic by 2026, or 20,000 papers published by 2030. It is questionable whether this enormous number of papers could be published as to do so would place enormous pressure on all the persons involved in the publication process, especially the editors and reviewers. Even now, it is hard to ensure a rigorous peer-review process as scientists are often overloaded with review requests. This situation is unsustainable in the long-term and indicates that we might expect important changes in the publication process of landslide-related scientific papers.

One possible step toward solving this situation is to better classify the type of paper published. Papers should be distinguished between novel, innovative papers and applications of existing techniques and methodologies (case studies). While the innovative papers are potentially highly citable, the case studies bring new examples. Thus, both types are important, but do nothing to decrease the overload faced by editors and reviewers.

Author Contributions: J.B. prepared the manuscript, M.J. and B.T. revised and edited the text. J.B., M.J. and B.T. were Guest Editors of this Special Issue and, therefore, reviewed all the submissions. All authors have read and agreed to the published version of the manuscript.

Funding: This research received no external funding.

Data Availability Statement: The data analyzed in this paper have been downloaded from "Web of Knowledge Core Collection" (https://www.webofscience.com/wos/woscc/advanced-search, on 2 November 2021).

Acknowledgments: The Guest Editors would like to acknowledge the external reviewers that assured the quality standards of the Applied Sciences journal were reached through their much-appreciated iterative review process. The authors, assistant editors and academic editors are also gratefully acknowledged for their interest in contributing to this Special Issue.

Conflicts of Interest: The authors declare no conflict of interest. The funders had no role in the design of the study; in the collection, analyses or interpretation of data; in the writing of the manuscript or in the decision to publish the results.

References

1. Thiery, Y.; Reninger, P.-A.; Nachbaur, A. Airborne Electromagnetics to Improve Landslide Knowledge in Tropical Volcanic Environments. *Appl. Sci.* **2021**, *11*, 3390. [CrossRef]
2. Gili, J.A.; Moya, J.; Corominas, J.; Crosetto, M.; Monserrat, O. Past, Present and Future Monitoring at the Vallcebre Landslide (Eastern Pyrenees, Spain). *Appl. Sci.* **2021**, *11*, 571. [CrossRef]
3. Fang, Y.M.; Chou, T.Y.; Hoang, T.V.; Bui, Q.T.; Nguyen, D.B.; Nguyen, Q.H. New Landslide Disaster Monitoring System: Case Study of Pingding Village. *Appl. Sci.* **2020**, *10*, 6718. [CrossRef]
4. Qiao, N.; Duan, Y.-L.; Shi, X.-M.; Wei, X.-F.; Feng, J.-M. Study on the Early Warning Methods of Dynamic Landslides of Large Abandoned Rockfill Slopes. *Appl. Sci.* **2020**, *10*, 6097. [CrossRef]
5. Dabiri, Z.; Hölbling, D.; Abad, L.; Helgason, J.K.; Sæmundsson, Þ.; Tiede, D. Assessment of Landslide-Induced Geomorphological Changes in Hítardalur Valley, Iceland, Using Sentinel-1 and Sentinel-2 Data. *Appl. Sci.* **2020**, *10*, 5848. [CrossRef]
6. Blahůt, J.; Balek, J.; Eliaš, M.; Meletlidis, S. 3D Dilatometer Time-Series Analysis for a Better Understanding of the Dynamics of a Giant Slow-Moving Landslide. *Appl. Sci.* **2020**, *10*, 5469. [CrossRef]
7. Troiani, F.; Martino, S.; Marmoni, G.M.; Menichetti, M.; Torre, D.; Iacobucci, G.; Piacentini, D. Integrated Field Surveying and Land Surface Quantitative Analysis to Assess Landslide Proneness in the Conero Promontory Rocky Coast (Italy). *Appl. Sci.* **2020**, *10*, 4793. [CrossRef]
8. Feng, C.; Tian, B.; Lu, X.; Beer, M.; Broggi, M.; Bi, S.; Xiong, B.; He, T. Bayesian Updating of Soil–Water Character Curve Parameters Based on the Monitor Data of a Large-Scale Landslide Model Experiment. *Appl. Sci.* **2020**, *10*, 5526. [CrossRef]
9. Tang, J.; Taro, U.; Huang, D.; Xie, J.; Tao, S. Physical Model Experiments on Water Infiltration and Failure Modes in Multi-Layered Slopes under Heavy Rainfall. *Appl. Sci.* **2020**, *10*, 3458. [CrossRef]
10. Fabbri, A.G.; Patera, A. Spatial Uncertainty of Target Patterns Generated by Different Prediction Models of Landslide Susceptibility. *Appl. Sci.* **2021**, *11*, 3341. [CrossRef]

11. Lai, J.-S. Separating Landslide Source and Runout Signatures with Topographic Attributes and Data Mining to Increase the Quality of Landslide Inventory. *Appl. Sci.* **2020**, *10*, 6652. [CrossRef]
12. Li, J.; Wang, W.; Han, Z.; Li, Y.; Chen, G. Exploring the Impact of Multitemporal DEM Data on the Susceptibility Mapping of Landslides. *Appl. Sci.* **2020**, *10*, 2518. [CrossRef]
13. Li, Y.; Liu, X.; Han, Z.; Dou, J. Spatial Proximity-Based Geographically Weighted Regression Model for Landslide Susceptibility Assessment: A Case Study of Qingchuan Area, China. *Appl. Sci.* **2020**, *10*, 1107. [CrossRef]
14. Jaboyedoff, M.; Del Gaudio, V.; Derron, M.-H.; Grandjean, G.; Jongmans, D. Characterizing and monitoring landslide processes using remote sensing and geophysics. *Eng. Geol.* **2019**, *259*, 105167. [CrossRef]
15. Definition of "Growth Rate" on the Webpage of The National Institute of Statistics and Economic Studies. Available online: https://www.insee.fr/en/metadonnees/definition/c1362 (accessed on 2 November 2021).

applied sciences

MDPI

Article

Airborne Electromagnetics to Improve Landslide Knowledge in Tropical Volcanic Environments

Yannick Thiery [1,*], Pierre-Alexandre Reninger [2] and Aude Nachbaur [3]

1 Risk and Prevention Direction, BRGM (French Geological Survey), 45100 Orléans, France
2 Geo-Resources Direction, BRGM (French Geological Survey), 45100 Orléans, France; pa.reninger@brgm.fr
3 Regional Division, BRGM (French Geological Survey), 97200 Martinique, France; a.nachbaur@brgm.fr
* Correspondence: y.thiery@brgm.fr; Tel.: +33-238-643-831

Abstract: Caribbean areas are particular volcanic territories in tropical environments. These territories juxtapose several landslide-prone areas with different predisposing factors (poorly consolidated volcanic materials, superimposition of healthy materials on highly weathered materials, high heterogeneity of thicknesses, etc.). In these environments, where rapid development of slopes and land use changes are noticeable, it is necessary to better characterize these unstable phenomena that cause damage to infrastructure and people. This characterization has to be carried out on the materials as well as on the initiation conditions of the phenomena and requires complementary investigations. This study, focusing on La Martinique, proposes a landslide analysis methodology that combines new information about landslide-prone materials acquired by an airborne electromagnetics survey with a physical-based model. Once the data are interpreted and compared with field observations and previous data, a geological model is produced and introduced into the physical model to test different instability scenarios. The results show that geophysical investigations (i) improve the knowledge of the internal structure of landslides and surficial formations, (ii) specify the spatial limits of the materials that are sensitive to landslides, and (iii) give a better understanding of landslide initiation conditions, particularly hydrogeological triggering conditions.

Keywords: airborne electromagnetics; landslide; physical-based modeling; tropical volcanic environment; La Martinique

Citation: Thiery, Y.; Reninger, P.-A.; Nachbaur, A. Airborne Electromagnetics to Improve Landslide Knowledge in Tropical Volcanic Environments. *Appl. Sci.* **2021**, *11*, 3390. https://doi.org/10.3390/app11083390

Academic Editor: Jan Blahut

Received: 28 February 2021
Accepted: 6 April 2021
Published: 9 April 2021

1. Introduction

Landslides are ubiquitous phenomena in the Caribbean [1–6], particularly in La Martinique [7–9]. With more than 600 events [10,11], this territory is the most affected area in the French Caribbean islands [8]. Phenomena can be shallow, deep, rotational, translational, or complex. The many landslides in the Caribbean are mainly due to the following reasons:

(i) A complex lithology with heterogeneous volcanic materials that are sometimes poorly consolidated and superimposition of healthy materials on highly weathered materials [12–16];
(ii) Hurricane and tropical storms associated with intense end heavy rainfalls [1,11,16,17].

Landslides regularly strike the coasts and the hinterlands, and because the island has much built-up land, resulting in anarchic development of the slopes [10,18], landslides can generate damage to the population and infrastructure, creating high rehabilitation costs [8]. Among the most remarkable recent events was (i) the Bellefontaine collapse (vol. = 15×10^4 m^3) in 1991 [18], which required €7 million in work to rehabilitate the slope after the event, and (ii) the Morne Callebasse landslide (vol. = 2×10^5 m^3) in 2011, which destroyed more than 20 buildings and the road 'RD 48′, bringing 75 expulsions and more than €17.1 million in works [15,18]. Therefore, anticipating landslides and improving their prevention in this French overseas territory has become a major challenge [8,15].

The first step in achieving these actions consists of assessing hazards [19–21]. Several methods that are more or less complex and range from qualitative to quantitative can be used. A large review of their uses can be found in the scientific literature [19–23], and some of them are particularly suited to the French regulatory context [24]. Among them, methodologies using physical-based models (PBMs) at the site scale (<1:5000) are the only methods designed for works [25] or that modified and revised existing regulatory hazard maps, taking into account triggering factors [24]. At this scale of work, it is possible to reach a good understanding of the mechanisms and probabilities of failure or to analyze the runout phenomena by taking into account physical processes. However, in Martinique, the parametrization of PBMs can be difficult because of a lack of information about the nature and depth of materials. Indeed, the island presents a very complex geology with different types of lavas deposited from the Oligocene to the present and exhibits extreme spatial variability and heterogeneity with a high degree of weathering [26–30].

Geophysical investigations can be an alternative to classical field investigations (i.e., geotechnical surveys; [31,32]) to obtain quick information about grounds in large areas. These investigations presuppose correlations between measured properties and physical or geotechnical characteristics [33,34]. In the case of landslides, it is possible to delineate the body of the moving mass of stable grounds due to the changing geophysical parameters [34,35]. However, with heterogeneous grounds, the indirect information that is provided can be biased or difficult to interpret. Therefore, calibrating them with direct observations is indispensable [31], thereby limiting investigations in time and space for very large areas [32].

Airborne electromagnetics (AEM) provides information on lithology and regolith over large surfaces and/or hard-to-access areas [36–40]. It is able to provide plentiful pseudo-3D information about geological structures reaching a few hundred meters [41,42] and has been successfully used in many environmental studies [30,43–47] particularly in volcanic settings. AEM is therefore useful to analyze the internal structure of large landslides [48–51].

In 2013, an AEM survey was conducted over Martinique Island [52]. The SkyTEM system was designed for mapping geological structures and for hydrogeological and environmental investigations [53]. AEM data, acquired along flight lines, provide information to a depth of 150 m and allow continuous and homogeneous imagery of resistivity variations [30]. If the subsurface resistivity has an indirect relationship with the soil characteristics [31], it can also provide relevant information on both formation thicknesses and their spatialization [32]. To be fully relevant, correlation with direct observations or independent geological datasets (outcrop, borehole geological log, etc.; [31,32]) remains mandatory. Recently, in the framework of future development [15], there has been demonstrated interest in AEM to delineate landslide-prone areas and to improve landslide hazard assessment. New information brought by AEM has allowed a landslide hazard map to be modified by combining a 3D geological model derived from a joint analysis of field observations and AEM results and a PBM. This study, for regulatory purposes, focused on only one type of island environment and on shallow (failure < 2 m) and moderately deep-seated landslides (failure from 2 m to 10 m). Therefore, deeper landslides (failure > 10 m) and associated formations inducing recurrent and very costly damage were not assessed. Considering these results and despite an exhaustive inventory and landslide hazard maps available for the entire island, there is still a lack of knowledge about (i) the nature of certain landslide-prone materials and (ii) the failure mechanisms and triggering conditions for landslides with different failure depths. This work suggests a landslide analysis methodology that combines information about regolith and bedrock derived from AEM results with a spatialized PBM (SPBM) adapted for different types and depths of failures.

Two different areas that are well known for different recurrent slope instabilities, are typical of the island and have caused significant damage for more than 20 years, were chosen to challenge the methodology. The latter is divided into 3 main steps:

1. Definition of the internal structure of grounds by coupling field observations, geological data and AEM results;
2. Building of a geological model;
3. Integration of the topography and geological model in a SPBM and analysis of slope stability following a specific protocol taking field observations into account.

The test sites have benefited from previous field, geophysical and geotechnical studies [54–65]. These studies enable (i) a criticism of the AEM results and their interpretation and (ii) the development of realistic failure scenarios based on historical observations and measurements.

2. Martinique and Study Sites

2.1. Generalities

Martinique is part of the French West Indies or the Lesser Antilles (Figure 1a) and results from the westward subduction of the Atlantic plate under the Caribbean plate [26,27]. As the largest island of the archipelago (i.e., 1080 km^2), it can be divided into two parts: that with mountainous relief in the north (with Pelée Mountain, a.s.l. 1397 m) and that with a gentler slope in the south (Figure 1a). Orographic effects control rainfall; for instance, the average annual precipitation in the northern part ranges from 5000 to 6500 mm.yr^{-1} at the highest elevations and from 1200 to 1500 mm.yr^{-1} in the southern part (Figure 1c). Because the climate is characteristic of a humid tropical climate, a humid season from July to November and a dry season from January to April can be delineated. The annual temperature varies between 18 °C and 32 °C at Fort de France.

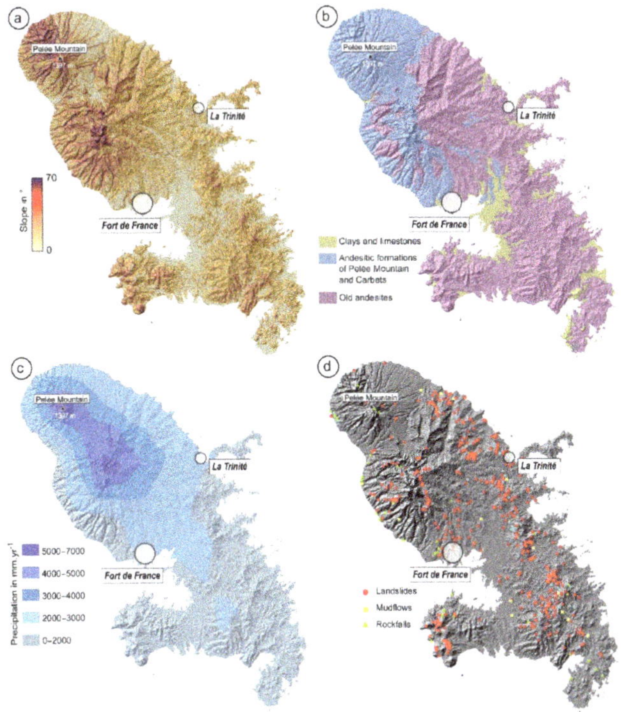

Figure 1. Location of La Martinique (the names of main cities are given in italics); (**a**) slope map; (**b**) simplified geological map; (**c**) precipitation map (mean annual rainfall per year); (**d**) landslides: type and locations (from French national landslide database: BD-MVT, https://www.georisques.gouv.fr/ accessed on the 18 May 2020).

2.1.1. Geology

In terms of the geological history, this island is unique because it is located between the old volcanic arc to the east and the recent volcanic arc to the west. Due to this specific position, it is possible to study the chronology of the volcanic front in detail [29]. Starting more than 25 Ma, the volcanic history is complex with alternating eruptive and marine sedimentation phases [26–29]. These successions allowed the edification of volcanic complexes, which were weathered and dismantled by erosion [30]. Figure 1b gives an overview of the old and recent volcanic deposits.

2.1.2. Landslides

More than 600 landslides have been inventoried during the last 20 years (Figure 2a). They are reported in the French National database (i.e., BD-MVT; https://www.georisques.gouv.fr/ accessed on the 18 May 2020). Each phenomenon is recorded with a minimum set of information, such as the date, the location of the event (centroid in the local geodesic system), the type of formation involved (lithology and/or regolith), and the associated damage, if any. Three types of landslides can be depicted: (i) landslides (i.e., debris slides, rotational and translational slides), (ii) mudflows, and (iii) rockfalls. Figure 1d gives an overview of the location of each phenomenon. Among the different landslides, seven are the subject of special attention because they have generated some damage, and they may continue to generate damage despite some engineering works [8,10].

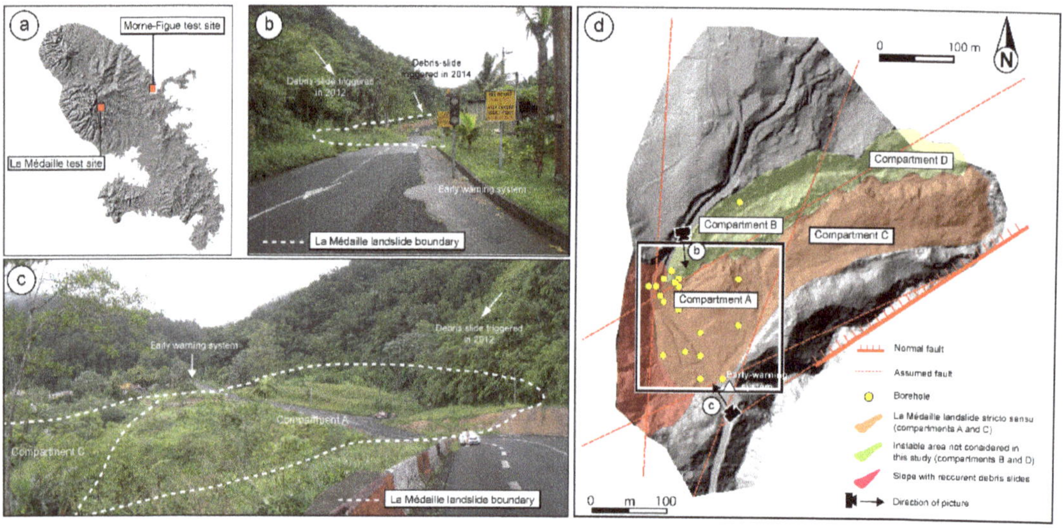

Figure 2. La Médaille landslide; (**a**) location of test site; (**b**) picture of the early warning system to block road traffic in case of landslide activity; (**c**) panorama of compartment A of the landslide; (**d**) structural scheme of the landslide (map is produced with the hillshade from Helimap DTM, 2013).

2.2. Study Sites: Presentation and Previous Works

The two selected sites, which are located in the municipalities of Fort-de-France and La Trinité (Figures 2a and 3), are characteristic of the observed landslides over 20 years and still regularly generate damage despite works and monitoring systems. The two sites have benefited from various geomorphological, geophysical and geotechnical studies that can help guide the various interpretations resulting from airborne electromagnetic surveys.

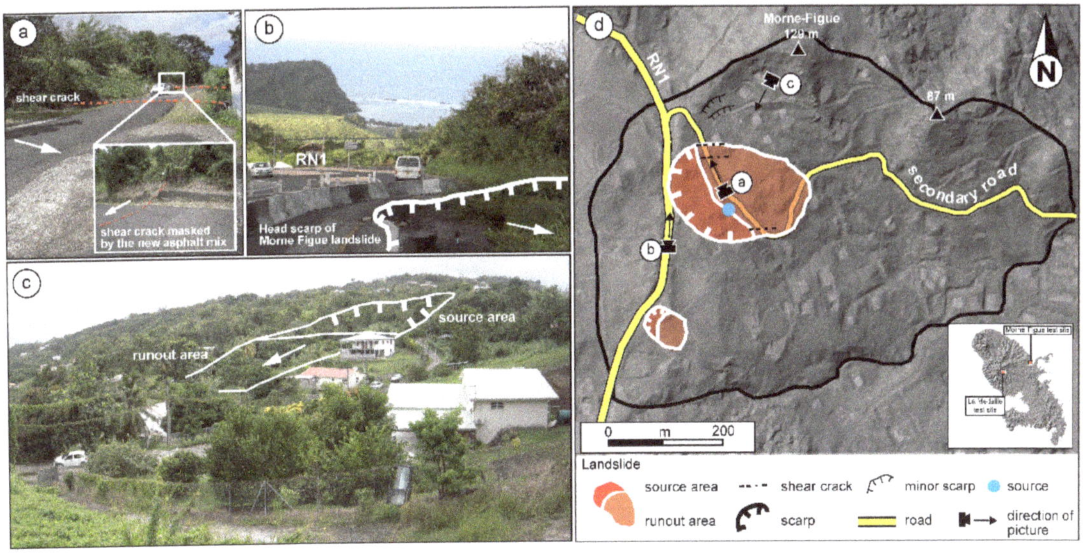

Figure 3. Morne-Figue area; (**a**) picture of the limit north of the Morne-Figue landslide (the bumps on the road are produced by the shear cracks of the landslide); (**b**) main scarp of the Morne-Figue landslide in 1989; (**c**) panorama of the Morne-Figue area; (**d**) main morphological landslide features (the map was produced with an orthophoto from IGN, 2015, and hillshade map from Litto3D DTM, IGN, 2010).

2.2.1. La Médaille Landslide

The La Médaille landslide is located in Morne Balthazar to the north of the Fort-de-France Municipality (Figures 2a and 3a). With a volume of approximately 260,000 m^3 and an area of approximately 8 ha (Table 1), the landslide occurred in 1916 [54,55], removing the old village of La Médaille and causing five casualties. The landslide is bounded to the north and south by two water courses marking two parallel faults and by a cliff of approximately 110 m representing a normal fault in the west. It is probably located on a paleolandslide (rock avalanche) represented by dacitic deposits in the small-scale lithological map [26,27]. Since 1916, several peaks of activity have been recorded (i.e., 1958, 1966, and 1993; [54–62]), each time with the rehabilitation or the relocation of road RN3 due to its crossing the moving mass. It is a slow-moving landslide but is likely to experience accelerations due to GWL variations. More recently, a debris slide in weathered andesitic formations occurred in the upper part of the cliff overhanging the landslide (Figure 2b,c).

Since the 1960s, several investigations (i.e., field surveys, geotechnical studies, and monitoring; Table 1) were engaged to improve the knowledge of the landslide [54–63], especially for the upper part (compartment A). Indeed, the topography of the lower part (compartment C) of the landslide is chaotic and complex, and the very dense tropical vegetation prevents the deployment of different investigation devices. The different field surveys allowed delineation of four main compartments functioning more or less independently. The lower compartment (C) seems to have an influence on the upper compartment (A), while the compartments on the edges are considered independent of the main landslide stricto sensu [58]. All drilling campaigns (five campaigns between 1967 and 1996) were located in compartment A of the landslide, which presents a high blockage potential (Figure 2). Thus, this compartment benefited from 14 boreholes reaching a depth of approximately 20 m. These campaigns located the faults bordering the landslide (Figure 2d).

Table 1. Investigations and monitoring of the two sites since the 1960s.

	La Médaille Landslide		Morne-Figue Area	
	Number (n)	Year	Number (n)	Year
Boreholes	15	1966, 1968, 1974, 1981, 1986, 1996	9	1981, 1982 1989, 1997
Geotechnical characterization	2 campaigns	1981, 1983	1 campaign	1989
Geophysical investigations	1 campaign (AEM)	2013	2 campaigns (electrical and AEM)	2007, 2013
Hydrogeological investigations	2 periods	Precipitation= since 1969 Piezometer = 2000 and 2002	1 period	Precipitation 1988–1989 Piezometer = 1988–1989
Field investigations	9 campaigns	1966, 1968, 1974, 1981, 1986, 1996, 2002, 2014, 2016	7 campaigns	1981, 1982 1989, 1997, 2007, 2016, 2017

The first campaign in 1967 revealed a more or less clayey upper layer, up to 3 m thick, followed by approximately 19 m of dacitic screes upstream and 3 m of dacitic screes downstream of compartment A (Figure S1). Under this layer, a thin clay layer of a thickness of 0.5 to 0.9 m was observed. Finally, the lower layer was interpreted as andesite bedrock that was more or less fractured [54,55]. Hazmoune et al. [57] questioned this statement following new observations based on new boreholes. Indeed, the last layer would be a mixture of weathered andesites and breccias and could be part of the landslide, which would then be composed of two superimposed bodies. This information corroborates the conceptual scheme of landslide development established in [56,57,63], which mentioned the implementation of a rock avalanche in andesites. This layer would have been weathered and would then have been covered by a thick layer of dacite debris (breccias), which would have also been weathered.

The different boreholes defined a groundwater level between 1967 and 1968 in the upper part of the landslide with a maximum piezometric level of approximately 10 m under the topography. The fault system and a nearby spring probably feed this water level [57]. Finally, a series of rainfall and piezometric records attempted to prove the relationship between precipitation, groundwater level (GWL) and landslide activity. Unfortunately, despite the punctual implementation of a measurement network, shortcomings in the precipitation series or in displacement monitoring did not allow this relationship to be clearly proven. Nevertheless, when the GWL is high, the activity of landslides increases [54,55,57].

2.2.2. Morne-Figue Area

The Morne-Figue area is located in the commune of La Trinité on the east coast of the island (Figure 3b). The study site covers an area of 0.36 km^2 and is limited to the north and west by the Morne-Figue (129 m a.s.l.) and the Morne-Congo (232 m a.s.l.) areas, respectively, which are carved in compact andesite, to the south by the Gué stream and to the east by Crosmy Bay. The site is hilly with elevations varying between 10 m and 100 m a.s.l. and both steep and gentle slopes varying between 5° and 45°. The site, which is very anthropized (approximately thirty houses), is bordered to the west by national road RN1 and is crossed by a small road connecting the neighborhood to the center of the municipality (Figure 2c,d). Since 1977, the east-facing slope has been subject to landslides (rotational and shallow translational landslides). The main phenomena is located below RN1. This landslide, which is approximately 2.99 ha and was triggered in 1988 [64], is associated with rotational failure in the upper part and a translational component in the lower part ([64]; Figure 3d). This phenomenon has two very active periods: from 1987 to 1988 and in 2004. Since 2004, new geomorphological features such as cracks (i.e., traction and compression), small scarps and new tension cracks on the road were observed intermittently (Figure 3).

One drilling campaign [64] and geophysical investigations allowed the involved materials and their thicknesses to be defined (Figure S2). Four types of formations from the topographic surface were defined: (i) an upper layer composed of clays or backfill with a thickness between 1 m and 3 m; (ii) a second layer with a thickness from 8 m to 10 m and represented by weathered very clayey materials; (iii) a third layer with a thickness between 10 m and 12 m composed of weathered basalts (Pré-vert basalts); and (iv) a weak horizon of basalt whose thickness is not known. A piezometer was implemented for the period of 1987–1988. During the humid period (from May to October), the GWL lies between 1.4 m and 0.5 m below the topography from upstream to downstream. In the dry season, the GWL is approximately 8 to 9 m below the topography. The piezometric levels and displacement measurements were implemented for a short period of one year, showing a correlation between rainfall and landslide activity [64]. Consequently, the landslide was more active during the humid period, with a maximum displacement of approximately 1.4 cm for the period of 1987–1988. In 2016, the geological knowledge of the surroundings was refined by field observations along cross-sections. Beyond the acquisition of new geological information, two units within the landslide were delineated with a very active unit upstream and a latent unit downstream. New shallow parallel landslides were observed within the site.

3. Materials and Methods

The suggested methodology is based on a transdisciplinary approach in three steps. This approach reduced the uncertainties raised by previous studies, particularly concerning the internal structure of the La Médaille landslide and the different material types and thicknesses of the Morne-Figue area. The interconnected steps are (i) the identification and improvement of the knowledge of involved materials; (ii) the production of a geological model for each site, including the different regolith thicknesses and/or internal structures of landslides; and (iii) landslide susceptibility analysis by a spatialized physically based model based on limit equilibrium equations with research of best fitting geotechnical parameters. Step (iii) integrates empirical triggering factors according to field observations.

3.1. AEM Data

From 29 January to 16 March 2013, SkyTEM ApS conducted a survey over Martinique Island. This survey, supervised by the BRGM (French Geological Survey) and totaling 4233 line-kilometers, was flown along the N–S direction with a 400-m spacing; locally, this spacing was refined to 200 m over areas of interest for the hydrogeology or risk assessment [52]. Along each line, EM measurements were spaced approximately 30 m apart, with an average ground clearance of approximately 64 m due to the sharp topography of the island. Figure 4 shows the location of the flight lines flowing over the two studied sites.

SkyTEM is an AEM system developed by the Hydro Geophysics Group of Aarhus (Denmark) for hydroenvironmental investigations [53]. This system is composed of (1) a transmitter coil exciting the subsurface, (2) a receiver coil to measure the ground response, (3) a generator as a power source, and (4) several navigation instruments, such as GPS, tiltmeters, and laser altimeters, to locate the loop in space. The SkyTEM system operates in dual transmitter mode. The low moment, with a magnetic moment of approximately 2826 Am^2 and time gates from 11 µs to 1 ms for the present survey, provides early time data for shallow imaging, and the high moment, with a magnetic moment reaching 144,440 Am^2 and time gates from 109 µs to 8.9 ms, allows measuring later time data for deeper imaging. Locally, the depth of investigation of the method depends on the emitted magnetic moment, the bandwidth used, the subsurface electrical conductivity, and the signal/noise ratio.

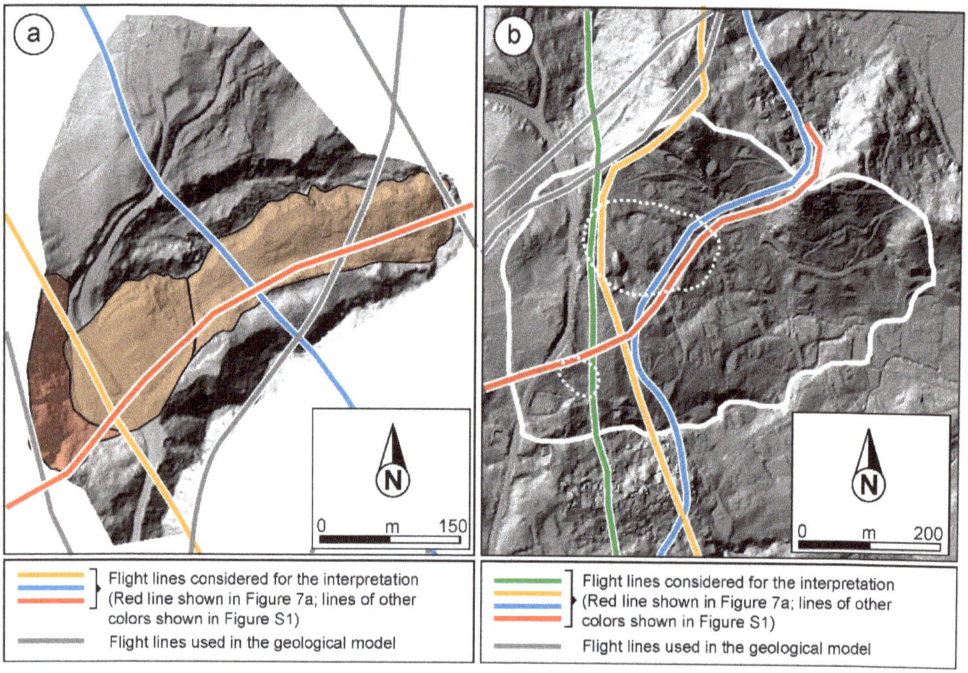

Figure 4. Location of the Airborne electromagnetics (AEM) flight lines. (**a**) La Médaille landslide (hillshade map is produced with Helimap DTM, 2013); (**b**) Morne-Figue area (hillshade map is produced with Litto3D DTM, IGN, 2010).

The AEM method allows imaging of the conductivity contrasts of the subsurface. To obtain usable measurements, several processes are applied to remove couplings with man-made installations and ambient noise from the signal. The processing scheme used is described in [39,40] and is based on singular value decomposition [39,40]. Data are then inverted using the spatially constrained inversion algorithm (SCI) [44]: (i) each usable AEM measurement is translated into a 1D (EM sounding) model divided into n layers, each defined by its thickness and resistivity, and the resistivity variations are displayed according to depth; during the inversion, constraints are applied vertically and spatially between nearby soundings (independently of flight lines). (ii) The ground clearance of the AEM system is also inverted, and the depth of investigation (DOI) is assessed as a final step in the inversion [66]. The results were obtained by running a smooth inversion for 25 layers from 0 m to 300 m deep. Each layer has a fixed and equal thickness. Only the resistivity values vary within a layer [39]. This approach is effective for imaging complex geological structures with the lowest dependency on the starting model, but it only displays a smoothed view of the subsurface. At this step, each flight line can be displayed as a resistivity profile composed of all the associated EM soundings. More 2D information can be obtained by interpolating, on raster grids, the resistivity of layers falling into a depth or elevation range. This interpolation is generally repeated to obtain slices over the entire range of investigations. Slices can then be merged to obtain a 3D resistivity model [45], (i) drawing profiles in any direction for confronting geological data (maps, boreholes, and field observations) and (ii) deriving interfaces for each imaged horizon.

3.2. Interpretations and Conceptualization

3.2.1. Confrontation with Independent Data

To interpret the different imaged horizons for defining interfaces of interest, a comparison with boreholes and field observations is carried out (Figures 2 and 3) by projecting the geological data on selected resistivity profiles. The distances between the boreholes and AEM lines vary from 2 to 10 m and 3 to 12 m for the La Médaille landslide and Morne-Figue area, respectively.

3.2.2. Geological Modeling

Based on interpreted resistivity profiles, boreholes, geological maps and field observations, a geological model is produced for each site [67,68]. AEM results are useful to constrain the model for different environments [37,38,41]. The principle is based on the first two steps of the approach described in [67] and the different positions of geological limits (contacts between two geological formations or lithologies). They are used to guide the interpolation by the kriging method as regular grids representing the bottom of each type of identified formation.

3.3. *Landslide Modeling*

3.3.1. ALICE Presentation

Landside modeling is performed with ALICE® (Assessment of Landslide Induced by Climatic Events) developed by the French Geological Survey (BRGM, [69]). This tool supports landslide susceptibility mapping for areas ranging from local sites (catchments) to large areas (several municipalities; [15,69–71]). Developed in a GIS environment (MAPINFO®), it is a SPBM described in [69] and summarized below (Figure 5).

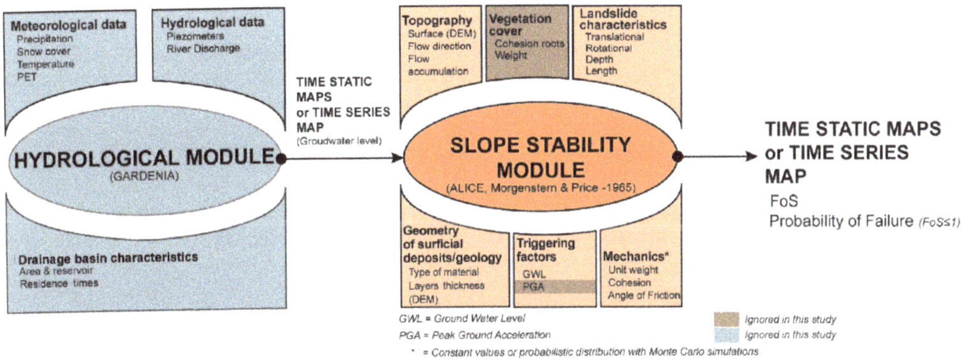

Figure 5. ALICE® concept (adapted from [69]).

The geometry of the studied area is introduced in raster format with (i) the topography and (ii) the geometry of different material layers. Geomechanical characteristics, cohesion (c), friction angle (φ) and specific bulk unit weight (γ), are given for each lithology and surficial formation. These parameters can be implemented by a constant value or by probabilistic distributions to take into account environmental variability and uncertainties [69,70]. The tool supports different landslide geometries and failures (i.e., rotational, translational, and complex with different lengths and depths). Two triggering factors can be used: (i) the groundwater level (GWL) or (ii) seismic acceleration. The GWL represents the saturation ratio ($m = h/z$), where h is the height of the water table and z is the depth of the material(s) taken into account. The GWL can be implemented empirically in one or several formations by increasing the saturation level from 0 (dry conditions) to 1 (saturated conditions) or with the help of a hydrogeological model taking into account the effective rainfalls. Seismic acceleration is represented by the peak ground acceleration (PGA).

The slope stability computation is based on a limit equilibrium method (LEM) and slice theory described in [72] and in Figure S1. The iteration process is based on the concept of reducing the number of iterations about the interslice function and therefore the computation time [73]. The hypothetical failure surface is divided into *n* vertical slices, and each slice I is subject to the normal shear interslice forces and to the shear resistance (Figure S3), where:

$$R_i = [w_i cos\alpha_i - u_i b_i \sec\alpha_i] tan\varphi'_i + c'_i b_i \sec\alpha_i$$

and the moving forces:

$$T_i = w_{i\sin\alpha_i}$$

where W_i: weight; α_i: base inclination; u_i: average water pressure; b_i: width of the slice; φ_i: effective friction angle; c'_i: cohesion along the base; and R_i: sum of the shear resistances, except the normal shear interslice forces. Ti is the component tending to cause instability. The potential failure is expressed by the factor of safety (FoS). If the FoS is below 1, the slope (i.e., the computation cell) is considered instable. The slope stability assessment is performed on regularly spaced 2D profiles automatically produced over the whole area and based on maximum gradient lines from the DTM. Two types of computation are possible, (i) the computation of the FoS or (ii) the computation of the probability, to obtain an FoS below 1 for each cell. In the second case, the tool performs a random selection of each geotechnical value following probabilistic distributions and Monte Carlo simulations [69–71].

3.3.2. Landslide Modeling Protocol

Landslide modeling, split into three steps, consists of defining the best set of representative parameters for each site.

(i) The first step focuses on the geotechnical characteristics acquired during the drilling campaigns at each site. The goal is to reduce the broad spectrum of geotechnical values because their range is quite large. Indeed, retaining geotechnical values giving recurrent instability under any saturation conditions would not be representative of real conditions, with landslides being episodically unstable. Therefore, a sensitivity study is undertaken by several iterations computing the FoS in dry and fully saturated conditions. Computations are carried out in 2D along a representative cross-section of the study site. Each geotechnical characteristic (cohesion, angle of friction, and bulk weight density) is modified iteratively by increasing its value and keeping constant values for the other two. If the results show recurrent instability under any condition (dry/saturated), the value is excluded. If the results show a recurring stability of the sites under any condition (dry/saturated), then this value is also excluded. Thus, by this sensitivity analysis, the values (cohesion, angle of friction, bulk unit weight) producing a factor of safety lower than one at a high GWL are preserved, and those producing a factor of safety greater than one for a null GWL are also preserved.

(ii) The second step consists of defining the optimum cell size for ALICE® to obtain reliable results. Indeed, for the spatialized model, in raster format, the cell size can generate errors that propagate in the results [74,75]. Thus, for the two sites, simulations are carried out with cell sizes of 10 m and 5 m to observe if discrepancies can be noted between the simulations. It is not necessary to decrease the resolution because with a lower cell size, the generated profiles are too small, and it is impossible to correctly compute failures shorter than 10 m ([62,69]). Conversely, performing calculations with a larger cell size results in a loss of spatial precision, sometimes generating areas with a low or high failure probability that is not characteristic of the field reality [62]. Computations are performed with the geological models developed with the new data from the AEM data and interpretation. For this step, the best geotechnical values defined previously are used. The GWL introduced in ALICE® is constant and corresponds to full saturation conditions.

(iii) The third step allows checking the influence of the GWL on the destabilization of the ground. This step takes into account the GWL recorded for the two sites from 1967 to 1968 and from 1987 to 1988 for the La Médaille landslide and Morne-Figue area, respectively. Therefore, following the different records, the maximum GWL is established with a maximum fixed at −0.5 m from the topography surface for the La Médaille landslide, −0.5 m from the topography surface for the moderately deep rotational landslide and a GWL at the level of topography for the translational shallow landslide for the Morne-Figue area. The different saturation ratios increased iteratively from dry conditions (GWL = 0) to full saturation conditions measured for each site (GWL = 1).

For each step, the failure geometries were fixed following the field observations (Table 2). The validation of simulations is performed by comparison with (i) an expert map derived from field observations and DTM derivatives for the La Médaille landslide [62,76] and (ii) the landslide inventory performed in 2016 for the Morne-Figue area [62]. Different classical statistical tests are computed (relative error and ROC-AUC; [69–71]) to validate the results. These tests were completed by expert (qualitative) verification.

Table 2. Landslide characteristics used for modeling.

	La Médaille	Morne-Figue	
Landslide type	Deep	Moderately deep	Shallow
Failure length (m)	45–50	45–50	10–15
Failure depth (m)	20–40	9–10	2–3

4. Results

4.1. Identification of Involved Materials and Geological Models

For each site, resistivity profiles and grids (at different depths) were combined with boreholes, field observations and geological maps (Figures 6 and 7). Interfaces of interest were then extracted to constrain the geological model.

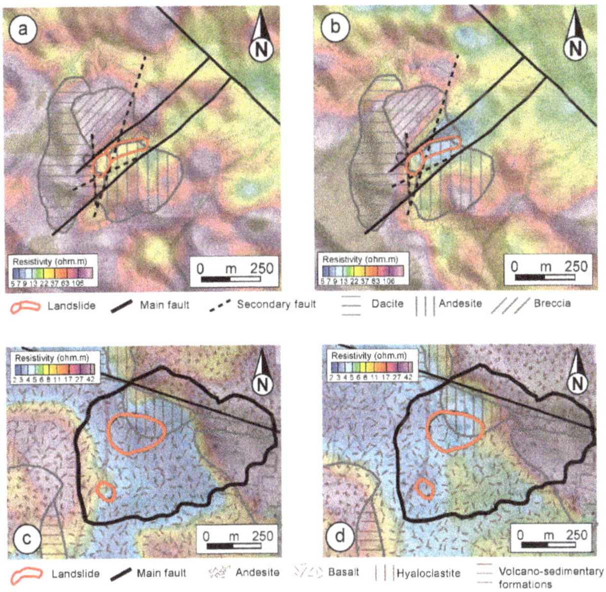

Figure 6. Resistivity grids for the two sites, with landslide areas and main lithological formations from geological maps. (**a**) Resistivity map between a depth of 4 m and 7 m for the La Médaille landslide; (**b**) resistivity map between a depth of 38 m and 47 m for the La Médaille landslide; (**c**) resistivity map between a depth of 2 m and 4 m for the Morne-Figue area; (**d**) resistivity map between a depth of 19 m and 24 m for the Morne-Figue area.

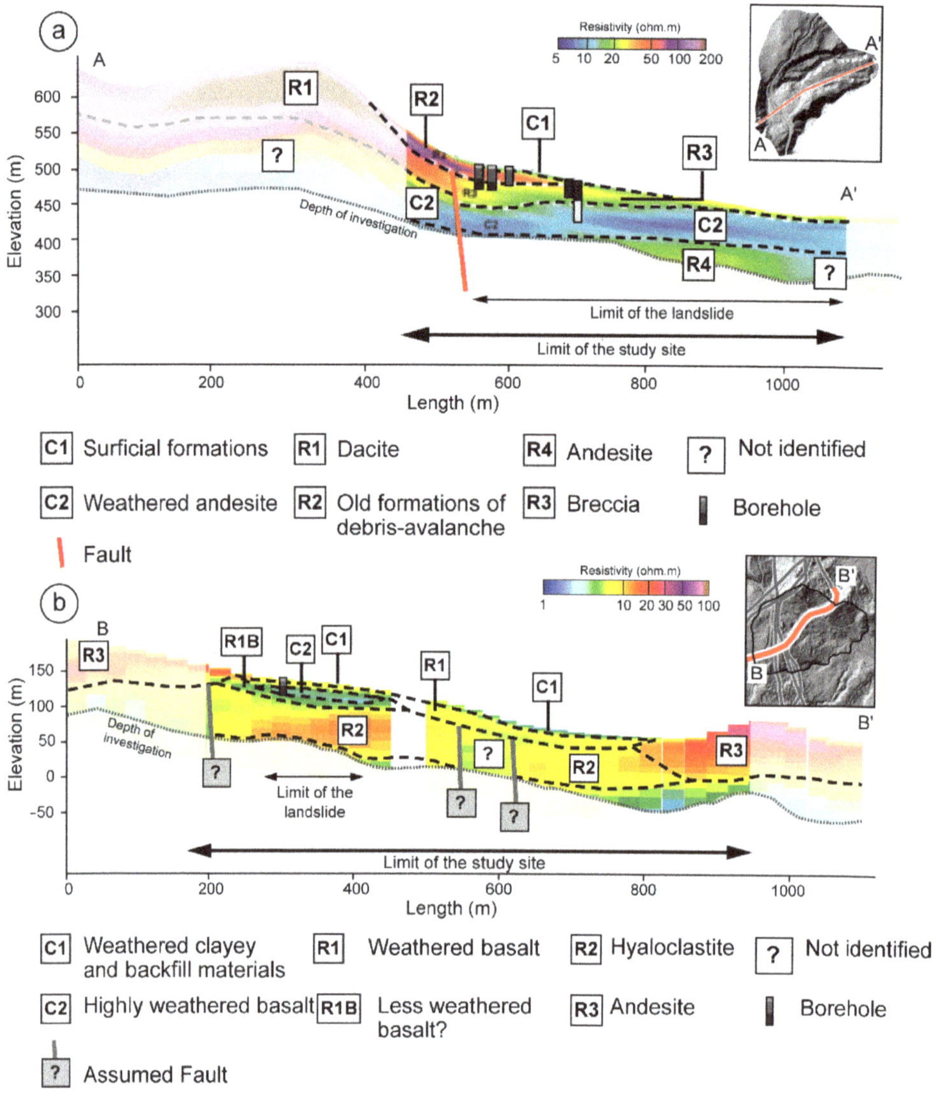

Figure 7. Interpreted resistivity profiles. (**a**) La Médaille landslide (hillshade map is produced with Helimap DTM, 2013); (**b**) Morne-Figue area (hillshade map is produced with Litto3D DTM, IGN, 2010).

4.1.1. La Médaille

Figure 6a,b shows two resistivity grids at two different depths, between 4 and 7 m and between 48 and 47 m deep. For each, the geological formations and the landslide were reported. The dacitic formations (breccias and bedrock) show high resistivity values (>65 Ω.m, Figure 6a). It is possible to observe them in the western part of the landslide. At the level of the landslide, the resistivity values are between 20 Ω.m and 35 Ω.m, corresponding to the dacitic screes observed in the boreholes and described in [58,59] and [61]. Deeper (Figure 6b), high resistivity values (>50 Ω.m) are imaged to the west and northwest of the landslide and correspond to the dacitic formations, whereas the landslide area is characterized by a low resistivity, below 15 Ω.m. A resistivity profile is shown in Figure 7a and is completed in Figure S1. Nearby boreholes are also projected on this profile to identify the different materials (Figure 7a). At the surface, a thin conductive layer (C1) is visible with a resistivity of approximately 15 Ω.m. It would represent the superior clayey layer described in the different boreholes. Under it, a more resistant layer (R2 ~50 Ω.m) can be observed. It is assimilated by the dacitic formations of the old debris avalanche described in [54]. Below R2, a thick layer (R3) with a lower resistivity (from 15 Ω.m to 30 Ω.m) is imaged. This layer is the one observed in Figure 6a and was interpreted as breccias or fossilized screes in [54,55]. The deep conductive thick layer (C2) would correspond to weathered andesite. This conductive layer, present along the landslide (Figure 6b, Figure 7a and Figure S1) and not revealed with the older boreholes, would be 30 m thick and would overcome the bedrock composed of andesite (R4). Table 3 details each formation.

Table 3. Characteristics of the different formations observed in the La Médaille landslide.

Name	Resistivity (Ω.m)	Thickness (m)	Description
R1	>50	>20	Very deep layer with high resistivity values corresponding to dacite
C1	15	1–3	Surficial layers corresponding to recent weathered materials rich in clay
R2	50–55	4–20	Layer with high resistivity values corresponding to dacitic screes and debris implemented by an ancient debris avalanche
R3	15–30	20–30	Layer under R2 with lower resistivity values corresponding to mixed materials (breccias or fossilized screes)
C2	<15	25–30	Layer with low resistivity values and a thickness of approximately 30 m corresponding to weathered andesite at the bottom, which was probably caused by hydrothermal water
R4	20–25	not identified	Formation with high resistivity values and likely corresponding to the bedrock (andesite)

4.1.2. Morne-Figue

Figure 6c,d show two resistivity grids at two different depths, between 2 and 4 m and between 19 and 24 m. The resistivity range is globally low (<50 Ω.m). At the surface, the highest resistivity corresponds to andesites and volcanic-sedimentary formations. Weathered basalts have a lower resistivity (< 8 Ω.m). Landslides occur in the majority of these materials described in [62,63]. In between, it is possible to define the hyaloclastites. Resistivity profiles are displayed in Figure 7b and Figure S2. Nearby boreholes are also shown. Six horizons are delineated: a conductive layer (C1; <5 Ω.m) corresponding to a very weathered clayey layer and backfill materials that is more or less saturated following the groundwater level, and a more resistive layer (R1; from 5 Ω.m to 8 Ω.m) corresponding to less weathered clay materials. Locally, this layer disappears in lieu of a more conductive layer C2 (from 2 Ω m to 5 Ω m).

According to borehole data (Figure 7b), this layer (C2) would correspond to highly weathered lavas, where water was observed [63]. C2 disappears progressively downstream of the slope and is less weathered at depth (R4) in favor of a few resistive layers (R1B and R1). Under R1, C1 and R1B, R2 shows higher resistivity values (between 8 Ω.m and 11 Ω.m). This layer, located 25 m below the topographic surface, would correspond to hyaloclastites. However, it remains difficult to know the exact nature of this layer due to the lack of deep boreholes. Finally, a more resistive layer (R3), with values higher than 20 Ω.m, corresponds to andesite formations overlying the hyaloclastite formations in some places. It should be noted that two sectors, where resistivity values drop sharply, probably correspond to faults delimiting the topographic depression at the center of the study site. Table 4 gives the different characteristics of each formation.

Table 4. Characteristics of the different formations observed in the la Morne-Figue area.

Name	Resistivity (Ω.m)	Thickness (m)	Description
C1	<5	2–4	Weathered clayey and backfill materials
C2	1–5	5–10	Highly weathered basalt lavas with preferential water circulation
R1	5–8	3–10	Basalt lavas that are more or less weathered
R1B	5–8	3–10	Basalt lavas that are more or less weathered that are replaced laterally and at depth C2
R2	8–11	15–20	Hyaloclastite formations
R3	>20	>20	Andesite formations that are more or less weathered

4.1.3. Geotechnical Models

From the interpreted resistivity models, different interfaces were derived for each site. Figures 8 and 9 illustrate the different geological models for the two sites.

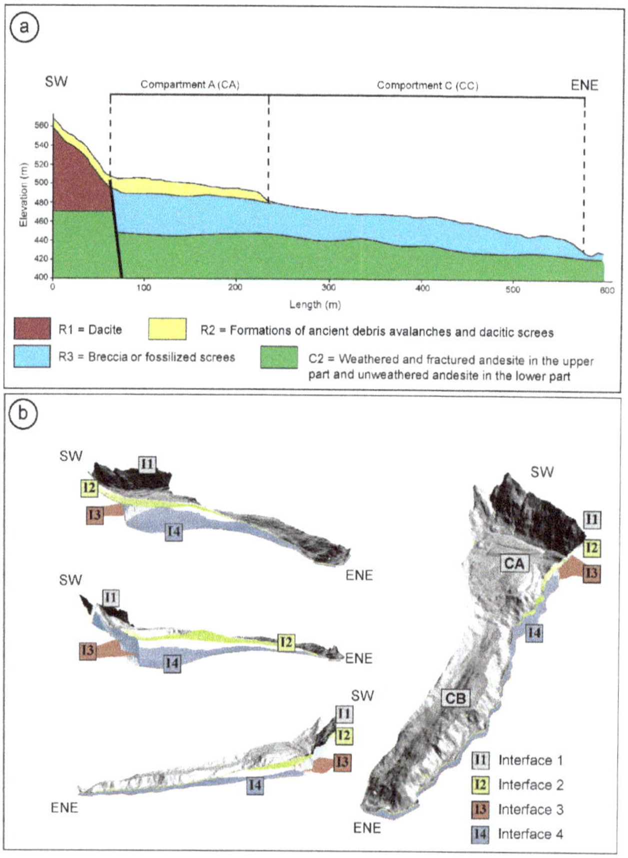

Figure 8. Geological model for the La Médaille landslide. (**a**) A 2D cross-section representing the different materials selected for physical-based modeling; (**b**) interfaces introduced for ALICE® (hillshade map is produced with Helimap DTM, 2013).

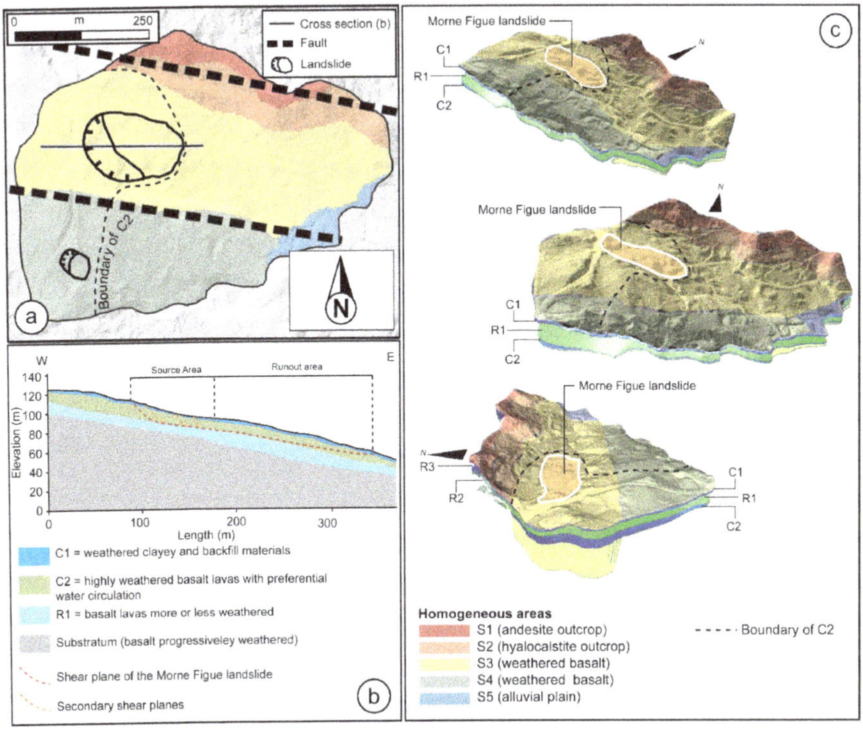

Figure 9. Geological model of the Morne-Figue area. (**a**) Location of homogeneous areas identified by resistivity data analysis, field observations and boreholes; (**b**) 2D cross-section representing the different materials selected for physical-based modeling; (**c**) interfaces introduced for ALICE® (hillshade map is produced with Litto3D DTM, IGN, 2010).

For the La Médaille landslide, four layers were defined (Figure 8a): (i) a first layer corresponding to dacitic screes (R2) upstream of the landslide on steep slopes and on compartment A of the landslide. The thickness varies from a few decimeters to approximately 5 m upstream of compartment A. On compartment A of the landslide, the thickness varies from 5 m to 20 m. (ii) A second layer corresponds to andesite screes (R3) with a thickness between 40 m under compartment A and a few meters downstream of compartment C. Therefore, R2 rests on R3 along compartment A and forms compartment C alone. (iii) A third layer corresponds to dacite (R1) located under R2 up to the concave slope failure corresponding to the upstream limit of the landslide. It is limited downstream by the north-south fault. Its thickness varies from 20 to 90 m. (iv) A fourth layer corresponds to andesite (C2) forming the bedrock. The fault observed in the resistivity profile (Figure 7a) and documented in the geological map [26,27] is not integrated in the geotechnical model. Figure 8a shows the failure hypothesis described in [54] and questioned in [57]. The new information on formations and their thicknesses introduced in the geotechnical model and numerical simulations under ALICE® should help to better understand how destabilization occurs.

For the Morne-Figue landslide, the area was divided into five homogeneous areas, defined by the main formations identified previously, the structure and the geomorphology (Figure 9a). Thus, S1 corresponds to the hyaloclastite formation (R2) resting on andesite formations (C1); this area corresponds to the main relief. S2 is composed of hyaloclastite (R2) marking steeper slopes. S2 is separated from S3 by a fault identified in the resistivity model and identified in [26,27]. S3 marks the beginning of the topographic depression of the site. It is characterized by a succession of conductive and more resistive layers (C1, R1, C2, and R1B), including the C2 layer over almost its entire surface. S4, in the southern part of the site, differs from S3 by an absence of layer C2 on the entire downstream and eastern part. Therefore, downstream of S4, only C1 and R1 are identified. Area S4 is probably limited to the north by an assumed fault separating it from S3. Finally, S5 represents the downstream of the site with gentle slopes covered by alluvial deposits. Faults in this area are not integrated into the geotechnical model. Numerical simulations with ALICE® constrained by the new information brought by the AEM (nature and extension of the formations and their thicknesses) allowed spatializing landslide hazards at the study site.

4.2. Slope Instability Analyses

For stability analyses, the best geotechnical parameters to be introduced in the models should be chosen. Moreover, this step tests the contribution of the information derived from the AEM and uses it to build the conceptual models on both the different materials involved and the different destabilization conditions.

4.2.1. Identification of the Best Geotechnical Parameters

The sensitivity analysis along the cross-sections (Figures 8a and 9b) must reduce the range of values of geotechnical parameters and define the consistency as the best fitting intrinsic for the involved materials. A set of 50 model iterations for each cross-section were carried out (i) with independent variation of each parameter within the range of values and (ii) in dry and fully saturated conditions to obtain the best range of combinations. Tables 5 and 6 give the selected formation and the retained boundaries to perform calculations. For the two sites, cohesion appears to be the most influential parameter in the safety factor modeling followed by the internal friction angle, while the weight bulk density appears to be the least influential, which is often noticed in this type of analysis.

Table 5. Geotechnical values selected after sensitivity analysis for the La Médaille landslide. IV = initial values; SV = selected values; γ = bulk unit weight; c = cohesion; φ = angle of friction; italic values are introduced as triangular probability distributions in ALICE®.

Formation	Thickness (m)	γ (kN.m^{-3})		c (KPa)		φ (°)	
		IV	SV	IV	SV	IV	SV
R1	>20	25–30	30	40–50	40	30–50	40
R2	2–20	10–17	*11–16*	10–37	*12–35*	5–30	*5–25*
R3	20–30	7–17	*8–16*	6–50	*6–40*	8–35	*10–30*
C2	25–30	25–35	28	35–40	37	28–47	40

Table 6. Geotechnical values selected after sensitivity analysis for the Morne-Figue area. IV = initial values; SV = selected values; γ = bulk unit weight; c = cohesion; φ = angle of friction; italic values are introduced as triangular probability distributions in ALICE®.

Homogeneous Area	Formation	Thickness (m)	γ (kN.m^{-3})		c (KPa)		φ (°)	
			IV	SV	IV	SV	IV	SV
S1	R3	>20	10–30	*15–20*	8–16	*10–15*	5–40	*10–30*
	Substratum (andesite)	infinite	26–30	30	40	40	40	40
S2	R2	15–20	05–29	*10–15*	5–20	*15–20*	1–35	*10–18*
	Substratum (hyaloclastite)	infinite	29–30	29	31–35	35	31–35	35
S3 and S4	C1	1–3	12–17	*12–17*	5–15	*5–10*	10–22	*10–20*
	R1	3–10 m	10–18	*12–18*	1–20	*10–20*	16–25	*17–22*
	C2	5–12 m	10–18	*12–18*	5–25	*10–20*	18–25	*18–25*
	R1B	3–10 m	10–18	*12–18*	5–25	*10–20*	16–25	*17–22*
	Substratum (basalt)	infinite	25–30/ 29–30	29	50–66/ 31–35	66	30–38/ 31–35	38

Therefore, for the La Médaille landslide, the results for rotational failure with a depth of 20 m show that the cohesion must not be below 12 KPa and 6 KPa for layers R2 and R3, respectively, especially if the angle of friction is low for each layer. If the value is below one of these values, the computed FoS remains under 1, corresponding to a recurrent instability that does not correspond to reality.

For the Morne-Figue area, there are two cases: shallow and moderately deep-seated landslides. For shallow landslides occurring in the R1 materials, if the angle of friction is higher than 15°, then the slope remains stable. For cohesion, when it is greater than 10 KPa, the slope remains stable. On the other hand, when cohesion is close to 5 KPa, the slope switches from stable to unstable according to the saturation scenario. For moderately deep-seated landslides, the situation is more complex, regardless of the conditions, and the slope is close to instability even with high cohesion values and/or high angle of friction values. When the GWL is low and the saturation conditions are null, then the FoS is high, but not higher than 1.5. Therefore, when the angle of friction is less than 10°, the stability remains low under any condition. If the angle of friction is higher than 20°, the slope is stable. For cohesion under 5 KPa, the slope is recurrently unstable (FoS < 1) regardless of the angle of friction. If the angle is greater than 10°, then according to the angle of friction, the stability increases.

Tables 5 and 6 show the initial and retained values to be integrated in the spatial modeling for the two case studies. Values in italics are used in ALICE® simulations. Values in italics and bold are used with a probability distribution necessary to compute failure probabilities with ALICE®. The probability distributions taken into account are triangular (Figure 10). This shape of the probability distribution is classically selected for ALICE®, as explained in [69,70], and offers the best compromise for testing new hypotheses of slope destabilizations.

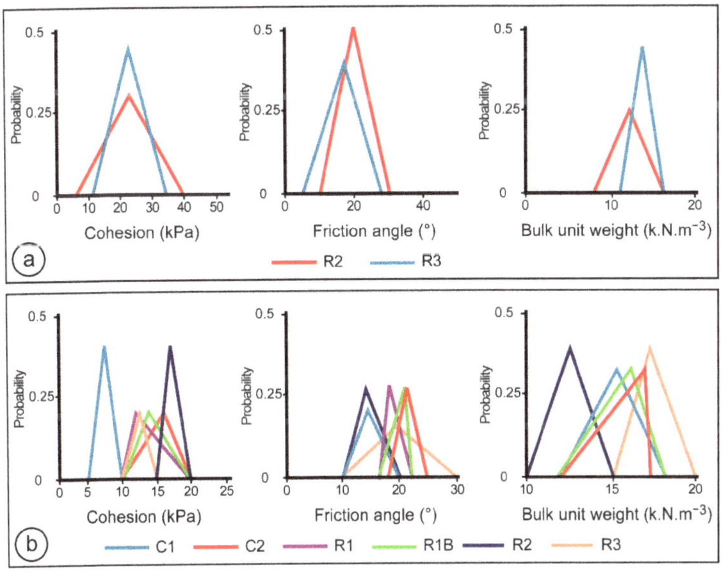

Figure 10. Representation of probability distributions of different materials. (**a**) La Médaille landslide; (**b**) Morne-Figue area.

4.2.2. Identification of the Optimum Cell Size

For this step, computations are performed with 10 m and 5 m cell sizes. The geotechnical dataset is the best previously defined for each site. For the La Médaille site, two computations are carried out with rotational failures with a depth of 20 m. For the Morne-Figue site, four computations are performed (two for shallow landslides with a maximum depth of 3 m and two for deep rotational landslides with a depth of 10 m). The goal was to define the best cell size given the best results, and it was decided to perform the computations taking into account the failure conditions observed in the field (i.e., with a high GWL corresponding to the saturation of the materials, GWL = 1). The comparison and validation of the results is carried out by two statistical tests and expert verification. The tests are classical: (i) the relative error analysis (performed with the observed failure identified) and (ii) the analysis of the area under an ROC curve (ROC-AUC). For the La Médaille site, failures correspond to the main scarps of the landslide (i.e., the scarp between compartments A and C and the scarps identified downstream). For the Morne-Figue, failures correspond to scarps of phenomena.

Figures 11 and 12 depict the results for computations carried out with the two cell sizes for each site. For the La Médaille test site, the different scarps are well identified by the models with high probability values of failure in the different scarps. The relative errors are very low, with 0.25 and 0.21 for the 10 m cell size and 5 m cell size, respectively. The two computed ROC-AUCs show that the different models have a high degree of fit with values of 0.89 and 0.91 for a 10 m cell size and a 5 m cell size, respectively. For the Morne-Figue site, for shallow landslides, the different models computed with cell sizes of 10 m and 5 m have a low degree of fit. Indeed, the relative error and ROC-AUC have values of 0.89 and 0.95 and 0.59 and 0.58 for 10 m and 5 m cell sizes, respectively. These results indicate a low degree of model fit and low model representativeness for translational shallow landslides. In contrast, for moderately deep rotational landslides, the results are very good, with low relative errors and high ROC-AUCs of 0.13 and 0.15 and 0.89 and 0.87 for a 10 m or 5 m cell size, respectively. The Morne-Figue failure area is well represented with high probability values from 0.1 and 1, indicating, according to the equations, high failure probabilities when the materials are saturated.

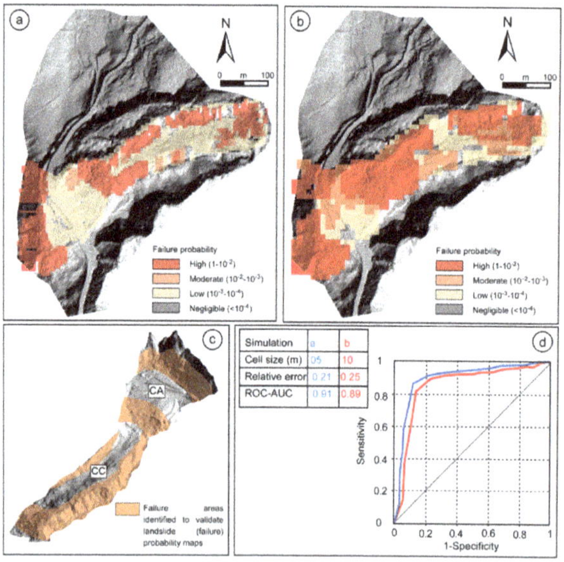

Figure 11. Comparison of results with cell sizes of 5 m and 10 m for the La Médaille landslide. (**a**) Map computed with a cell size of 5 m; (**b**) map computed with a cell size of 10 m. (**c**) Expert map of potential failures; (**d**) statistical tests for (**a**,**b**). For each computation for (**a**,**b**), the GWL = 1.

Hillshade maps were produced with the Helimap DTM (2013).

For both cases, the results few differ between computations with a cell size of 5 m or 10 m, either statistically or visually on the computed maps. Thus, for the two sites, a cell size of 10 m appears to be the best compromise for the next phase. For the Morne-Figue site, given the low representativeness of the models, it is necessary to have other information to better adjust the computations. Therefore, the influence of the GWL is not tested for shallow translational landslides.

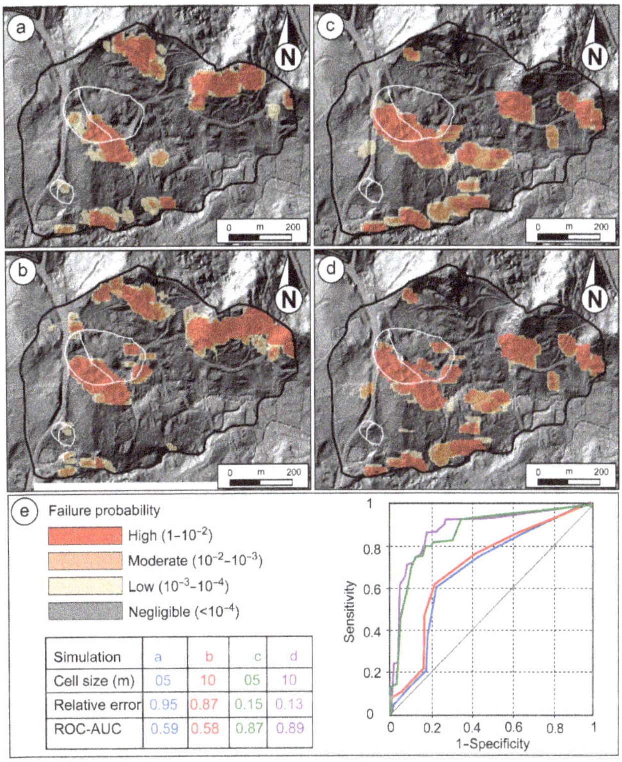

Simulation	a	b	c	d
Cell size (m)	05	10	05	10
Relative error	0.95	0.87	0.15	0.13
ROC-AUC	0.59	0.58	0.87	0.89

Figure 12. Comparison of results with cell sizes of 5 m and 10 m for the Morne-Figue area. (**a**) Map computed for shallow translational landslides with a cell size of 10 m. (**b**) Map computed for shallow translational landslides with a cell size of 5 m. (**c**) Map computed for moderately deep rotational landslides with a cell size of 10 m. (**d**) Map computed for moderately deep rotational landslides with a cell size of 5 m. (**e**) Statistical tests for the four computed maps. For each map the GWL = 1. Hillshade maps were produced with the Litto3D DTM (IGN, 2010).

4.2.3. Influence of the GWL

The initial hypothesis following antecedent monitoring and observations is that the two case studies are controlled by the rise of the GWL corresponding to the material saturation. Therefore, several scenarios were tested by gradually increasing the GWL. The scenarios take into account the best calculation cell size (i.e., 10 m), and the best set of geotechnical data. For the La Médaille landslide, the goal is to find the most favorable materials prone to failures, confirming the landslide-triggered hypotheses. Thus, several scenarios were tested with computations of failures (i) for R2 in dry and completely saturated situations (i.e., GWL = 0 and GWL = 1), (ii) for R3 in dry and completely saturated situations (i.e., GWL = 0 and GWL = 1), and finally, (iii) for both types of materials in dry and completely saturated situations (i.e., GWL = 0 and GWL = 1). Once the materials to be considered have been defined, the influence of the GWL is tested by gradually increasing its level from 0 to 1 (with a maximum at 0.5 m below the topography). For the Morne-Figue area, considering the stakes, the main goal is to improve the knowledge about the material of the compartment face and the rise of the GWL to best understand future landslide-prone areas. The maximum depth of the retained GWL takes into account the observations (i.e., the GWL reaches a maximum at 0.5 m below the topography for moderately deep rotational landslides).

For the La Médaille area, the results (Figure 13) show that in dry conditions (i.e., with a very low GWL), the landslide is computed as stable with a failure probability under the low class threshold. Only the dacitic screes (R2) on very steep slopes are computed with a high probability of failure. For the same R2 materials located on gentle slopes, by taking into account these materials alone, then the computed failure probabilities are low, even if the GWL = 1 (Figure 13a). Only a very thin area of the upper part and the northern boundary of compartment A are computed with high failure probabilities. The relative error and ROC-AUC are not satisfactory, with values of 0.73 and 0.55, respectively. By taking into account R3 formations, it is possible to better delineate some failure areas (Figure 13b). The tests are better than previous tests (Figure 13a) but not optimum, with a relative error of 0.32 and an ROC-AUC equal to 0.73 (Figure 13b). Nevertheless, the results are not exactly consistent with field observations, particularly the failure area observed upstream of compartment A. Finally, by taking into account both types of materials (R2 + R3), the failure areas observed in the field are well identified (Figure 13c). The relative error and ROC-AUC are acceptable, with values of 0.25 and 0.89, respectively.

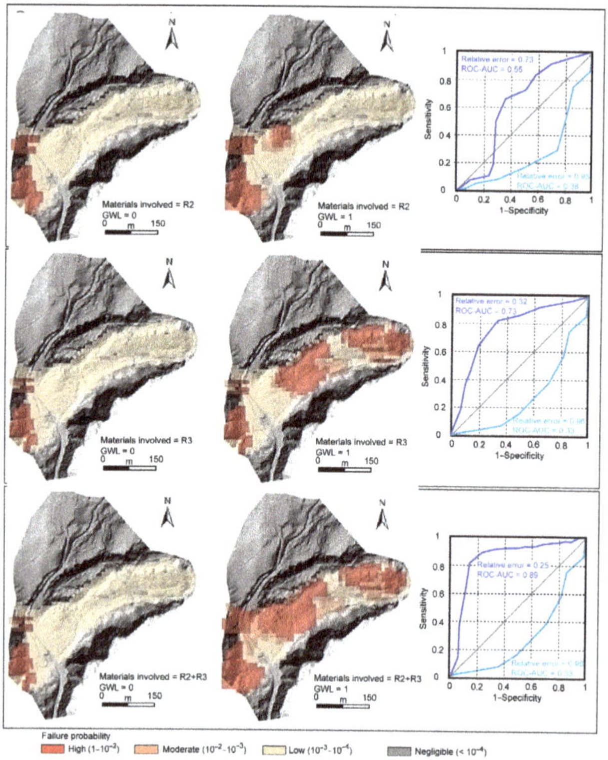

Figure 13. Influence of the type of materials introduced to compute failure probabilities with ALICE® for the La Médaille landslide. (**a**) Computations with R2; (**b**) computations with R3; (**c**) computations with R2+R3. Hillshade maps were produced with Helimap DTM (2013).

Figure 14 illustrates the sensitivity of materials when the GWL varies inside of those materials. Therefore, water appears as the engine of recurrent landslide destabilization; the more the GWL increases, the more the number of computed cells with high failure probabilities increase. Nevertheless, the different computations show that under a GWL of 0.9, the landslide remains stable with a high probability of failure located in the foot of the sliding mass. From a GWL level equivalent to 0.9, it is possible to notice that the areas with high probabilities of failure correspond to the main location of scarps (Figure 14) with a relative error and an ROC-AUC of 0.27 and 0.89, respectively.

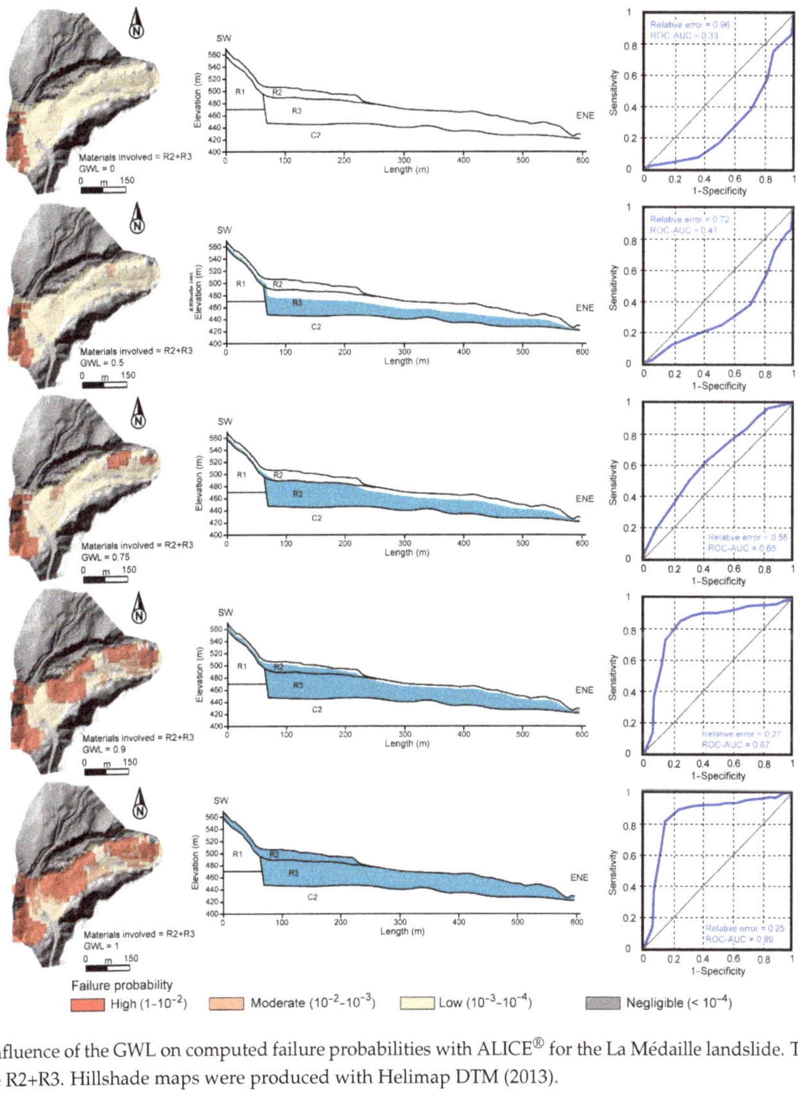

Figure 14. Influence of the GWL on computed failure probabilities with ALICE® for the La Médaille landslide. The materials involved are R2+R3. Hillshade maps were produced with Helimap DTM (2013).

For the Morne-Figue area, Figure 15 illustrates numerical simulations for moderately deep rotational landslides with a GWL variation ranging from 0 to 1 (i.e., reaching 0.5 m from the topography). With a GWL equivalent to 0.5, the slopes remain relatively stable, and the failure probabilities are equivalent to 0, except near the Morne-Figue slope, where few cells are computed with failure probabilities between 0.001 and 0.0001 (i.e., moderate failure probability). When the GWL value exceeds 0.7, the initiation area of the Morne-Figue landslide is computed with high probability values from 0.7 to 0.01, indicating a high hazard level following the JTC-1 classification of hazards [22]. A significant increase in probability values is noted for a GWL above 0.8. From this level, the model computes high probabilities of failure with a good recognition of the initiation area of the Morne-Figue landslide represented by values from 0.9 to 0.01. The northern hillside of the area, with slopes greater than 25°, appears to be more susceptible, with probabilities greater than 0.3, indicating a very high probability of material mobilization by landslides. Finally, when the GWL is near the maximum, the Morne-Figue landslide failure area is clearly identified with failure probabilities between 0.9 and 1, indicating a very strong probability of landslide hazards. The northern hillside is also simulated with very high failure probabilities exceeding 0.4. Note that failure is possible along the Gué stream south of the study site.

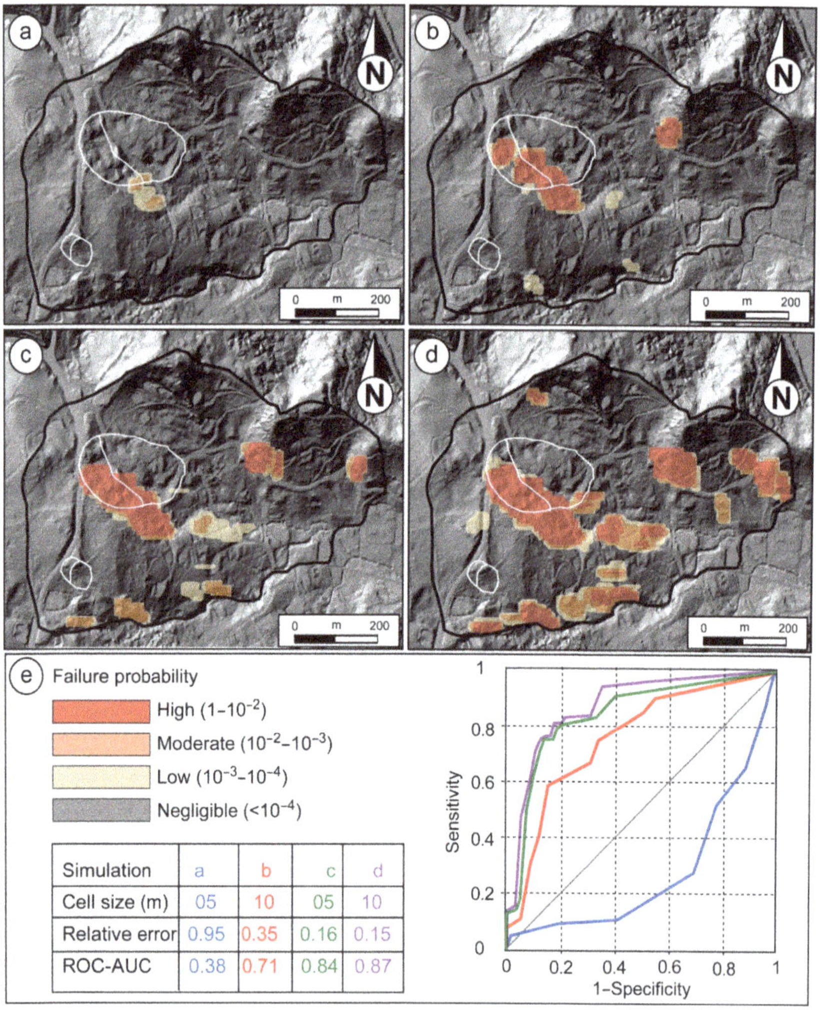

Simulation	a	b	c	d
Cell size (m)	05	10	05	10
Relative error	0.95	0.35	0.16	0.15
ROC-AUC	0.38	0.71	0.84	0.87

Figure 15. Influence of the type of materials introduced to compute failure probabilities with ALICE® for the Morne-Figue area. (**a**) Map computed with a GWL = 0.5; (**b**) map computed with a GWL = 0.7; (**c**) map computed with a GWL = 0.8; (**d**) map computed with a GWL = 1; (**e**) statistical tests for the four computed maps. Hillshade maps were produced with the Litto3D DTM (IGN, 2010).

5. Discussion

AEM data have already been used to study large landslides in temperate alpine environments or in Japan [38,48–51]. Few studies have been carried out on multiple types of landslides simultaneously in complex tropical environments with superimposed lavas that are more or less weathered at depth. Despite this lithological and structural complexity, through this study, AEM data appear to be of primary interest to assess these phenomena generating damages and losses in the West Indies [1]. In the following paragraphs, the AEM method, the role of water in slope failures and the differences with regulatory landslide hazard maps are discussed.

Thus, first, the AEM data allowed a better definition of the internal structure and the materials prone to landslides for both the La Médaille and Morne-Figue sites. For the La Médaille site, some hypotheses [57–60] that were difficult to validate due to the lack of deep boreholes were confirmed, such as the existence of two superimposed main bodies (one associated with a very low resistivity <5 Ω.m) composing the landslide body. The extracted information also allows building a specific geotechnical model integrated into ALICE®. Figure 16 gives an overview of the conceptualization of the La Médaille landslide gathering ancient information with the information derived from the AEM data. Without this latter, the geological model should be different (Figure 16), with downstream and upstream compartments (A and C) defined in [54,55] and located on one type of thinner material. Using this model (Figure 16a), the results of the numerical failure simulations would probably show that only the upper part of the landslide is likely to be unstable. Conversely, taking into account the interpretations derived from AEM data, it is possible to show that the shear surface identified in [54,55] and questioned in [56,57] turns out to be true concerning failure probability computations.

For the Morne-Figue site, the data from the AEM allowed us to clarify the geological structure of the site: (i) the hyaloclastite footprint and (ii) the different successive low and high resistive horizons in the basalts with likely different weathered levels. The derived geological model was able to reproduce the unstable behavior of highly saturated materials, especially for moderately deep rotational slides. The results obtained for shallow landslides are less conclusive. However, a previous study conducted in [15] showed that AEM data and modeling under ALICE® could give satisfactory results for shallow landslides. Here, this is not the case, and we decided to stop the modeling of shallow phenomena because of the mediocrity of the results. To improve the results for this type of landslide, two points should be emphasized: (i) the shallow landslide is located on a preferential location of water circulation mentioned in [77,78]. The use of a static GWL in ALICE® did not allow faithful reproduction of the real influence of water at this location. (ii) The landslide was triggered in complex materials mixing road embankments and likely weathered lavas. This type of material was not investigated during the geotechnical laboratory tests, and at this location, the simplification of the model meant that these specific materials were not included in the study.

Second, beyond these considerations, the results of physical numerical modeling show that water (i.e., introduced by the GWL) remains the main factor of slope destabilization in this environment. For both sites, the GWL has to be high to obtain high failure probabilities. This statement confirms the field observations and the punctual measurements and monitoring during the 1960s and 1980s for the La Médaille landslide and Morne-Figue area, respectively. However, the question of the origin of this water remains. For example, for the La Médaille landslide, if the results seem conclusive, there are still uncertainties, such as the role of the probable aquifer of the C2 formations and the probable upwelling of water along the discontinuities in C2. The measured levels during the monitoring of the landslide activity show that the water supply to the landslide from the source is not sufficient to activate it.

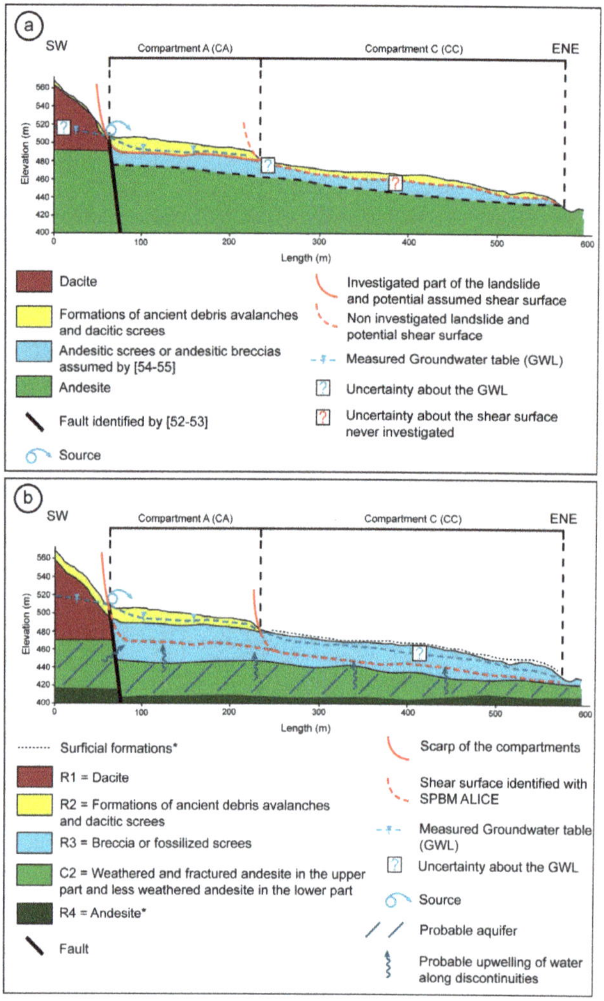

Figure 16. La Médaille conceptualization. (**a**) Before this study; (**b**) after this study (* data not used in the computations of failure probabilities with ALICE®).

Careful consideration of the AEM results shows low resistivity bodies (C2) with values of approximately 5 Ω.m under each unstable site for both sites. This low resistivity may come from (i) highly mineralized water; (ii) preferential flow in more or less weathered and fractured materials; and (iii) very weathered materials with likely a high proportion of smectite [30]. Taking into account the recent works in Martinique [12,30], it is possible to retain hypotheses (i) and (ii) with deep water upwelling in fractured materials feeding shallow water tables in landslide-prone materials, as illustrated in Figure 16b. However, to obtain such a resistivity value, this water must be highly mineralized, and currently, it is difficult to know the mineralization processes [30]. Therefore, this hypothesis needs to be confirmed by deep drilling and less punctual monitoring than those previously undertaken at these sites.

Third, the new geological interpretation and physical failure modeling allowed refinement of the landslide susceptibility maps, especially for Morne-Figue. Figure 17 shows the areas identified by the expert approach in 2004 at the 1:20,000 scale of work for all landslide types and the results of numerical failure simulations for an extreme GWL for only moderately deep rotational landslides. The obtained maps are different, particularly (i) the area between the two landslides at Morne-Figue and the 1977 landslide and (ii) the area along the Gué stream or (iii) the areas around the Mont Morne-Figue carved in the andesite formations. One important point to highlight is that Figure 17a integrates all landslide types, and Figure 17b shows only one type of event. At this stage, if the numerical failure modeling resulting from the geotechnical model can help experts create some failure scenarios and better identify some potentially unstable sectors, it is necessary to further model shallow landslides, especially the triggering factors and the material thickness involved. There is a certain amount of uncertainty in the results of the shallow failure probabilities during the calibration phase. Additional investigations must be envisaged, probably based on the work of [15]. The new results of simulated failure probability maps allow better definition of the boundaries of high-hazard areas and improvement of the future final landslide hazard map. This will be achieved during a project to revise the landslide hazard regulatory maps between 2021 and 2022. The goal is not to reject one approach and its results compared to another. Indeed, they are complementary; the expert approach is based on the subjectivity of the expert who, through his or her experience, can intrinsically bring elements not taken into account in the modeling [19,24]. In addition, the geotechnical model may contain some uncertainties (i.e., interpretation, oversimplification of the GWL, and material not considered) and generate errors in the stability computations under ALICE®. Therefore, this approach based on interdisciplinarity is in accordance with the landslide hazard mapping strategy proposed in [24] for France with local revisions based on numerical spatialization tools able to provide new information on slope stability quickly for decision support. Therefore, this study complements the study conducted in 2017 on a nearby site for shallow landslides.

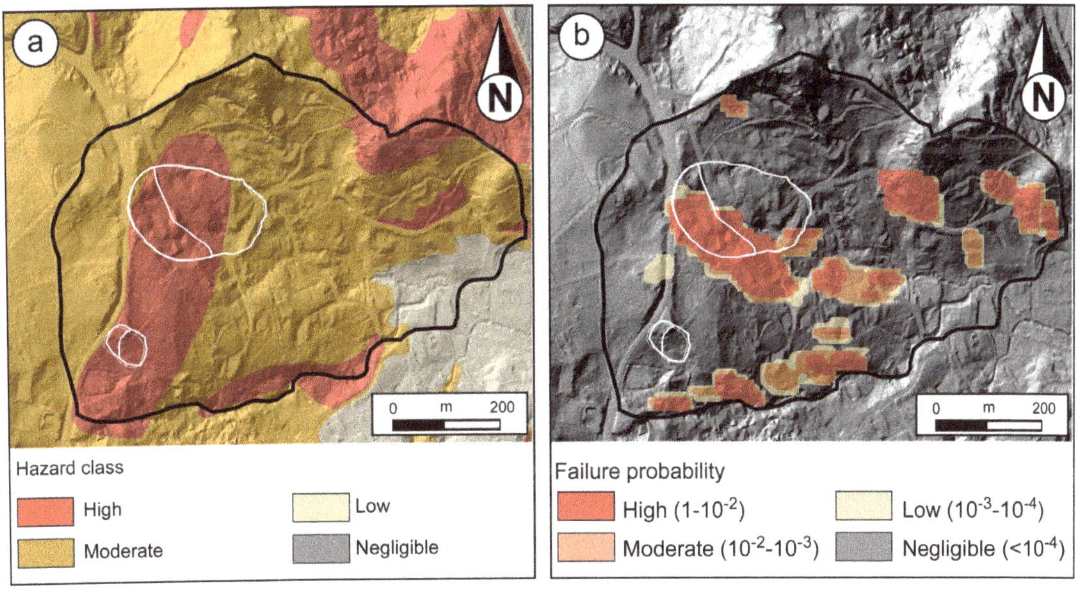

Figure 17. Landslide hazard maps for the Morne-Figue area. (**a**) Expert landslide hazard map obtained by expert approach and carried out at the 1:25,000 scale of work for all landslide types; (**b**) failure probability map obtained with ALICE® for moderately deep rotational landslides; map computed with GWL = 1; hillshade maps were produced with the Litto3D DTM (IGN, 2010).

6. Conclusions

AEM data are powerful in revealing in-depth resistivity contrasts in a complex tropical environment [30]. Once the results are compared with in situ investigations, such as boreholes and field observations, it is possible to better delineate the different superimposed lithologies and to understand the internal structure of various grounds. This study illustrates an additional possibility to obtain reliable data on the first 50 m by this noninvasive geophysical method. The results are fundamental for landslide studies in these environments with complex accesses limiting the use of classical field investigation techniques. Thus, it would be essential to verify low-resistance bodies located at approximately 5 Ω.m between a depth of 10 and 50 m. The two examples of the La Médaille landslide and the Morne-Figue area show that under landslides, some low resistivity formations could play the role of aquifers concentrating underground flows, punctually recharging the upper layers and influencing the activity of slope instabilities. The study of these bodies over the whole island, in correlation with recent hydrological studies [12,30] using AEM data, must be considered to better define this role on landslides [46]. It may be possible to refine the knowledge of unstable slopes from this information. From an operational point of view, the data acquired over the whole island could thus be exploited to (i) refine the knowledge of large landslides that regularly generate material damage and (ii) improve the spatialization of hazards and thus regulatory maps of landslide risks. Beyond the island of Martinique, the AEM method seems to be one of the most cost-effective methods to obtain geophysical information on shallow and deep formations, particularly in complex environments (topography, vegetation, and complex geological structure), such as the Caribbean. Thus, with the information acquired by this study and in [15], new perspectives to improve landslide hazard analysis and mapping for Martinique and other Caribbean islands can be considered.

Supplementary Materials: The following are available online at https://www.mdpi.com/article/10.3390/app11083390/s1, Figure S1: EM profiles for the La Médaille landslide. Figure S2: EM profiles for Morne-Figue area. Figure S3: Concept of slope stability computations with a slice method.

Author Contributions: Conceptualization, Y.T. and P.-A.R.; AEM data processing, P.-A.R.; methodology, Y.T.; software, Y.T. and P.-A.R.; validation, Y.T. and A.N.; formal analysis, Y.T.; investigation, Y.T., P.-A.R. and A.N.; writing—original draft preparation, Y.T., P.-A.R., A.N.; writing—review and editing, Y.T.; supervision, Y.T.; project administration, Y.T. All authors have read and agreed to the published version of the manuscript.

Funding: This research was supported by the Risk and Prevention General Direction of the French Ministry of Ecological Transition and the French Geological Survey (BRGM), contracts PS14DRP041, PS15DRP017 and AP16DRP017. The heliborne geophysical survey, called "MartEM" program [53], was co-funded by BRGM, the FEDER funds for Martinique, the Regional Office for Environment Planning and Housing (DEAL), the Regional Council and the Water Office of Martinique (ODE).

Institutional Review Board Statement: Not applicable.

Data Availability Statement: Some results of the study are available at http://infoterre.brgm.fr/rechercher/switch.htm?scope=9 (accessed on 4 April 2021); http://ficheinfoterre.brgm.fr/document/RP-66605-FR (accessed on 4 April 2021); http://ficheinfoterre.brgm.fr/document/RP-65407-FR (accessed on 4 April 2021).

Acknowledgments: We acknowledge the useful comments and suggestions from the three anonymous referees who helped us to enhance the manuscript. We also acknowledge our colleagues from the BRGM, Gilles Grandjean, Rosalie Vandromme and Severine Bernardie, for the constructive discussion about the approach.

Conflicts of Interest: The authors declare no conflict of interest. The funders had no role in the design of the study; in the collection, analyses, or interpretation of data; in the writing of the manuscript, or in the decision to publish the results.

References

1. CHARIM. Available online: http://www.charim.net (accessed on 18 November 2020).
2. Van Westen, C.J. National Scale Landslide Susceptibility Assessment for Grenada. Caribbean Handbook on Risk Information Management, World Bank GFDRR, ACP-EU Natural Disaster Risk Reduction. Available online: http://www.charim.net (accessed on 18 November 2020).
3. van Westen, C.J. National Scale Landslide Susceptibility Assessment for Dominica. Caribbean Handbook on Risk Information Management, World Bank GFDRR, ACP-EU Natural Disaster Risk Reduction. Available online: http://www.charim.net (accessed on 18 November 2020).
4. Van Westen, C.J. National Scale Landslide Susceptibility Assessment for Saint Vincent and the Grenadines. Caribbean Handbook on Risk Information Management, World Bank GFDRR, ACP-EU Natural Disaster Risk Reduction. Available online: http://www.charim.net (accessed on 18 November 2020).
5. Van Westen, C.J. National Scale Landslide Susceptibility Assessment for Saint Lucia. Caribbean Handbook on Risk Information Management, World Bank GFDRR, ACP-EU Natural Disaster Risk Reduction. Available online: http://www.charim.net (accessed on 18 November 2020).
6. Van Westen, C.J. National Scale Landslide Susceptibility Assessment for Belize. Caribbean Handbook on Risk Information Management, World Bank GFDRR, ACP-EU Natural Disaster Risk Reduction. Available online: http://www.charim.net (accessed on 18 November 2020).
7. Momtpelat, J.M. Unités Cartographiques et Évaluation de L'aléa Mouvements de Terrain en Guadeloupe (Antilles Françaises). Ph.D. Thesis, Institut de Physique du Globe de Paris, Paris, France, 1994.
8. Peronet, L. Les Mouvements de Terrain Dans les Petites Antilles: Contribution à L'analyse Comparative de Leur Gestion de Crise. Ph.D. Thesis, Universités des Antilles, Antilles, France, 2015.
9. Thiery, Y.; Reninger, P.A.; Vandromme, R.; Nachbaur, A. *Contribution of Heliborne Electro-Magnetic Survey for Landslide Prediction: Application to La Martinique (West Indies, France)*; EGU: Vienne, Austria, 2017.
10. Nachbaur, A. Les Mouvements de Terrain Majeurs en Martinique, Rapport Final, BRGM/RP-59250-FR. Available online: http://infoterre.brgm.fr/rapports//RP-59250-FR.pdf (accessed on 18 November 2020).
11. Degraff, J.V.; Bryce, R.; Jibson, R.W.; Mora, S.; Rogers, C.T. Landslides: Their extent and significance in the Caribbean. In *Landslides: Extent and Economic Significance*; Brabb, E.E., Harro, B.L., Eds.; A.A. Balkema: Rooterdam, The Netherlands, 1989; pp. 51–80.
12. Vittecoq, B.; Deparis, J.; Violette, S.; Jaouën, T.; Lacquement, F. Influence of successive phases of volcanic construction and erosion on Mayotte Island's hydrogeological functioning as determined from a helicopter-borne resistivity survey correlated with borehole geological and permeability data. *J. Hydrol.* **2013**, *509*, 519–538. [CrossRef]
13. Rogers, C.T. Landslide hazard data for watershed management and development planning, St Lucia, West Indies. In *Natural Hazards and Hazard Management in the Greater Caribbean and Latin America, Unit for Disasters Studies*; Ahmad, R., Ed.; The University of the West Indies: Kingston, Jamaica, 1988; pp. 150–164.
14. Vittecoq, B.; Reninger, P.A.; Violette, S.; Martelet, G.; Dewandel, B.; Audru, J.C. Heterogeneity of hydrodynamic properties and groundwater circulation of a coastal andesitic volcanic aquifer controlled by tectonic induced faults and rock fracturing—Martinique island (Lesser Antilles—FWI). *J. Hydrol.* **2015**, *529*, 1041–1059. [CrossRef]
15. Thiery, Y.; Reninger, P.A.; Lacquement, F.; Raingeard, A.; Lombard, M.; Nachbaur, A. Analysis of slope sensitivity to landslides by a transdisciplinary approach in the framework of future development: The case of la trinité in martinique (French West Indies). *Geosciences* **2017**, *7*, 135. [CrossRef]
16. Sepúlveda, S.A.; Petley, D.N. Regional trends and controlling factors of fatal landslides in Latin America and the Caribbean. *Nat. Hazards Earth Syst. Sci.* **2015**, *15*, 1821–1833. [CrossRef]
17. Nachbaur, A.; Legendre, Y.; Lombard, M.; Dewez, T. *Caractérisation Géologique et Identification des Mécanismes d'Instabilité de la Falaise Samperre, Rapport Final. BRGM/RP-68564-FR*; BRGM: Paris, France, 2019.
18. Durville, J.L.; Rivière, D. *Les Risques de Mouvements de Terrain sur le Site de Morne Callebase à Fort de France (Martinique)*; Conseil Général de l'Environnement et du Développement Durable: Paris, France, 2013.
19. van Westen, C.J.; van Asch, T.W.J.; Soeters, R. Landslide hazard and risk zonation—Why is it still so difficult? *Bull. Eng. Geol. Environ.* **2005**, *65*, 167–184. [CrossRef]
20. Fell, R.; Corominas, J.; Bonnard, C.; Cascini, L.; Leroi, E.; Savage, W.Z. Guidelines for landslide susceptibility, hazard and risk zoning for land use planning. *Eng. Geol.* **2008**, *102*, 85–98. [CrossRef]
21. Corominas, J.; van Westen, C.; Frattini, P.; Cascini, L.; Malet, J.P.; Fotopoulou, S.; Catani, F.; Van Den Eeckhaut, M.; Mavrouli, O.; Agliardi, F.; et al. Recommendations for the quantitative analysis of landslide risk. *Bull. Eng. Geol. Environ.* **2014**, *73*, 209–263. [CrossRef]
22. Reichenbach, P.; Rossi, M.; Malamud, B.D.; Mihir, M.; Guzzetti, F. A review of statistically-based landslide susceptibility models. *Earth-Sci. Rev.* **2018**, *180*, 60–91. [CrossRef]
23. Lombardo, L.; Opitz, T.; Ardizzone, F.; Guzzetti, F.; Huser, R. Space-time landslide predictive modelling. *Earth-Sci. Rev.* **2020**, *209*, 103318. [CrossRef]
24. Thiery, Y.; Terrier, M.; Colas, B.; Fressard, M.; Maquaire, O.; Grandjean, G.; Gourdier, S. Improvement of landslide hazard assessments for regulatory zoning in France: STATE–OF–THE-ART perspectives and considerations. *Int. J. Disaster Risk Reduct.* **2020**, *47*, 101562. [CrossRef]

25. Soeters, R.; Van Westen, C.J. Slope instability, recognition, analysis, and zonation. In *Landslides Investigation and Mitigation; Transportation Research Board, Special Report 247*; Turner, A.K., Schuster, R.L., Eds.; National Research Council: Washington, DC, USA, 1996; pp. 129–177.
26. Westercamp, D.; Andreieff, P.; Bouysse, P.; Cottez, S.; Battistini, R. *Notice Explicative de la Carte Géologique À 1: 50000ème de la Martinique, éditions*; BRGM: Paris, France, 1989.
27. Westercamp, D.; Pelletier, B.; Thibault, P.M.; Traineau, H. *Carte Géol. France (1: 50000ème), Feuille Martinique*; Bureau de Recherches Géologiques et Minières: Orléans, France, 1990.
28. Germa, A.; Quidelleur, X.; Labanieh, S.; Lahitte, P.; Chauvel, C. The eruptive history of Morne Jacob volcano (Martinique Island, French West Indies): Geochronology, geomorphology and geochemistry of the earliest volcanism in the recent Lesser Antilles arc. *J. Volcanol. Geotherm. Res.* **2010**, *198*, 297–310. [CrossRef]
29. Germa, A.; Quidelleur, X.; Labanieh, S.; Chauvel, C.; Lahitte, P. The volcanic evolution of Martinique Island: Insights from K–Ar dating into the Lesser Antilles arc migration since the Oligocene. *J. Volcanol. Geotherm. Res.* **2011**, *208*, 122–135. [CrossRef]
30. Vittecoq, B.; Reninger, P.A.; Lacquement, F.; Martelet, G.; Violette, S. Hydrogeological conceptual model of andesitic watersheds revealed by high-resolution heliborne geophysics. *Hydrol. Earth Syst. Sci.* **2019**, *23*, 2321–2338. [CrossRef]
31. Jongmans, D.; Garambois, S.P. Geophysical investigation of landslides: A review. *Bull. Soc. Géol. Fr.* **2007**, *178*, 101–112. [CrossRef]
32. Whiteley, J.S.; Chambers, J.E.; Uhlemann, S.; Wilkinson, P.B.; Kendall, J.M. Geophysical monitoring of moisture-induced landslides: A review. *Rev. Geophys.* **2019**, *57*, 106–145. [CrossRef]
33. Grandjean, G.; Hibert, C.; Mathieu, F.; Garel, E.; Malet, J.P. Monitoring water flow in a clay-shale hillslope from geophysical data fusion based on a fuzzy logic approach. *C. R. Geosci.* **2009**, *341*, 937–948. [CrossRef]
34. Pazzi, V.; Morelli, S.; Fanti, R. A review of the advantages and limitations of geophysical investigations in landslide studies. *Int. J. Geophys.* **2019**, *2019*, 1–27. [CrossRef]
35. Hack, R. Geophysics for slope stability. *Surv. Geophys.* **2000**, *21*, 423–448. [CrossRef]
36. Baroň, I.; Supper, R.; Winkler, E.; Motschka, K.; Ahl, A.; Čarman, M.; Kumelj, Š. Airborne geophysical survey of the catastrophic landslide at Stože, Log pod Mangrtom, as a test of an innovative approach for landslide mapping in steep alpine terrains. *Nat. Hazards Earth Syst. Sci.* **2013**, *13*, 2543–2550. [CrossRef]
37. Siemon, B.; Christiansen, A.V.; Auken, E. A review of helicopter-borne electromagnetic methods for groundwater exploration. *Near Surf. Geophys.* **2009**, *7*, 629–646. [CrossRef]
38. Auken, E.; Boesen, T.; Christiansen, A.V. A review of airborne electromagnetics methods with focus on geotechnical and hydrological applications from 2007 to 2017. *Adv. Geophys.* **2017**, *58*, 47–93. [CrossRef]
39. Reninger, P.A.; Martelet, G.; Deparis, J.; Perrin, J.; Chen, Y. Singular value decomposition as a denoising tool for airborne time domain electromagnetic data. *J. Appl. Geophys.* **2011**, *75*, 264–276. [CrossRef]
40. Reninger, P.-A.; Martelet, G.; Perrin, J.; Dumont, M. Processing methodology for regional AEM surveys and local implications. *Explor. Geophys.* **2019**, *51*, 143–154. [CrossRef]
41. Dumont, M.; Peltier, A.; Roblin, E.; Reninger, P.-A.; Barde-Cabusson, S.; Finizola, A.; Ferrazzini, V. Imagery of internal structure and destabilization features of active volcano by 3D high resolution airborne electromagnetism. *Sci. Rep.* **2019**, *9*, 18280. [CrossRef]
42. Dumont, M.; Reninger, P.A.; Pryet, A.; Martelet, G.; Aunay, B.; Join, J.L. Agglomerative hierarchical clustering of airborne electromagnetic data for multi-scale geological studies. *J. Appl. Geophys.* **2018**, *157*, 1–9. [CrossRef]
43. Auken, E.; Jørgensen, F.; Sørensen, K.I. Large-scale TEM investigation for groundwater. *Explor. Geophys.* **2003**, *34*, 188–194. [CrossRef]
44. Viezzoli, A.; Christiansen, A.V.; Auken, E.; Sørensen, K. Quasi-3D modeling of airborne TEM data by spatially constrained inversion. *Geophysics* **2008**, *73*, 105–113. [CrossRef]
45. Schamper, C.; Pedersen, J.B.; Auken, E.; Christiansen, A.V.; Vittecoq, B.; Deparis, J.; Jaouen, T.; Lacquement, F.; Nehlig, P.; Perrin, J.; et al. Airborne transient EM methods and their applications for coastal groundwater investigations. In *Groundwater in the Coastal Zones of Asia-Pacific*; Wetzelhuetter, C., Ed.; Springer Science: Berlin, Germany, 2013; pp. 121–153.
46. Supper, R.; Baroň, I.; Ottowitz, D.; Motschka, K.; Gruber, S.; Winkler, E.; Jochum, B.; Römer, A. Airborne geophysical mapping as an innovative methodology for landslide investigation: Evaluation of results from the Gschliefgraben landslide, Austria. *Nat. Hazards Earth Syst. Sci.* **2013**, *13*, 3313–3328. [CrossRef]
47. d'Ozouville, N.; Auken, E.; Sorensen, K.; Violette, S.; de Marsily, G.; Deffontaines, B.; Merlen, G. Extensive perched aquifer and structural implications revealed by 3D resistivity mapping in a Galapagos volcano. *Earth Planet. Sci. Lett.* **2008**, *269*, 517–521. [CrossRef]
48. Nakazato, H.; Kuroda, S.; Okuyama, T.; Sasaki, Y. Improving an Airborne Electromagnetic Method and Measuring the 3D Resistivity Distribution of a Landslide Region. *NIRE Res. Top.* **2004**, 91–92. [CrossRef]
49. Pfaffhuber, A.A.; Grimstad, E.; Domaas, U.; Auken, E.; Foged, N.; Halkjær, M. Airborne EM mapping of rockslides and tunneling hazards. *Lead. Edge* **2010**, *29*, 956–959. [CrossRef]
50. Sasaki, Y.; Nakazato, H. Inversion of airborne EM data accounting for terrain and inaccurate flight height. *SEG Tech. Program Expand. Abstr.* **2004**, 648–651. [CrossRef]
51. Nakazato, H.; Konishi, N. Subsurface structure exploration of wide landslide area by Aerial electromagnetic exploration. *Landslides* **2005**, *2*, 165–169. [CrossRef]

52. Deparis, J.; Reninger, P.A.; Perrin, J.; Martelet, G.; Audru, J.C. *Acquisition Géophysique Héliportée de la Martinique*; Openfile BRGM Report RP-62428-FR; BRGM: Paris, France, 2020. [CrossRef]
53. Sorensen, K.I.; Auken, E. SkyTEM–A new high-resolution helicopter transient electromagnetic system. *Explor. Geophys.* **2004**, *35*, 194–202. [CrossRef]
54. Deneufbourg, G. *Compte Rendu de la Reconnaissance de Terrain au Lieu-Dit "La Médaille"—Fort de France (Martinique)—15 Novembre 1966*; BRGM: Paris, France, 1967.
55. Deneufbourg, G. *Etude Géologique des Glissements de la Trace (R.N. 3) Entre Camp de Balata et le Quartier Propreté (Martinique). Décembre 1968. BRGM 68 RME 024 ANT*; BRGM: Paris, France, 1968.
56. Humbert, M. *Mouvement de Terrain Sécurité et Aménagement du Quartier de La Médaille (Commune de Fort de France). BRGM 86 MQE 035*; BRGM: Paris, France, 1986.
57. Hazmoune, A.; Closset, L.; La Fata, P. *Glissement de Terrain au Lieu-Dit "La Médaille"—N3—Fort de France—Synthèse des Connaissances et Bilan du Suivi Topographique 2000–2001.BRGM/RP-51496-FR, 5 fig.; 7 tabl., 1 pl.h.t., 8 ann*; BRGM: Paris, France, 2002.
58. Allard, J.F. *Glissement Nord de La Médaille—Fort de France, Étude Géotechnique. Note BRGM. 81. MQE. 15*; BRGM: Paris, France, 1981.
59. Chargueron, C.; Comte, J.P.; Mathon, C.; Périan, G. *Glissement de Terrain de « La Médaille »—RN3-(Commune de Fort de Rance) Phase 3: 2002–2003. Suivi Topographique, Piézométrique et Instrumentation. Rapport BRGM/RP-52893-FR, 42 Figure 6 tabl. 8 ph., 3 ann*; BRGM: Paris, France, 2004.
60. Belz, H.; Chalivat, P.R.N. 3, *Lieu-dit "La Médaille", Fort de France, Martinique—Reconnaissances Géologiques et Géotechniques dans la Partie Nord du Glissement de Terrain. Synthèse des Connaissances et Recommandations. Rapport BRGM N 2342, Septembre 1996, 2 fig., 3 tabl., 5 ann*; Natural Resources Canada, ESS, Scientific and Technical Publishing Services: Ottawa, ON, Canada, 1996.
61. ANTEA. *RN 3 Glissement de Terrain au Lieu-Dit « La Médaille »—Fort de France. Synthèse Géotechnique—A21824. 2 ann*; ANTEA: St. Paul, MN, USA, 2000.
62. Thiery, Y.; Reninger, P.A.; Nachbaur, A.; Vandromme, R. *Exploitation de Levés D'Électromagnétisme Héliporté Dans une Perspective de Réévaluation des Cartes d'aléa «Mouvement de Terrain» en Milieu Volcanique Tropical. Application aux Antilles—Phase 2. Rapport Intermédiaire. BRGM/RP-65407-FR, ill. 31*; BRGM: Paris, France, 2015.
63. Atlan, Y.; Besson, J.C. *Evolution des Reliefs en Liaison avec la Sécurité et l'Aménagement du Territoire en Martinique. BRGM 83. ANT. 016*; BRGM: Paris, France, 1983.
64. Allard, J.F. *RN1 Carrefour Nord de Trinité—Glissement du Morne Figue—Trinité—Martinique, Etude Géotechnique. Rapport n 89 MTQ 082*; BRGM: Paris, France, 1989.
65. Barras, A.V.; Ollagnier, S. *Glissement de Terrain dans le Secteur de Morne Figue Trinité—Martinique: Etude Hydrogéologique et Campagne Géophysique. Rapport n BRGM/RP-55469-FR*; BRGM: Paris, France, 2007.
66. Vest Christiansen, A.; Auken, E. A global measure for depth of investigation. *Geophysics* **2012**, *77*, WB171–WB177. [CrossRef]
67. Martelet, G.; Calcagno, P.; Gumiaux, C.; Truffert, C.; Bitri, A.; Gapais, D.; Brun, J.P. Integrated 3D geophysical and geological modelling of the Hercynian Suture Zone in the Champtoceaux area (south Brittany, France). *Tectonophysics* **2004**, *382*, 117–128. [CrossRef]
68. Calcagno, P.; Chilès, J.P.; Courrioux, G.; Guillen, A. Geological modelling from field data and geological knowledge: Part I. Modelling method coupling 3D potential-field interpolation and geological rules. *Phys. Earth Planet. Inter.* **2008**, *171*, 147–157. [CrossRef]
69. Vandromme, R.; Thiery, Y.; Bernardie, S.; Sedan, O. ALICE (Assessment of Landslides Induced by Climatic Events): A single tool to integrate shallow and deep landslides for susceptibility and hazard assessment. *Geomorphology* **2020**, *367*, 107307. [CrossRef]
70. Sedan, O. *Logiciel ALICE Version 7-Guide d'Utilisateur, Technical Report, RP-60004*; BRGM: Orléans, France, 2013.
71. Thiery, Y.; Vandromme, R.; Maquaire, O.; Bernardie, S. Landslide susceptibility assessment by EPBM (Expert physically based model): Strategy of calibration in complex environment. In *Proceedings of Advancing Culture of Living with Landslides, Volume 2 Advances in Landslide Science, Proceedings of the 4th World Landslide Forum, Ljubljana, Slovenia, 29 May–2 June 2017*; Springer: Berlin, Germany, 2017; pp. 917–926.
72. Morgenstern, N.R.; Price, V.E. The analysis of the stability of general slip surfaces. *Géotechnique* **1965**, *15*, 79–93. [CrossRef]
73. Zhu, D.Y.; Lee, C.F.; Qian, Q.H.; Chen, G.R. A concise algorithm for computing the factor of safety using the Morgenstern–Price method. *Can. Geotech. J.* **2005**, *42*, 272–278. [CrossRef]
74. Fressard, M.; Maquaire, O.; Thiery, Y.; Davidson, R.; Lissak, C. Multi-method characterisation of an active landslide: Case study in the Pays d'Auge plateau (Normandy, France). *Geomorphology* **2016**, *270*, 22–39. [CrossRef]
75. Viet, T.T.; Lee, G.; Thu, T.M.; An, H.U. Effect of digital elevation model resolution on shallow landslide modeling using TRIGRS. *Nat. Hazards Rev.* **2016**, *18*, 04016011. [CrossRef]
76. Glenn, N.F.; Streutker, D.R.; Chadwick, D.J.; Thackray, G.D.; Dorsch, S.J. Analysis of LiDAR-derived topographic information for characterizing and differentiating landslide morphology and activity. *Geomorphology* **2006**, *73*, 131–148. [CrossRef]
77. Dubois, L. *Mission «Mouvement de Terrain». Les 20 et 21 Mars 2003. Rapport CETE Normandie-Centre pour la Direction Départementale de la Martinique. Affaire n 8482*; BRGM: Paris, France, 2003.
78. Allard, J.F.; Bove, Y. *CD25 BIS. Etude Géotechnique du Glissement Survenu au Quartier de Morne Congo à Trinité (Martinique). BRGM/78-ANT-004*; BRGM: Paris, France, 1978.

applied sciences

MDPI

Article

Spatial Uncertainty of Target Patterns Generated by Different Prediction Models of Landslide Susceptibility

Andrea G. Fabbri [1,*] and Antonio Patera [2]

1 DISAT, Università di Milano-Bicocca, 20126 Milan, Italy
2 Istituto Nazionale di Geofisica e Vulcanologia, 00143 Rome, Italy; antonio.patera@ingv.it
* Correspondence: andrea.fabbri@unimib.it; Tel.: +39-3338313289

Abstract: This contribution exposes the relative uncertainties associated with prediction patterns of landslide susceptibility. The patterns are based on relationships between direct and indirect spatial evidence of landslide occurrences. In a spatial database constructed for the modeling, direct evidence is the presence of landslide trigger areas, while indirect evidence is the presence of corresponding multivariate context in the form of digital maps. Five mathematical modeling functions are applied to capture and integrate evidence, indirect and direct, for separating landslide-presence areas from the areas of landslide assumed absence. Empirical likelihood ratios are used first to represent the spatial relationships. These are then combined by the models into prediction scores, ordered, equal-area ranked, displayed, and synthesized as prediction-rate curves. A critical task is assessing how uncertainty levels vary across the different prediction patterns, i.e., the modeling results visualized as fixed, colored groups of ranks. This is obtained by a strategy of iterative cross validation that uses only part of the direct evidence to model the pattern and the rest to validate it as a predictor. The conducted experiments in a mountainous area in northern Italy point at a research challenge that can now be confronted with relative rank-based statistics and iterative cross-validation processes. The uncertainty properties of prediction patterns are mostly unknown nevertheless they are critical for interpreting and justifying prediction results.

Keywords: landslide susceptibility; ranking; cross validation; prediction model; prediction pattern; target pattern; uncertainty pattern

Citation: Fabbri, A.G.; Patera, A. Spatial Uncertainty of Target Patterns Generated by Different Prediction Models of Landslide Susceptibility. *Appl. Sci.* **2021**, *11*, 3341. https://doi.org/10.3390/app11083341

Academic Editor: Jan Blahut

Received: 29 January 2021
Accepted: 5 April 2021
Published: 8 April 2021

1. Introduction

Predicting landslide susceptibility of a region or study area has become a critical necessity with the continuing expansion of urbanizations across hazardous landscapes, increasing soil deterioration and the extensive damages inflicted by landslides [1]. Regional approaches to quantitative predictions have developed following the availability of thematic maps in digital form, including that of interpreted aerial photography and remotely sensed images. Examples of themes related to mass movement processes are expressions of various aspects of the rocks, soil, land cover, land use, soil permeability, groundwater table, and topographic surfaces [2,3].

In any application, the themes are hopefully representing the gravity-induced physical context of the process of slope failure. For this reason, the landslides are considered spatially related with the thematic map units or values over the study area of concern. These spatial relationships are established and integrated by mathematical models that transform raw data from specific landslide distribution maps and contextual maps into prediction maps or maps of the likelihood of future landslide occurrence [4].

Obviously, the detail and accuracy of the digital maps used for modeling must be congruous and the statistical relationships significant to obtain convincing likelihood maps of landslide occurrences that, hopefully, are more informative than what is already known. The term "convincing" must reflect the interpretable quality of the integrated likelihood maps as spatiotemporal predictions and the certainty—or conversely, the uncertainty—associated with that quality.

The applications discussed in this contribution have the purpose of exposing aspects of spatial predictions that are commonly ignored—validation, robustness, and uncertainty of the resulting susceptibility levels expressed as relative ranks. They are the modeled values that make up the prediction patterns.

Analyzed are a database of landslide occurrences and their spatial context in a study area in northern Italy. The database has been the focus of experiments on comparison of models and analysis of modeling structure. Here, we will focus exclusively on the uncertainties affecting the ranks representing relative predicted levels of susceptibility to active landslides.

The following section deals with the predictive methodology, the terminology, and the analytical procedures applied. The next section describes the study area, its database for modeling, and previous research on landslide susceptibility. Experimental results follow on uncertainty identification and its relative measures. They consist of modeling prediction, target and uncertainty patterns, their interpretation, and comparison. Conclusions deem that uncertainty assessments are a desirable and necessary endeavor for making spatial prediction modeling a worthwhile practice.

2. Predictive Methodology, Terminology, and Analytical Procedures

A unified framework for modeling prediction patterns was proposed by Chung and Fabbri (1993) [5]. It was termed the "favorability function" and is used in the applications that are the focus of this contribution. The term favorability was selected for its generality in the meaning intended to be comprehensive and unified to cover a variety of mathematical interpretations. The following is a summary of the concepts in the modeling, the terminology used, and the processing strategy that enables us to assess spatial uncertainty.

A study area (SA) is first assumed to be selected by experts as sufficiently representative of the processes that are considered hazardous within it. Hopefully then, the statistics from spatial quantitative data available in the SA are suitable for information extraction by modeling. A second assumption is that we can have at least two subareas. First, an occurrence subarea is identified as affected by the hazardous landslide process, with the known presence of landslide trigger areas of a specific and congruous dynamic type. Then, another subarea is selected of non-occurrences in which landslides are known to be absent (or are unknown if present). The occurrence subarea is where the occurrences are located and characterized, and it is considered a training area (TA) for establishing the spatial relationships. Generally, the nonoccurrence subarea is the complement of the occurrence subarea within the SA. In some cases, however, the nonoccurrence subarea is only part of the SA or even outside the SA. This could also be the case with the occurrence area. Commonly, the occurrence subarea is a very small portion of the SA. A third assumption is that future occurrences within the SA will be similar in type and setting to the known ones.

To allow favorability function prediction modeling, a proposition is constructed as follows:

F_i: that a point i in the SA will be affected by a future landslide | the presence of spatial evidence. (1)

The symbol | indicates "given." The proposition is to be supported as true by means of the known occurrence distribution and their setting. The setting is considered spatial evidence. Therefore, the proposition links the occurrence locations (or their neighborhoods) and their setting as multivariate context. To satisfy the proposition as a mathematical statement, both the digital images representing the occurrence locations and their setting have to be transformed into spatial relationships. These new images have been termed the direct supporting pattern (DSP). They are the indexed occurrence locations. Indirect supporting patterns of the proposition (ISPs) are termed the images of their setting. The term "pattern" is used to indicate the results of this functional transformation into spatial relationships.

The landslide occurrences are converted into clusters of adjacent picture elements or pixels with a numerical value sequentially identifying each occurrence. The image pixel values of the settings are transformed into normalized frequencies if derived from categorical maps such as lithology or land use. Continuous field digital images, such as elevation or slope angles, are transformed into density functions. The ratios of the normalized frequencies and the density functions for the TA are divided by those of the nonoccurrence subareas within the SA. These ratios are termed empirical likelihood ratios (ELRs). We could also apply other types of normalization; nevertheless, the end results would have the same purpose and functionality. ELRs, as described, range in value between zero and infinity and provide a measure of support of the proposition in (1).

Many different mathematical models can be used to convert and integrate the ratios into prediction patterns that classify the SA into levels of likelihood of future landslide occurrence. We will be using five different models in our applications but the favorability function modeling and the processing structure are independent of the mathematical models that will be indicated later on.

In essence, the likelihood ratios enable us to separate the presence of occurrence from their supposed absence within the SA. This separation uses another assumption. While the normalized frequencies or the density functions in the TA (the occurrence subarea) are related to observed/mapped landslide occurrences, the ones in the nonoccurrence subarea relate to both the absence of occurrences and the unknown occurrences. For the latter, we assume that their setting, due to being relatively rare, is diluted within the setting of the absence of occurrences. This assumption allows considering the ratios as a form of contrast between areas with presence and areas with a presumed absence of occurrences. Commonly, the TA with respect to the SA is three or four orders of magnitude smaller.

A favorability function at each point of a SA should have two properties for modeling, i.e.: (1) should be able to measure a relative level of likelihood that a point i in the SA contains part of a future landslide of the given dynamic type and (2) should be able to provide a measure of uncertainty associated with the function by using only the part of all possible landslide occurrences present in the training area, TA.

Now suppose that we use a given prediction model, still unspecified at this point, that generates integrated scores as predicted values for every pixel of the SA. The scores are in some a-dimensional units or values between a minimum and a maximum. How do we establish the very high, high, intermediate, low, and very low likelihood of future landslide occurrence? This is not a simple question to answer, and it leads to further assumptions depending on the kind of data available for the SA, database of DSP and ISPs, and additional information.

Suppose, for instance, that we have temporal characterization for the many occurrences of the given specific dynamic type in the TA. We could use the occurrences from an older time interval to model a prediction pattern and then overlay it with the locations of the occurrences from the younger time interval. In this way, we obtain some statistics about their proportions within higher likelihood scores. Should some of them fall within the lower likelihood score areas, we would consider them as poorly predicted. Vice versa, well-predicted occurrences should fall on the higher likelihood score areas. We have termed this exercise as cross validation. In practice, this is a natural way to establishing how "good" our prediction pattern is as a "predictor of future occurrences."

Now let us suppose that the information on time partitioning of the occurrences, the DSP, is not available, as it is in most cases. How can we proceed with some other forms of cross validation? We could empirically pretend, for instance, not to know the existence of some of the occurrences, e.g., 25% of them, and use the remaining 75% to obtain a prediction pattern and then cross-validate it with the 25% we pretended not to know and that was not used as DSP for modeling. In this case, the "future" landslide occurrences for cross validation are the "next" 25%. Furthermore, we could operationally devise iterative strategies, depending on the number of occurrences available as DSP, such as (1) sequential exclusion of an arbitrary number of occurrences to be used for cross validation of the pattern produced with the remaining ones, (2) sequential selection of a number of occurrences for modeling and using the remainder for cross validation, or (3) random selection of a number of occurrences repeated a convenient number of times. All these strategies will provide different ways to predicting the next arbitrarily selected number or proportion of occurrences. Recall that we have not yet selected any particular mathematical model for predicting.

Cross validation is a strategy for assessing the quality of our prediction modeling and also for comparing different prediction patterns, produced either by varying the number of ISPs or the mathematical models. The results of cross validations are tables of prediction scores for numbers or proportions of occurrence pixels in the SA. How do we interpret these dimensionless integrated scores generated for each pixel of a SA and ranging between a minimum and a maximum?

Given the number of transformations from the original map unit names or continuous values used to compute the ISPs and their conversion into relative scores, these are considered here as impossible to interpret directly by recognizing systematic changes or breaks. Instead, the prediction scores are easily converted into equal-area ranks after sequencing them in decreasing order. It was found convenient to obtain 200 ranks each corresponding to 0.5% of the SA. By equal-area ranking the prediction scores, it becomes practical and simple to display the prediction patterns and generate cross-validation tables of proportions of validation occurrences for each rank and cumulative tables of these proportions to be represented as curves.

A consequence of iterative cross validation is that a different prediction pattern is generated each time so that the set of patterns can be used to further characterize the initial prediction pattern obtained using all the occurrences in the DSP. We can then pile up the patterns and, if their numbers are sufficiently large, apply some form of statistics to obtain average and variance for each stuck of pixels in the SA. We have defined "target pattern" as what we wish to have—a representation of all past and future landslide occurrences (as susceptibility scores and associated uncertainty scores). In our case, some form of averaging the ranks of the prediction patterns from the iterations. We have defined "uncertainty patterns" as the expression of deviations from these averages. From practice, we have found that a very robust means for generating a target pattern is selecting the median rank for each pixel from the iteration prediction patterns, and for the uncertainty pattern, the rank of the ranges of deviations from the median ranks.

Furthermore, three assumptions must be reasonably satisfied [6], namely, (a1) the known landslide occurrences, the DSPs, are a "random selection" of all existing ones, known and unknown (allowing to extend the favorability function from the TA to the entire SA); (a2) the ISPs are correlated with the target pattern (allowing to estimate the function using the known part of the target pattern in the TA); and (a3) the process of slope failure is not random and follows a certain rule (allowing to model the favorability function).

Target and uncertainty patterns have opened the way to further characterize prediction patterns. Some of these will be discussed in Section 4. Let us now consider some favorability modeling functions of empirical likelihood ratios commonly applied and used here in our analyses. They are fuzzy set membership function [7], linear and logistic regression functions [8,9], empirical likelihood ratio function [10–12], and Bayesian predictive discriminant function [13]. We will abbreviate them as FZ, LI, LO, LR, and BP, respectively. The modeling functions imply different representation and combination rules of spatial relationships [5,7,14].

We will not discuss theoretical aspects of the modeling functions here because they were amply dealt with in the above-mentioned contributions [5,7,10–14]. The focus of our study is the characterization and interpretation of prediction patterns, independently of the particular prediction models applied. Our concern is that their applications to the same input data generate prediction patterns whose scores are in entirely different units, and these are considered as not interpretable or comparable except in terms of equal-area ranking. Moreover, they require entirely different combination rules. They imply diverse assumed interpretations of the spatial relationships expressed by the empirical likelihood ratios, and each interpretation combines ISPs with assumptions of conditional independence (between categorical, continuous field ISPs and for integration of the two types) [12]. In geomorphology, and geosciences in general, maps are often spatially correlated so that such independence seldom exists or can be hypothesized. Nevertheless, it has been found that the existence of a correlation between ISPs causes minor alterations to equal-area rankings, mostly to the lower ranks [12,15].

3. Study Area, Database, and Previous Research

The Tirano South study area, whose location is shown in Figure 1, occupies the southern half of a community area termed "Comunità Montana Valtellina di Tirano" in the Province of Sondrio, Lombardy Region, in northern Italy. It was established in 1971 for socioeconomic and environmental protection (www.cmtirano.so.it; accessed on 7 April 2021). The geomorphology and landslide processes in the area were described by Sangalli (2008) [16], who compiled the available information and constructed a database for an initial landslide susceptibility analysis. Part of the data is being used in this contribution that also covers the same study area.

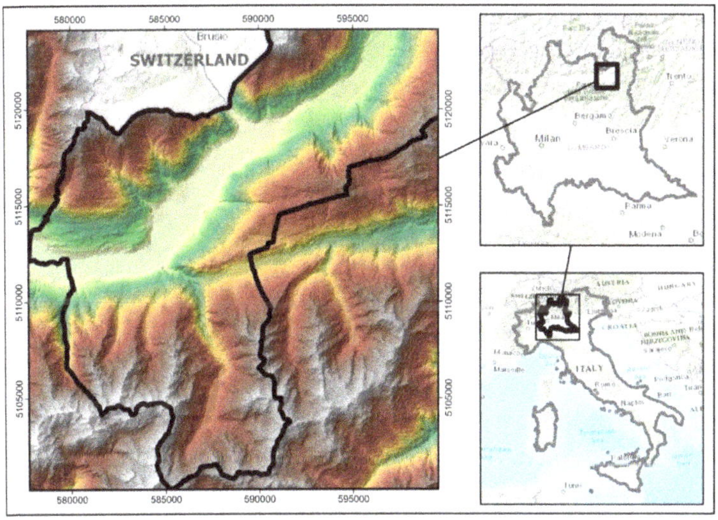

Figure 1. Locations of the Tirano South study area, in the Lombardy Region, northern Italy.

Elevations in this young Alpine geology area range between 300 and 3000 m a.s.l. and rainfall from 700 to 1900 mm, thus providing different climatic conditions. Vegetation consists of broad-leaved forests at lower elevations and coniferous forests at higher elevations. The geomorphology is controlled mostly by glacier activity, but anthropic activity significantly affects land use at valley bottoms with tourism, agriculture, and industries. Glacial and torrent erosion, together with intense rainfalls, are the triggers of slope instabilities. Out of a variety of landslide phenomena in the area, 70 active landslides were identified from published inventories. However, they did not contain sufficient information for separating rotational from translational dynamic types. In addition to the inventories used to generate digital images of landslide trigger area locations, various cartographies related to slope instability were available for compilation and digitization into the database, as shown in Figure 2. The database offers opportunities for experimenting on spatial prediction modeling of landslide susceptibilities, hazards, and risks.

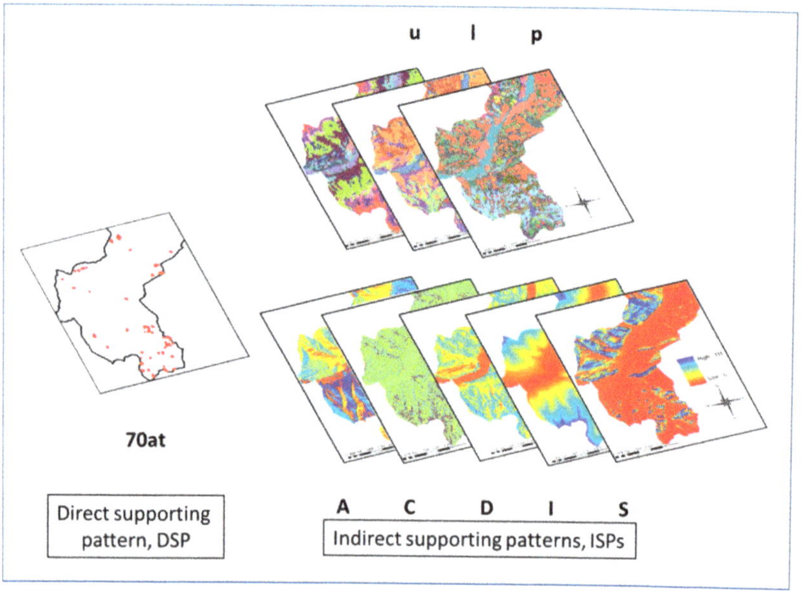

Figure 2. The Tirano South database is shown that will be converted into a direct supporting pattern (DSP) the 70 active landslide trigger areas, **70at,** and the three categorical and five continuous field digital maps to be converted into indirect supporting patterns (ISPs), **ulp** and **ACDIS**. The explanation is in the text.

The Tirano South database consists of digital images within a raster of 1090 pixels by 1194 lines with 20 m resolution. The study area proper within the raster covers 646,091 pixels corresponding approximately to 258 km^2. The trigger zones of the 70 active landslides, converted into DSP and abbreviated as **70at**, cover 697 pixels, i.e., as a training area (TA) a little less than 0.11% of the study area (SA). Eight cartographies are used to generate, at the same 20 m resolution, the ISPs consisting of maps with 23 land use classes, $\mathbf{u}_{1\text{–}23}$, with 51 geologic (lithologic) units, $\mathbf{l}_{1\text{–}51}$, and with eight permeability classes, $\mathbf{p}_{1\text{–}8}$ (three categorical maps), in addition to aspect (0° to 359°), **A**, topographic curvature (−32 to +29), **C**, topographic digital elevation (350–2906 m), **D**, internal relief (0–111 m for 3 × 3 pixels), **I**, and slope (0° to 61°), **S**, all derived from a 5 × 5 m resolution digital elevation model, DTM, resampled to 20 × 20 m (5 continuous field maps).

Previous research by Poli and Sterlacchini (2007) [17] on landslide susceptibility in the Tirano area used an earlier and simplified database with 28 active complex landslides and five binary and binarized factor maps applying the weights-of-evidence mathematical model and generating arbitrary thresholds of prediction ranks. That and similar applications of spatial prediction modeling were criticized [18] so that in joint contributions that used the same database, kindly shared by those authors, cross-validation procedures were preferred. The purpose was to assess the relative quality of the modeling results with an alternative model and different analytical processes that exposed the uncertainties associated with the prediction patterns [19,20]. A further study with a wealth of controlled information on Alpine landslides and three national–regional mapping initiatives provided new data for the construction of a high-quality database for the Tirano study area. It contained landslide inventories in addition to geological and soil–land use cartographies. For the northern part of the Tirano study area, prediction patterns were thus generated using both active and quiescent translational–rotational landslide scarps, the empirical likelihood ratio function model, and iterative cross-validation strategies based on blind tests [21].

Indeed, the Tirano study area had become the focus of much more research on landslide susceptibility modeling and other types of landslides and associated floods due to the wealth of information available and certified. Blahut et al. (2010a) [22] used two sets of aerial photographic coverages, 20 years apart, to select natural-condition debris flows for the estimation of the spatiotemporal probability of hazard initiation. The resulting distribution map was then used to model runout zone limitations using an empirical GIS-based simulation tool. The experiments indicated that the spatial variability observed needed validation tests for satisfactory interpretation. Furthermore, another study was carried out by those authors [23] of the degree of spatial pattern agreement between different landslide susceptibility maps that showed similar debris-flow predictive power. Their database contained the distribution of 573 scarps, and one half was used for modeling a susceptibility map, while the other half was used to validate the map. The similarity of predictive power, however, did not necessarily produce similar spatial configurations of susceptibility ranks, a fact that pointed at the variability of ranks and related spatial uncertainty.

Two more activities in the Tirano study area make it an exemplary instance of supporting background for natural hazard and risk studies. Blahut et al. (2010b) [24] constructed a reliable operational inventory of landslides from three available official inventories. Their tasks were (1) to prepare debris flow and factor maps inventories for susceptibility modeling, (2) generate indices of accountability/reliability and selection of factor maps, (3) evaluate/validate susceptibility maps, and (4) compare the results of different maps for combining them into an integrated susceptibility representation. Various methods of subdivision of debris flows and factor maps were obtained, including the separation of the Tirano study area into three more congruous subareas—northern, central, and southern. Again, high spatial variability of the susceptibility ranks was observed as related to the different combinations of factor maps. This made it difficult to delimit the ranks into susceptibility classes. Blahut et al. (2012) [25] studied the historical information on landslides and floods in the Tirano area for hazard estimation and definition of tentative risk scenarios. They developed a case study to exemplify the usefulness in their database of 489 records of damaging events (from the years 1600 to 2001), for realistic scenario generation, producing damage classification maps of territorial threats.

Recent analyses of the Tirano South database [16] focused on (1) credibility analysis of a fuzzy set modeled prediction pattern of landslide susceptibility and separation of well predicting from poorly predicting landslide occurrences [26] and (2) a generalized procedural strategy for comparisons of prediction patterns of active and dormant landslides by different models [27]. This contribution wants to expand that procedure and strategy attempting to interpret the uncertainties associated with target patterns and their consequences for understanding the prediction patterns generated by the application of those very same models.

4. Experimental Results

The Tirano South database, shown in Figure 2, was reanalyzed in order to expose the uncertainties associated with prediction patterns. The stepwise procedure for generating and assessing prediction patterns, proposed by Fabbri et al. (2017) [27], was also followed here, i.e., (1) use models to obtain prediction patterns from likelihood ratios, (2) cross-validate the patterns, (3) interpret the cross-validation results, and (4) obtain and compare the target and uncertainty patterns via equal-area ranks. Furthermore, step (5) was added, namely, analyze target, uncertainty, and 50% combination pattern relationships.

Table 1 shows the likelihood ratios for the ISPs computed for the Tirano South study area. For the ISP units and value ranges, only the ratios ≥ 2 or ≈ 2 are shown in the table. A ratio of 2 represents a normalized frequency in the presence of occurrences that is twice that in their absence. The arbitrary value of 2 was used as an empirical rule of thumb to separate the supportive ISP units or value ranges from the unsupportive ones of the proposition in (1). A ratio of 1 represents that the frequency in the presence of occurrences is the same as the one in their absence, i.e., null support of the proposition. Only the maximum values of the supportive ratios and respective units and value ranges are shown in Table 1: two land-use classes (2.29 and 4.48); eight lithology units (5.19 to 5.52); three permeability classes (1.92, 2.06 and 1.91), (categorical ISPs); one range of aspect angles (maximum ratio 2.07); two curvature ranges (11.07 and 5.62); two elevation ranges (4.24 and 2.31); one internal relief range (5.97); and one slope angle range (4.26). All prediction patterns generated by the five models have been based on the supports provided by all the ratio values of which, for simplicity, only the most supportive are listed in Table 1.

Table 1. Subset is listed of categorical and continuous field ISPs in the Tirano South study area and their respective empirical likelihood ratio values. They will be used for predictions using as DSP the distribution of the set of landslides, **70at** with **ulpACDIS** as ISPs. Abbreviations for categorical ISPs are $\mathbf{u}_{1\text{–}23}$, land-use classes; $\mathbf{l}_{1\text{–}51}$, lithology units; and $\mathbf{p}_{1\text{–}8}$, permeability classes. For the continuous field ISPs, A, C, D, I, and S are aspect, curvature, digital elevation, internal relief, and slope, respectively. Values are bold if the ELR $\geq$ 2.00. In Italics is the corresponding range of values, with the maximum ratio in brackets. In this reduced version, only ratios $\geq$2 or $\approx$2 are shown (table modified after Tables 1 and 2 in Fabbri et al., 2017 [27]).

Supporting Categorical Units or Classes Converted to ISPs for DSP 70at	
Land use $\mathbf{u}_{1\text{–}23}$	$\mathbf{u}_2$, Rock and scree vegetation; $\mathbf{u}_3$, Bedrock outcrops and surficial deposits.
Lithology $\mathbf{l}_{1\text{–}51}$	$\mathbf{l}_6$, Sandstones; $\mathbf{l}_{20}$, Non-vegetated deposit; $\mathbf{l}_{26}$, Active non-vegetated scree slope; $\mathbf{l}_{40}$, Outcropping quartzites; $\mathbf{l}_{44}$, Outcropping ypo-abyssal rocks; $\mathbf{l}_{45}$, Intrusive rocks; $\mathbf{l}_{46}$, Outcropping intrusive rocks; $\mathbf{l}_{49}$, Serpentinites.
Permeability $\mathbf{p}_{1\text{–}8}$	$\mathbf{p}_2$, Cohesive units with low permeability; $\mathbf{p}_3$, Cohesive units with very low permeability; $\mathbf{p}_4$, Non-cohesive units with high-medium permeability.
Categorical ISPs with ELR values$\geq$2 or$\approx$2	
Land use $\mathbf{u}_{1\text{–}23}$ Lithology $\mathbf{l}_{1\text{–}51}$ Permeability $\mathbf{p}_{1\text{–}8}$	$\mathbf{u}_2$ = **2.29**, $\mathbf{u}_3$ = **4.48**; $\mathbf{l}_6$ = **5.19**, $\mathbf{l}_{20}$ = **18.91**, $\mathbf{l}_{26}$ = **5.89**, $\mathbf{l}_{40}$ = **6.15**, $\mathbf{l}_{44}$ = **2.41**, $\mathbf{l}_{45}$ = **13.10**, $\mathbf{l}_{46}\mathbf{l}_{46}$ = **19.79**, $\mathbf{l}_{49}$ = **5.52**; $\mathbf{p}_2$ = **1.92**, $\mathbf{p}_3$ = **2.06**, $\mathbf{p}_4$ = **1.91**.
Continuous field ISPs with ELR values$\geq$ 2	
Aspect (0°-359°), **A** Curvature (−32–+29), **C** D. Elevation (350–2906m), **D** Int. Relief (0–111 m, 3 × 3 pixels), **I** Slope (0°–61°), **S**	$\geq$2: *168–198*, (max **2.07**); $\geq$2: *−24–−7*, (max 11.07); *+7–+17*, (max **5.62**); $\geq$2: *1737–2104*, (max 4.24); *2484–2629*, (max **2.31**); $\geq$2: *24–56*, (max **5.97**); $\geq$2: *37–57*, (max **4.26**)

4.1. *Use Different Mathematical Models to Obtain Prediction Patterns from Likelihood Ratios*

The empirical likelihood ratios represent the spatial relationships within the database. They were obtained using the trigger zones distribution image of the 70 active landslides, **70at**, as DSP, and the ratio transformed images of the three categorical and the five continuous-field maps, **ulpACDIS**, as ISPs. Each model was applied to generate a different prediction pattern. The models integrated the ratios for each ISP. The patterns, displayed in Figure 3, were generated by the five mathematical models each requiring different assumptions and providing incompatible scores computed from identical inputs of likelihood ratios. To identify the patterns and their inputs in the computations, a sequence of short names was used as follows: MODEL_DSP_ISPs, as in **FZ_70at_ulpACDIS** to **BP_70at_ulpACDIS**. The one-letter ISP abbreviations were used in analyses with subsets of them.

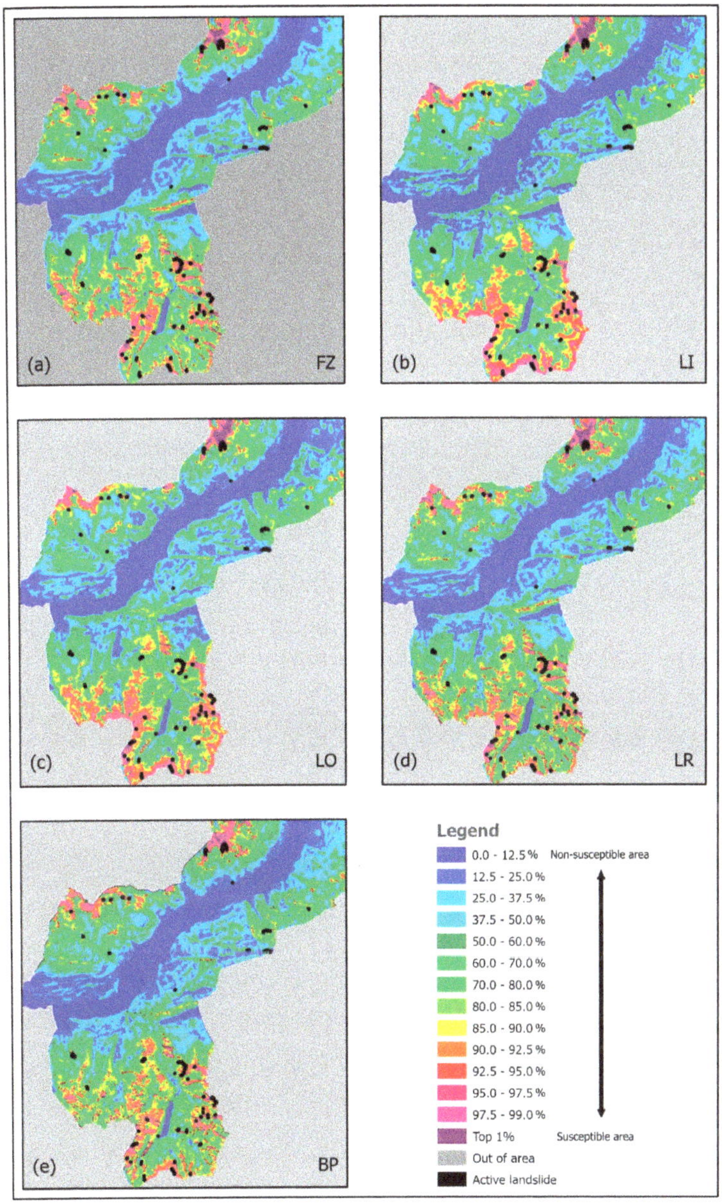

Figure 3. Prediction patterns using the different models: FZ, LI, LO, LR, and BP in (**a**–**e**), respectively, all using **70at_ulpACDIS**. Black patches are the oversized trigger zones of the 70 active landslides, **70at**. Colors in the legend indicate groups of ranks of % of the study area (SA) in ascending order.

A convenient way to interpret the scores resulting from the modeling is by converting them into 200 equal-area ranks after sequencing in descending order. Fixed recognizable groups of ranks are then associated with pseudo-colors, as shown in the legend of Figure 3. The 200 ranks are displayed in wider groups for lower ranks of lesser concern, and in successively narrower groups for higher ranks of greater concern, e.g., 12.5%, 10%, 5%, 2.5%, 1.5%, and top 1%, of the area of the SA.

The illustration shows the prediction patterns from the five models, all overlaid with the distribution as black polygons of the 70 active landslide occurrences. The patterns represent the likelihood of future landslide occurrences in the SA.

Comparing them, we can observe similarities among higher-ranking groups in the northern part of the study area but strong differences are found in the southern part. Wider patches with high values are in Figure 3b,c, from the LI and LO models, respectively. Discontinuous patches are in Figure 3a,c,d, for the FZ, LR, and BP models. Altogether, there is a similarity between Figure 3a,d,e), and some similarity between Figure 3b,c. Note that we have to compare the same classes, fixed groups of equal-area ranks, in zones of concern. They are the zones with relatively higher ranks but located far from the known occurrences. In particular, we can focus on the top 10% ranks (90–99% and top 1%) corresponding, for instance, to higher elevations (1740–2100 and 2480–2630 m a.s.l.), high slope angles (37–57°), and very low permeability, in addition to the particular types of land use and lithology (see Table 1). How "good predictors" are the prediction patterns? Which one is the best or preferable? For answering these questions, we will have to consider the prediction-rate curves in Figure 4, described in the next step. They will provide a measure of predictability of future landslide occurrences, i.e., in our case, the "next seven."

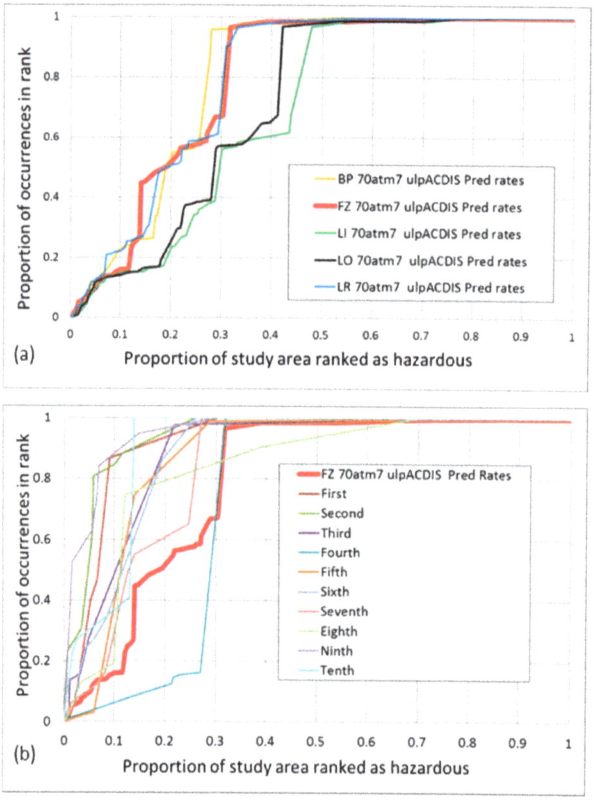

Figure 4. Prediction-rate curves are displayed generated by iterative cross validation using a sequential exclusion strategy of 7 out of 70 active landslides. (**a**) shows the aggregated iteration results for models FZ, LI, LO, LR, and BP, and (**b**) shows the individual curves for the 10 iterations of the FZ model and the aggregated curve as a thick red curve, corresponding to the red curve in (**a**).

4.2. Cross-Validate the Prediction Patterns

A cross-validation strategy was used to obtain and characterize the patterns as predictors. Iterative cross validation by the sequential exclusion of seven was tentatively selected out of the 70 active-landslide trigger zones in the DSP. It was computed to obtain the corresponding prediction-rate tables. The tables associate cumulative proportions of equal-area ranks of the prediction pattern with the corresponding cumulative proportions of validation occurrences. From the tables, cumulative curves are obtained, as shown in Figure 4. The illustration provides the prediction-rate curves from the iterative process using the five models. The horizontal axis shows the proportion of study area as cumulative equal-area ranks, each of 0.5% of the study area, i.e., ≈3230 pixels. The vertical axis shows the corresponding cumulative proportion of occurrence pixels in the cumulative class. The 697 occurrence pixels are distributed over the **70at** landslides so that the proportion of occurrence corresponds to the pixel numbers in each.

Note that the curves in Figure 4b represent proportions of seven occurrences, while the thick red curve represents the proportion of all 70 occurrences (curve **FZ_70atm7_ulpACDIS**, where m indicates minus).

The widespread distribution of the curves is indicative of the uncertainty affecting their aggregated thick red curve. What is predicted here is the likelihood of the "next" seven occurrences. In the iterations, they are assumed to be unknown. In all iterations, 63 occurrences are used to generate a prediction pattern that is then cross-validated as a predictor of the remaining seven occurrences. Note also that the proportions shown on the vertical axis are three orders of magnitude smaller than those on the horizontal axis (proportions of 697 pixels versus proportions of 646,091 pixels). As a reminder, the vertical axis was kept half the length of the horizontal axis.

We can observe that the prediction-rate curves are not particularly good. For instance, the thick red curve in Figure 4a shows that the top 10% ranks (0 to 0.1) contain about 14% of the occurrences for FZ, LI, and LO, and 22% for LR and BP; the top 20% ranks contain around 50% of the occurrences for FZ, LR, and BP, and about 26% for LI and LO; the top 30% ranks contain 96%, 91%, and 67% for BP, LR, and FZ, respectively, and about 57% for LI and LO.

These relative proportions represent the predictive capability of the modeling results: the proportion of predicted occurrences in the corresponding equal-area ranks. The shallow initial part of the curves is a sign of poor congruity of the setting of the landslide trigger areas used for cross validation. A good prediction pattern should provide an initially steep prediction-rate curve through cross validation in which most of the validation occurrences fall on higher ranks. The curves in Figure 4a show that the FZ, BP, and LR patterns of Figure 3 predict better than those for LI and LO. Figure 4b shows, in addition to the FZ prediction-rate curve identified with a thick red line (**FZ_70atm7_ulpACDIS**), the curves for each of the 10 iterations of the FZ model.

Note that a shallow initial curve is the one from the fourth iteration. Recall that in the iterations the cumulative proportions of occurrences in the diagram refers to the seven occurrences being cross-validated, while the proportions for the aggregated FZ curve refer to the 70 occurrences, validated into successive 10 groups of 7. What is being predicted here are the "next" 7 occurrences using the "previous" 63. This is the best approach when not having the time of the occurrences.

4.3. Interpret the Cross-Validated Results

The next step was to assemble sets of 10 prediction patterns to obtain via rank-based statistics the corresponding target and uncertainty patterns. The term "target" refers to what we are looking for as a validated prediction result. The term "uncertainty" refers to the stability of it. Indeed, the prediction pattern is the most informed prediction because it is using all information available, all the 70 active occurrences as DSP. However, we do not know its prediction capability. This is estimated by generating the target pattern via iterative cross validations using whichever strategy we can apply.

Figure 5 shows the uncertainty patterns obtained applying the five models to the database, the **70at** as DSP and **ulpACDIS** as ISPs. To obtain the target patterns, rank-based statistics was used to select for each pixel the corresponding median rank of the 10 prediction patterns generated by the iterative cross-validation process **70atm7**. The display of the target patterns is not provided here due to the extreme similarity with the corresponding prediction patterns in Figure 3 when using the same color legend. More informative are the uncertainty patterns in Figure 5, generated by ranking the ranges of ranks around the median ranks of the target pattern. The wider is the range the higher is the uncertainty of the corresponding target pattern (and consequently also of the prediction pattern).

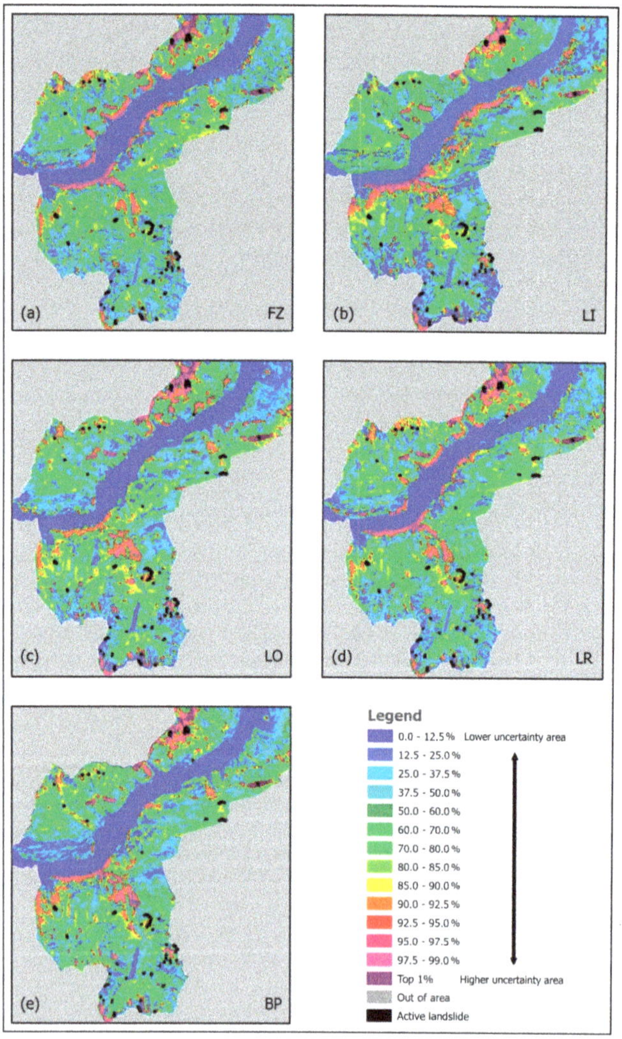

Figure 5. Uncertainty patterns, all using **70atm7_ulpACDIS** as cross-validation process, from the different models FZ, LI, LO, LR, and BP in (**a**–**e**), respectively. Black patches are the oversized trigger zones of the 70 active landslides. Colors in the legend indicate groups of ranks of % of SA in ascending order.

We can use the same color legend for the uncertainty ranks as for the prediction or target ranks. The uncertainty ranks relate to all target ranks from high to low. For uncertainty, however, the desirable ranks are the lower ones, indicating lower uncertainties. Obviously, if a high target rank corresponds to a high uncertainty rank, it is considered less credible than one corresponding to a low uncertainty rank. Figure 5 shows areas of high uncertainty, higher ranks, to the North in all the five uncertainty patterns. As to those along the valley edges, they are visible only in Figure 5a,b,d, for FZ, LI, and LR, respectively. Other areas of high uncertainty to the south are common to all the five patterns in the illustration.

At this point, it becomes important to study the relationships between target and corresponding uncertainty ranks with the 70 active landslide occurrences, **70at,** from the **70atm7_ulpACDIS** process of cross validation.

4.4. Obtain and Compare the Target and Uncertainty Patterns via Equal-Area Ranks

By cross-validating the target and uncertainty patterns with the 70 active landslides, **70atm7**, the relationship between target and uncertainty ranks were visualized, conveniently expressed in *1000 units (for instance, rank 900 corresponds to the top 10% equal-area rank). Figure 6 shows as an example the relationship between the target ranks, in descending order on the horizontal axis, and the respective uncertainty ranks, in ascending order on the vertical axis. It was obtained from the cross-validation process **FZ_70m7at_ulp_ACDIS.** In the illustration, the diagram shows that the 70 points are distributed so that the higher target ranks on the horizontal axis appear to correspond to relatively lower uncertainty ranks. It is worth noting that there are five encircled points corresponding to areas of relatively high uncertainty ranks and the lowest target ranks. They are occurrences that contribute to the shallow part of the prediction-rate curves in Figure 4. They could be considered outliers amongst the occurrences used as DSP.

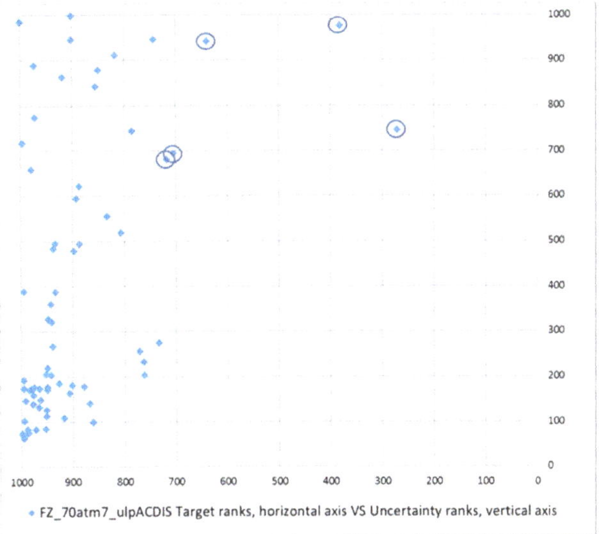

Figure 6. Target ranks in *1000 for the FZ model on the horizontal axis versus uncertainty ranks on the vertical axis, for the 70 active landslides, **70at**, in the Tirano South study area. The five encircled points have the lowest target ranks. See text for explanation.

Such a tendency for the patterns from the five models can also be visualized by sequencing the target ranks of the 70 occurrences in decreasing order, constructing histograms with pair of columns of target and corresponding uncertainty ranks. This is accomplished in Figure 7 that shows in blue the target ranks and in red the corresponding uncertainty ranks for each occurrence. For all models, the histograms express a preferential distribution of higher uncertainty ranks for lower target ranks.

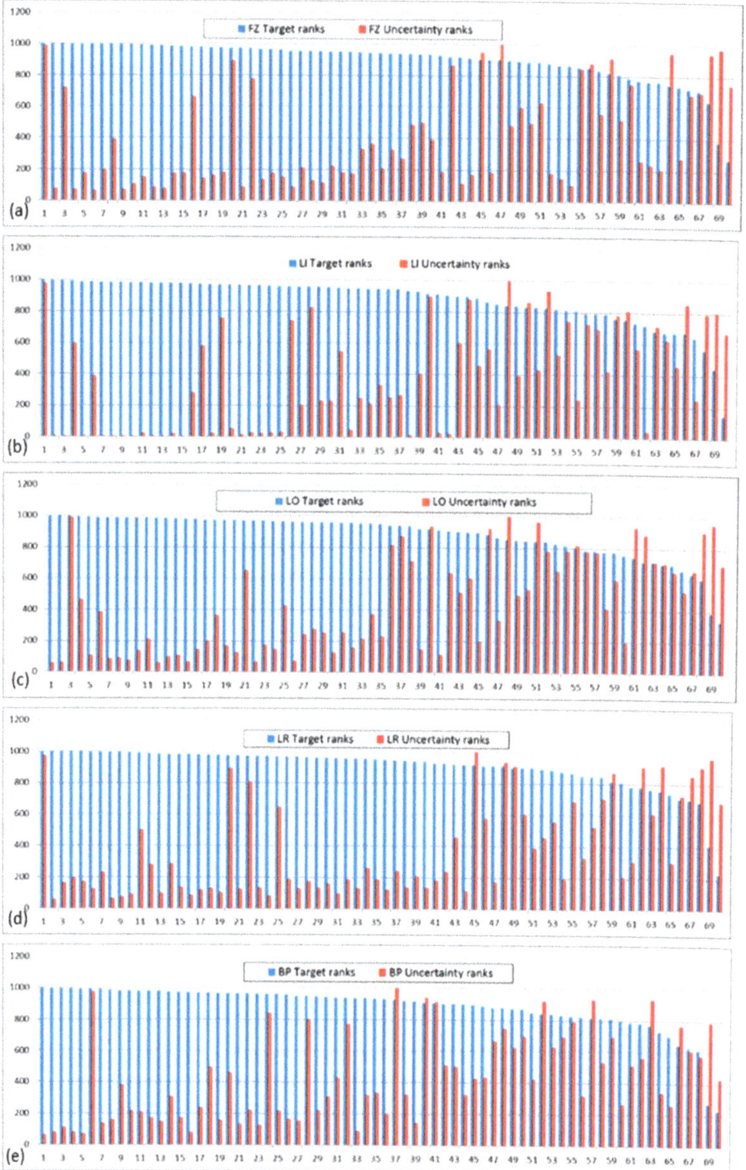

Figure 7. Histograms of decreasing target ranks (blue columns) and corresponding uncertainty ranks (red columns) for models FZ, LI, LO, LR, and BP from (**a**–**e**), respectively, using the cross-validation process **70atm7_ulpACDIS**.

We may wonder whether that tendency is visible also in the target patterns (or in the prediction patterns). This can be observed by generating combination patterns that relate uncertainty and target ranks by tentatively thresholding the uncertainty patterns. We have generated the 50% combination patterns for each target pattern, as shown in Figure 8. A threshold value was arbitrarily set at the lower 50% uncertainty ranks in the study area to select the corresponding target ranks (or alternatively, we could have used prediction ranks from the respective prediction patterns).

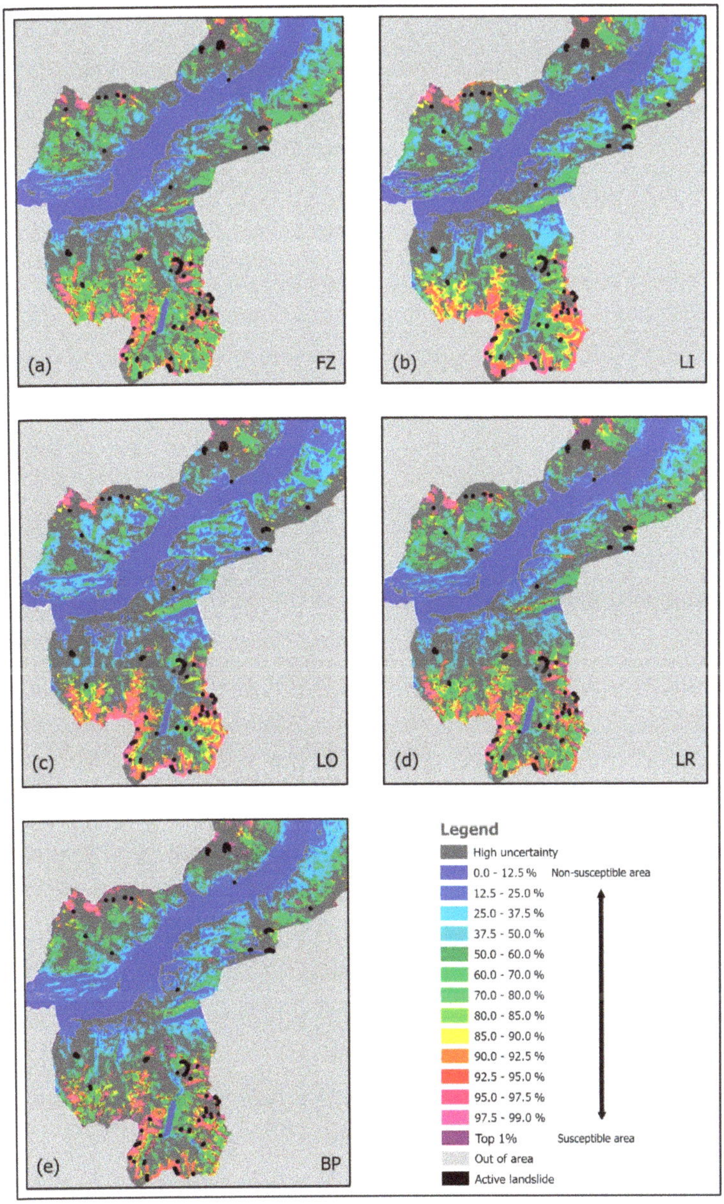

Figure 8. The 50% combination patterns, **70atm7**, using the different models FZ, LI, LO, LR, and BP in (**a–e**), respectively. Black patches are the oversized trigger zones of the 70 active landslides, **70at**. The dark-gray areas represent 50% of the study area with relatively higher uncertainty. The remaining 50% show the target pattern ranks for areas with lower uncertainty. Colors in the legend indicate groups of ranks of % of SA in ascending order.

By comparing the uncertainty pattern in Figure 5 with the prediction patterns in Figure 3, we can infer that the combination patterns in Figure 8 have retained high values in the southern part of the SA that shows low uncertainty (lower 50% uncertainty ranks). In the northeastern part, however, uncertainty is relatively high (upper 50% uncertainty ranks) so that the high target ranks are not visible in the combination patterns. What are then the characteristics of the 50% combination patterns? How is the uncertainty range threshold shaping them?

Let us consider some revealing details of the rank distribution in a small window of the 50% combination patterns, as shown in Figure 9. Seven landslide trigger areas are displayed as black contours in the illustration. Some are predicted as uncertain and fall on gray areas, in Figure 9b,c,e by LI, LO, and BP models. On the contrary, they have low uncertainty in Figure 9a,d by FZ and LR models. Topologically, similar dispersed patches are visible in Figure 9a,d,e, and more continuous ones in Figure 9b,c. The top 1% combination ranks, purple color, show the high variability of these clusters of pixels. These are the main characteristics of the patterns. They indicate weak robustness of the ranking and the uncertainty at the occurrence location.

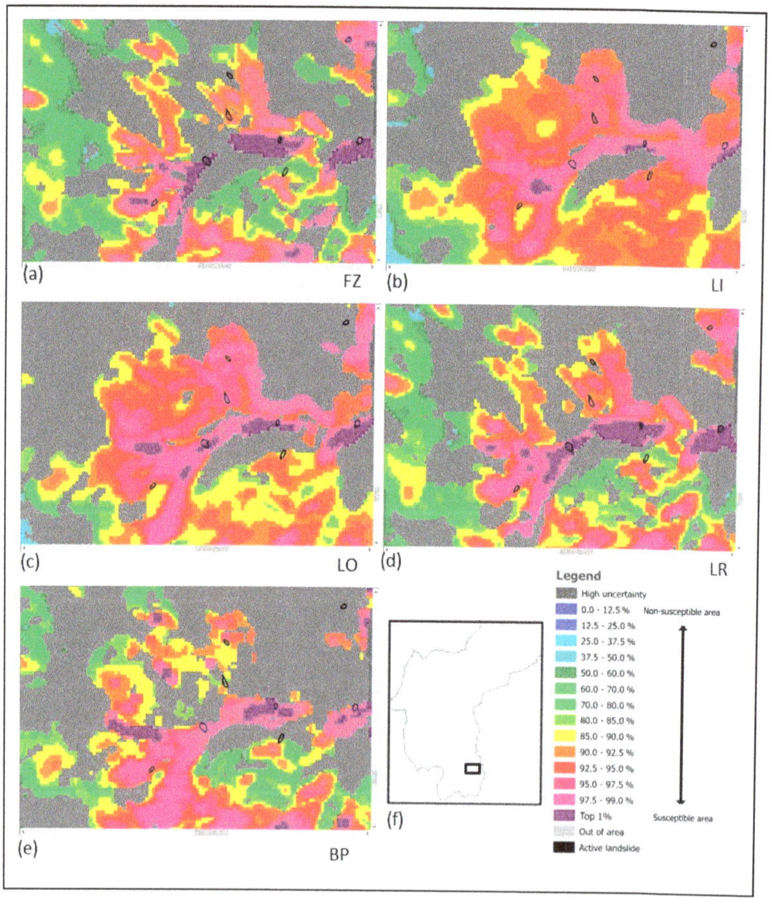

Figure 9. A particular small subarea is shown of 50% combination patterns with overlaid boundaries of the seven active landslides located in it. In (**a**,**b**) are the patterns from models FZ and LI; in (**c**–**e**) from models LO, LR and BP, respectively. In (**f**) is the location of the subarea. Colors in the legend indicate groups of ranks of % of SA in ascending order.

Next, we can derive a more general characterization of an entire 50% combination pattern by looking at low-uncertainty proportions or pixel numbers in the individual 200 ranks.

4.5. Analyze the Target, Uncertainty, and 50% Combination Pattern Relationships

Increasing uncertainty ranks with decreasing target ranks, as we have seen with the **70at** landslide trigger areas via the **70atm7** process in Figures 6 and 7, are also found in the target patterns. We can visualize their suspected higher uncertainty (hinted by the curves in Figure 4b) for lower prediction ranks. This is carried out by generating plots similar to the one in Figure 6 but with 646,091 points corresponding to the pixels in the SA, using pairs of target and uncertainty patterns. Instead, the following simpler visualization procedure was followed.

We have generated 200 equal-area ranks out of the prediction, target, and uncertainty ranks, each corresponding to 0.5% of the study area (≈3230 pixels). Being concerned mostly with the highest ranks, e.g., the higher 80%, we have displayed the numbers of pixels in each combination pattern that corresponded with the 50% lower uncertainty ranks, leaving out those with higher uncertainty ranks. We observed, therefore, the decrease in pixel number for each rank due to the elimination of higher uncertainty ranking pixels and detected which ranks were consequently more uncertain. These were the intermediate ranks. Figure 10 compares this decrease in the 50% combination patterns from the five models with the respective target ranks, i.e., the straight lines. In all cases, the intermediate ranks show higher uncertainty, i.e., higher "loss" of target pixels in the combination ranks, between rank 160th and rank 80th. The curves in Figure 10 have strongly variable configurations with one or two concavities and different higher pixel numbers in the vicinity of the 200th and the 40th ranks. Figure 10f contrasts the curves from models FZ and LO. Note the absence of a drop in pixel number for the top ranks in Figure 10e. It shows a sharper pick, as to be expected from the yellow prediction-rate curve from model BP in Figure 4a that appears to be less sensitive to the suspected outlier occurrences.

We may wonder whether the curves indicate some form of database or modeling signature. The curves in Figure 10 show how the combination patterns reach high uncertainties at intermediate ranks. The relative uncertainties increase from initial lows to reach one or two maximum values between rank 190th and rank 60th. Since this is observed for all the patterns from the five models, it appears as a database property. Observe the prediction-rate curves in Figure 4a and consider the curves in Figure 10 showing the number of target ranks corresponding to lower uncertainties close to the two extreme ranks displayed. Would we want to select the top 10% combination ranks, from 200th to 180th, as the acceptable part? Or alternatively, would we prefer the top 20% as more significant? The properties of uncertainty and combination patterns are still unknown and remain a research challenge.

4.6. Considerations on Prediction Patterns as Maps

What resembles a map, such as the display of a prediction pattern, is not necessarily a map but more a representation of information extraction through normalizations, conversions, assumptions, and integrations. At present, with what we know about a study area, we have to be satisfied by the higher part of the prediction-rate curve and by the corresponding combination pattern, perhaps fine-tuning the process. Otherwise, additional information must be used to choose a satisfactory part of the prediction pattern. Regarding the prediction patterns, would we want to discover instances of unmapped landslide trigger zones in the higher 10% ranked areas away from the known ones? Is the 50% combination pattern helpful in mapping more of what we know or that we do not know? Should we wait for the next seven landslides to see where they will appear? Where should we concentrate on more detailed mapping? How much of the study area should we consider unfit for particular land uses or developments? Would we opt for using cost/benefit or using safety criteria? Providing answers for decision making and subsequent risk analysis is the logical function of the prediction patterns of landslide susceptibility, as we have discussed.

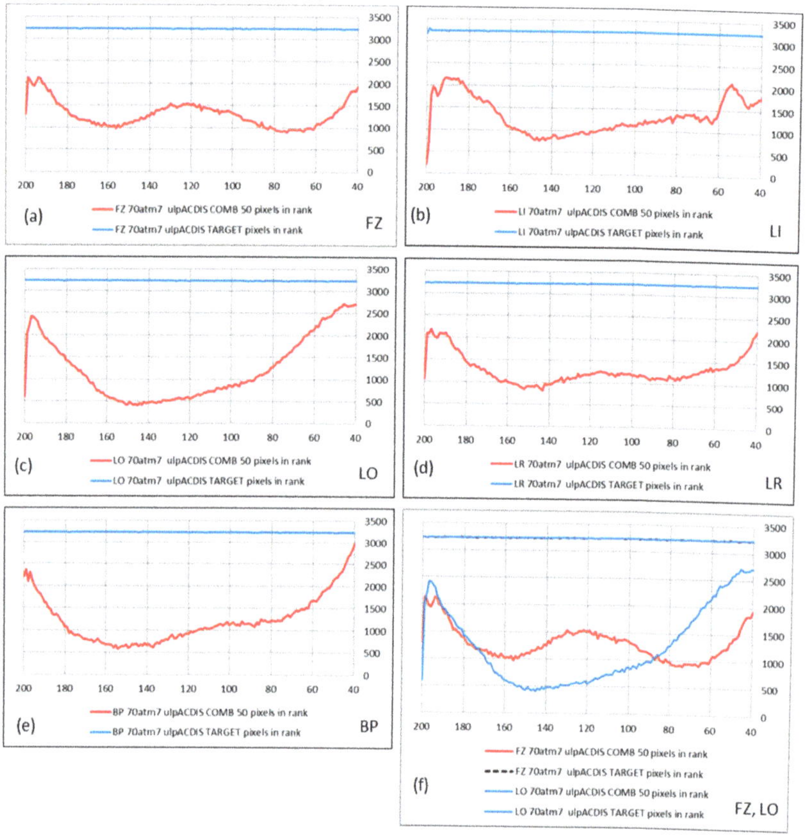

Figure 10. Comparison of pixel numbers in equal-area target ranks with the pixel numbers in the corresponding low uncertainty target ranks of the 50% combination pattern. Models FZ, LI, LO, LR, and BP are in (**a–e**). In (**f**) is the overlay of curves for models FZ and LO. The target equal-area ranks contain about 3230 pixels. Only the higher 160 ranks are shown here.

5. Concluding Remarks

Favorability function modeling was applied to the Tirano South database through five different models of spatial relationships. All the prediction patterns obtained were represented as relative ranks, including their derived target and uncertainty patterns. The 50% lower uncertainty ranks were tentatively used to extract the corresponding target ranks as 50% combination patterns. In them, for all models, the proportion of less uncertain pixels in the rank represents the level of confidence in the prediction. Possibly, the top 10% combination pixels are a significant part of the prediction pattern. We are trying to find answers to questions such as the following: What are the consequences of higher uncertainty for intermediate ranks in combination patterns? How are these diagrams providing a measure of the quality of the prediction? Are they characteristic of the modeling because of the model or the data? How much of the prediction pattern is reliable? How to evaluate the higher ranks in the prediction patterns? Should we select the top 10% or 20% combination ranks? What would our cost/benefit considerations and choices be? All we have generated from the modeling are "relative" integrated equal-area ranks that must be interpreted as susceptibility to land-sliding. The model preference needs to be a function of the interpretation of prediction patterns and their representation. In our case, FZ, BP, and LR are equally satisfactory but less so for LI and LO, as to the pattern of predictive performance. However, this may be just one of the criteria that might be used.

A five-step procedure is proposed for modeling prediction and uncertainty. The uncertainty ranks are considered important to properly select the susceptible part of the study area, i.e., susceptible to the "next seven" future" landslide occurrences. While various other strategies have also been applied for iterative cross validation to obtain slightly different prediction patterns, the sequential exclusion of seven occurrences appears to do justice to the properties of the database. The procedure is proposed as critical in spatial prediction modeling. Independently of the models used, a necessary research issue is in the interpretation of the uncertainty associated with prediction patterns.

We have described our analysis and modeling results to indicate a way to predict and the assumptions implied. Obviously, we suspect that our experiments on this particular database have a more general significance beyond the specific study area or the five mathematical models used. These considerations, we hope, will be useful to researchers and users of susceptibility maps. This contribution does not provide a solution but poses questions whose answers point at possible solutions.

Author Contributions: Investigation, A.G.F. and A.P. All authors have read and agreed to the published version of the manuscript.

Funding: This contribution was initially and partly supported by the European Commission Project "Mountain Risks: from Prediction to Management and Governance" (MRTN-CT-2006-035978, 2007–2010), Mountainrisk (2007).

Acknowledgments: The authors are grateful to four anonymous reviewers and to Jan Blahut (The Czech Academy of Sciences) for helping to improve this manuscript.

Conflicts of Interest: The authors declare no conflict of interest.

References

1. Schuster, R.L.; Highland, L.M. *Socioeconomic and Environmental Impacts of Landslides in the Western Hemisphere*; USGS Open-File Report 01-276; US Department of the Interior, US Geological Survey: Denver, CO, USA, 2001; 48p.
2. Kojima, H.; Chung, C.-J.; van Westen, C.J. Strategy on the landslide type analysis based on the expert knowledge and quantitative prediction model. *Int. Arch. Photogramm. Remote Sens.* **2000**, *33 Pt 7*, 701–708.
3. Shano, L.; Raghuvanshi, T.K.; Meten, M. Landslide susceptibility evaluation and hazard zonation techniques—A review. *Geoenviron. Disasters* **2020**, *7*, 1–19. [CrossRef]
4. Fabbri, A.G.; Chung, C.-J.; Cendrero, A.; Remondo, J. Is prediction of future landslide possible with a GIS? *Nat. Hazards* **2003**, *30*, 487–499. [CrossRef]
5. Chung, C.-J.; Fabbri, A.G. Representation of geoscience data for information integration. *Nat. Resour. Res.* **1993**, *2*, 122–139. [CrossRef]
6. Fabbri, A.G.; Cavallin, A.; Masetti, M.; Poli, S.; Sterlacchini, S.; Chung, C.-J. Spatial Uncertainty of Groundwater-vulnerability Predictions Assessed by a Cross-validation Strategy: An Application to Nitrate Concentrations in the Province of Milan, Northern Italy. In *Risk Analysis VII and Brownfields V*; Brebbia, C.A., Ed.; Wit Press: Southampton, UK, 2010; Volume 43, pp. 497–514. [CrossRef]
7. Chung, C.F.; Fabbri, A.G. Prediction Models for Landslide Hazard Using Fuzzy Set Approach. In *Geomorphology and Environ-Mental Impact Assessment*; Marchetti, M., Rivas, V., Eds.; Balkema: Rotterdam, The Netherlands, 2001; pp. 31–47.
8. Chung, C.F. *Computer Program for the Logistic Model to Estimate the Probability of Occurrence of Discrete Events*; GSC Paper 78-11; Geological Survey of Canada: Ottawa, ON, Canada, 1978; 23p.
9. Davis, J.C.; Chung, C.-J.; Ohlmacher, G.C. Two models for evaluating landslide hazards. *Comput. Geosci.* **2006**, *32*, 1120–1127. [CrossRef]
10. Chung, C.-J.F.; Fabbri, A.G. Validation of spatial prediction models for landslide hazard mapping. *Nat. Hazards* **2003**, *30*, 451–472. [CrossRef]
11. Chung, C.-J.; Fabbri, A.G. Systematic Procedures of Landslide Hazard Mapping for Risk Assessment Using Spatial Prediction Models. In *Landslide Hazard and Risk*; Glade, T., Anderson, M.G., Crozier, M.J., Eds.; Wiley: New York, NY, USA, 2004; pp. 139–174.
12. Chung, C.-J. Using likelihood ratio functions for modeling the conditional probability of occurrence of future landslides for risk assessment. *Comput. Geosci.* **2006**, *32*, 1052–1068. [CrossRef]

13. Chung, C.F.; Fabbri, A.G. Probabilistic prediction models for landslide hazard mapping. *Photogramm. Eng. Remote Sens.* **1999**, *65*, 1389–1399.
14. Chung, C.-J.F.; Moon, W.M. Combination rules of spatial geoscience data for mineral exploration. *Geoinformatics* **1991**, *2*, 159–169. [CrossRef]
15. Chung, C.-J.; Fabbri, A.G. Mineral occurrence target mapping: A general iterative strategy in prediction modeling for mineral exploration. *Nat. Resour. Res.* **2019**, *29*, 115–134. [CrossRef]
16. Studio Statistico Dell'area di Tirano Sud (Predisposition of Slopes to Mass Movements through Prediction-Rate Curves. Statistical Study of the Tirano South Area). Master of Sciences Thesis, University of Milano-Bicocca, Milan, Italy, 2008. (in Italian).
17. Poli, S.; Sterlacchini, S. Landslide representation strategies in susceptibility studies using weights-of-evidence modeling technique. *Nat. Resour. Res.* **2007**, *16*, 121–134. [CrossRef]
18. Fabbri, A.G.; Chung, C.F. On blind tests and spatial prediction models. *Nat. Resour. Res.* **2008**, *17*, 107–118. [CrossRef]
19. Fabbri, A.G.; Poli, S.; Sterlacchini, S.; Cavallin, A.; Chung, C.-J. Uncertainty of Class Membership in Spatial Prediction Modeling: Follow-up Study to an Application to Complex Landslides. In Proceedings of the International Association for Mathematical Geosciences—IAMG—Mathematical Geosciences at the Crossroad of Theory and Pratice, Salzburg, Austria, 5–9 September 2011. 12p. [CrossRef]
20. Fabbri, A.G.; Poli, S.; Patera, A.; Chung, C.-J. Estimation of Information Loss When Masking Conditional Dependence and Cate-gorizing Continuous Data: Further Experiments on a Database for Spatial Prediction Modeling in Northern Italy. In *Mathematics of Planet Earth*; Pardo-Igúzquiza, E., Guardiola-Albert, C., Heredia, J., Moreno-Merino, L., Durán, J.J., Vargas-Guzmán, J.A., Eds.; Springer: Heidelberg, Germany, 2014; pp. 291–294.
21. Fabbri, A.G.; Mauriello, M.R.; Chung, C.J. Using Blind-Tests for Database Evaluation in the Prediction of Rotational-Translational Landslides in the Tirano North Area, Northern Italy. In *Risk Analysis VIII*; Brebbia, C.A., Popov, V., Eds.; WIT Press: Southampton, UK, 2012; Volume 44, pp. 263–280. [CrossRef]
22. Blahut, J.; Horton, P.; Sterlacchini, S.; Jaboyedoff, M. Debris flow hazard modelling on medium scale: Valtellina di Tirano, Italy. *Nat. Hazard Earth Syst. Sci.* **2010**, *10*, 2379–2390. [CrossRef]
23. Sterlacchini, S.; Ballabio, C.; Blahut, J.; Masetti, M.; Sorichetta, A. Spatial agreement of predicted patterns in landslide susceptibility maps. *Geomorphology* **2011**, *125*, 51–61. [CrossRef]
24. Blahut, J.; Van Westen, C.J.; Sterlacchini, S. Analysis of landslide inventories for accurate prediction of debris-flow source areas. *Geomorphology* **2010**, *119*, 36–51. [CrossRef]
25. Blahut, J.; Poretti, I.; De Amicis, M.G.M.; Sterlacchini, S. Database of geo-hydrological disasters for civil protection purposes. *Nat. Hazards* **2011**, *60*, 1065–1083. [CrossRef]
26. Fabbri, A.G.; Chung, C.-J. How credible is my hazard map? Dissecting a prediction pattern of landslide susceptibility. *WIT Trans. Eng. Sci.* **2018**, *121*, 3–19. [CrossRef]
27. Fabbri, A.G.; Cavallin, A.; Patera, A.; Sangalli, L.; Chung, C.-J. Comparing Patterns of Spatial Relationships for Susceptibility Prediction of Landslide Occurrences. In *Advancing Culture of Living with Landslides*; Advances in Landslide Science Set, 2; Mikoš, M., Tiwari, B., Yin, Y., Sassa, K., Eds.; Springer International Publishing: Cham, Switzerland, 2017; Volume 2, pp. 1135–1144. [CrossRef]

applied sciences

MDPI

Article

Past, Present and Future Monitoring at the Vallcebre Landslide (Eastern Pyrenees, Spain)

Josep A. Gili [1,*], Jose Moya [1], Jordi Corominas [1], Michele Crosetto [2] and Oriol Monserrat [2]

1 Department of Civil and Environmental Engineering, Division of Geotechnical Engineering & Geosciences, Technical University Catalonia (UPC-BarcelonaTech), 08034 Barcelona, Spain; jose.moya@upc.edu (J.M.); jordi.corominas@upc.edu (J.C.)
2 Remote Sensing Department, Geomatics Division, Centre Tecnològic Telecomunicacions Catalunya (CTTC/CERCA), Castelldefels, 08860 Barcelona, Spain; michele.crosetto@cttc.es (M.C.); oriol.monserrat@cttc.es (O.M.)
* Correspondence: j.gili@upc.edu

Featured Application: Monitoring of very slow landslides with Classical and Novel Techniques.

Abstract: Works carried out to monitor the displacements of the Vallcebre landslide (Pyrenees range, NE of Spain) since 1987 are presented. The landslide, which extends over an area of about 0.8 km^2 and affects more than 20×10^6 m^3, has experienced displacements of up to one meter per year in some points and periods. It has been periodically monitored since 1987, using a wide range of surface and in-hole techniques: triangulation with theodolite, Terrestrial Photogrammetry, Electronic Distance Measurement, GNSS-GPS, inclinometers, wire extensometers, piezometers, DInSAR (satellite) and GBSAR (terrestrial). The results obtained using new techniques are compared with those obtained with GNSS-GPS and a wire extensometer, and checked against fixed stable points. From this comparison, we conclude that even though wire extensometers and inclinometers may have the highest precision, in practice, all systems play potentially valuable roles in providing meaningful data for monitoring at different study stages. In the near future, we envisage the installation of a Distributed Fiber Optic array to monitor the risk with a certain space and time continuity. After the evaluation of the precision and advantages of the different methods, the complementary use of some of them is strongly recommended.

Keywords: monitoring; landslides; photogrammetry; global positioning system; in-hole wire extensometer; DInSAR; GBSAR

Citation: Gili, J.A.; Moya, J.; Corominas, J.; Crosetto, M.; Monserrat, O. Past, Present and Future Monitoring at the Vallcebre Landslide (Eastern Pyrenees, Spain). *Appl. Sci.* **2021**, *11*, 571. https://doi.org/10.3390/app11020571

Received: 23 November 2020
Accepted: 4 January 2021
Published: 8 January 2021

1. Introduction and Site Description

The Vallcebre landslide is located in the Eastern Pyrenees, approximately 125 km north of Barcelona, Spain (Figure 1). Its situation, geological context and a complete geomorphological description can be found in [1,2]. The present paper is an expanded and updated version of a previous work [3] where we described the landslide as a translational slide with a stair-shape profile. As in most landslides, its structure and behavior are not simple. The landslide is 1200 m long and 600 m wide, involving an area of 0.8 km^2, which shows superficial cracking and distinct ground displacements (Figure 2). The mobilized material consists of a set of shale, gypsum and claystone layers gliding over a thick limestone bed. A geological cross-section is presented later in Section 2.4. The landslide consists of three main slide units, which show a decreasing thickness towards the landslide toe. Each unit is formed by a gentle slope surface bounded in its uphill edge by a scarp of a few tens of meters high. At the base of each scarp, an extension area develops in the form of a crack system and a graben [1]. This fact indicates that the lower units move more rapidly than upper ones, which has been repeatedly confirmed by an installed monitoring network [1,2].

Figure 1. Location of Vallcebre valley (red arrow) within the Pyrenees range, NE of Spain.

Figure 2. General view of the Vallcebre landslide from W. The unstable area (yellow contour) is about 0.8 km^2, has an average slope of only 17° (green line), and is sliding as a translational slide. The white arrows are the average slide directions of the upper, middle and lower units, which are separated by scarp zones (dashed yellow lines).

The average slopes of both the topography and the basal shear surface in the middle and lower unit are gentle, i.e., in the range between 6 and 10°. The basal shear surface depth is 42 m in the middle unit and 15 m in the lower one, being quite planar for most of the sliding surface.

Complete information on the geology and geotechnics of Vallcebre, and a stability analysis, are beyond the scope of this article, but can be found in [1–8].

Figure 3 shows a geomorphological sketch of the landslide and the location of the monitored points and boreholes. The most active area is the lower unit, whose toe is being eroded continuously by the Vallcebre torrent. Consequently, most of the monitored points were located there.

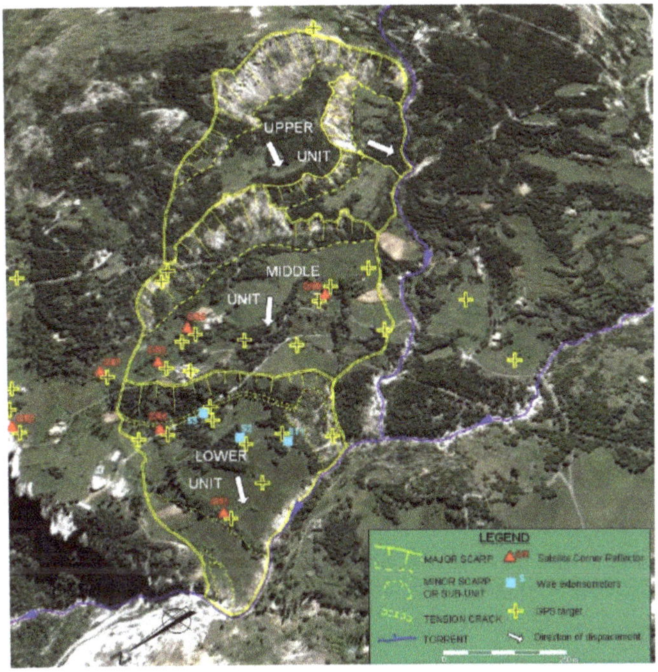

Figure 3. Geomorphological scheme of the Vallcebre landslide superimposed over a vertical aerial image. Several GPS benchmarks, corner reflectors and borehole extensometers have been highlighted with symbols.

The measurement of displacements is very often the simplest way to observe the evolution of a landslide and to analyze either the kinematics of the movement, the response to the triggering conditions (e.g., rainfall) or the efficiency of corrective measures.

A variety of measuring techniques have been developed to track the movements of unstable areas (see i.e., [9,10]). This second work, also known as the 'Red Book' of Geotechnical Instrumentation and Monitoring, is due to John Dunnicliff. Several of these methods have been used in Vallcebre since 1987, beginning with "classical" surveying and photogrammetry, and using Global Positioning Systems (GPS/GNSS) from 1995. In 1996, this site was included in the framework of the NEWTECH Project, funded by the European Commission. Between July 1996 and March 1997, 14 boreholes were drilled in the slope and equipped with inclinometers, wire extensometers and open standpipe piezometers (Figure 3).

In 2004–2005, multipoint piezometers were installed in three additional boreholes. Later, the Vallcebre landslide was included in an EU-FP7-funded research project known as SAFELAND. In 2007, seven artificial radar corner reflectors (CR) were installed to test the DInSAR monitoring capabilities. In 2010–2011, still within the SAFELAND project, the lower part of the landslide was monitored using a Ground-Based SAR (GBSAR). In this sense, the Vallcebre landslide can be considered to be a real-scale laboratory where the performance of different monitoring techniques can be assessed and compared. For instance, Laser Scanning (both Terrestrial or Aerial) have been disregarded because the ground displacements, mainly parallel to the topography, would be difficult to detect; also the vegetation would mask the results.

In the paper, Section 2 describes the measuring systems. In Section 3, some results are presented and compared in terms of precision and repeatability. Finally, some conclusions are outlined regarding the advantages and drawbacks of the systems tested in Vallcebre. The aforementioned comparison was made in terms of cost, ease of use, precision and continuity of the results.

2. Monitoring Methods Used in Vallcebre and Sample Results

A summary of the monitoring systems used consecutively from 1987 to the present in Vallcebre is given here. More details can be found in [1,4–7].

2.1. *Terrestrial Photogrammetry*

The first monitoring network established on the landslide was based on terrestrial (or "close-range") photogrammetry. A total of seven campaigns were performed at the landslide foot between 1987 and 1992, covering only a small area of about 100 × 50 m (Figure 4). Stereopairs were taken with a Wild P32 metric camera (Figure 4). Each campaign included three photograms that produced two photogrammetric models.

Figure 4. Left: example of a 6 × 8 cm glass plate photogram of landslide toe taken from the Vallcebre Torrent. **Right**: Wild P32 camera for terrestrial photogrammetry over a theodolite, which was used for the landslide toe monitoring between 1987 and 1992.

The results of each survey were the change in the coordinates of the main points (displacements). The precision ranges between 1 cm in well-defined points (pointing targets for instance) and 10 cm (rock blocks, trees and so on). Maps with contour lines were also produced (Figure 5). The interpretation of this kind of maps is not simple: the variations of the terrain surface are caused by the landslide movement, but also by the soil erosion during rainfalls and the eroding effect of the Vallcebre Torrent at the base of the toe. Although several blocks exhibit displacement of up to 8 m, the variation of the contour lines is very moderate as a whole (Figure 5).

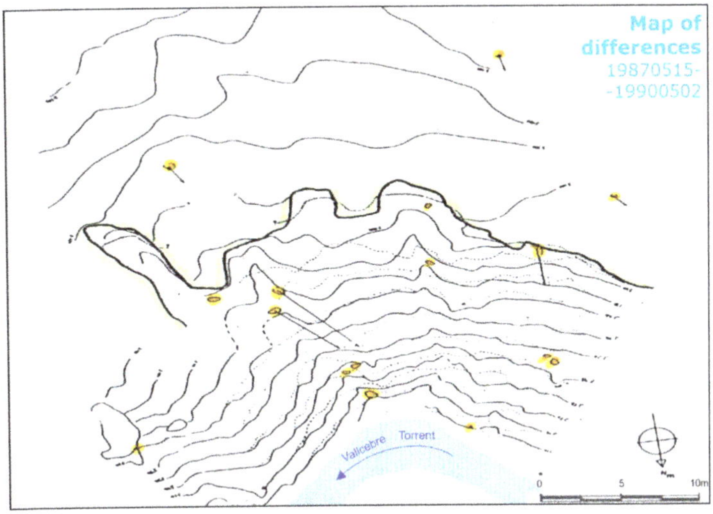

Figure 5. Terrestrial photogrammetry sample result: Map of differences between two campaigns (one-meter contour-line interval) at the landslide toe (Figure 4).

At that time, the use of terrestrial photogrammetry in landslides was not straightforward, because of the difficulty of creating a proper setup with an adequate view over the hill slope. Moreover, specialized and costly equipment had to be available for a precise stereocompilation. Although out of the scope of this paper, it is worth to say that the photogrammetry processing has evolved in the last decade towards automation, ease of use and low cost, thanks to the so-called Structure from Motion (SfM) concept [11–13]. In parallel, the use of drones (or UAS) as aerial photogrammetry platforms permits to overcome the point-of-view issues during acquisition [12–17] In this way, photogrammetry is once again a powerful landslide monitoring technique, providing information that is almost continuous in space, but discontinuous in time.

2.2. Triangulation and Electronic Distance Measurement

From 1987 to 1995, geodetic measurements with theodolite and EDM (Electronic Distance Measurement) were carried out. Between 1987 and 1992, up to 17 points were triangulated in the toe of the lower unit, mainly for orientation of the photogrammetric models (Figure 6a). In the period 1988–1994, three additional points in the middle unit were monitored with single distance variation measurements (Figure 6b). During 1994 and 1995, a "triangulateration" from new base benchmarks E1 and E2 (Figure 6c) was extended to 16 points spread out through the whole landslide. The angle measurements were carried out with a Wild T2 theodolite (Wild, Heerbrugg, Switzerland, 1975), and the EDM with a Wild DIOR 3002S (Wild, Heerbrugg, Switzerland, 1988).

The rate of movement measured during the period 1987–1995 was strongly dependent on the rainfall and the target position within the landslide. Rates up to 4 m per year were observed at certain points near the toe in the rainy years, while almost no displacement occurred during periods of drought (Figure 7). In the middle landslide unit (Figure 3), the rate of displacement was significantly smaller, in the range of 1 to 30 cm/year.

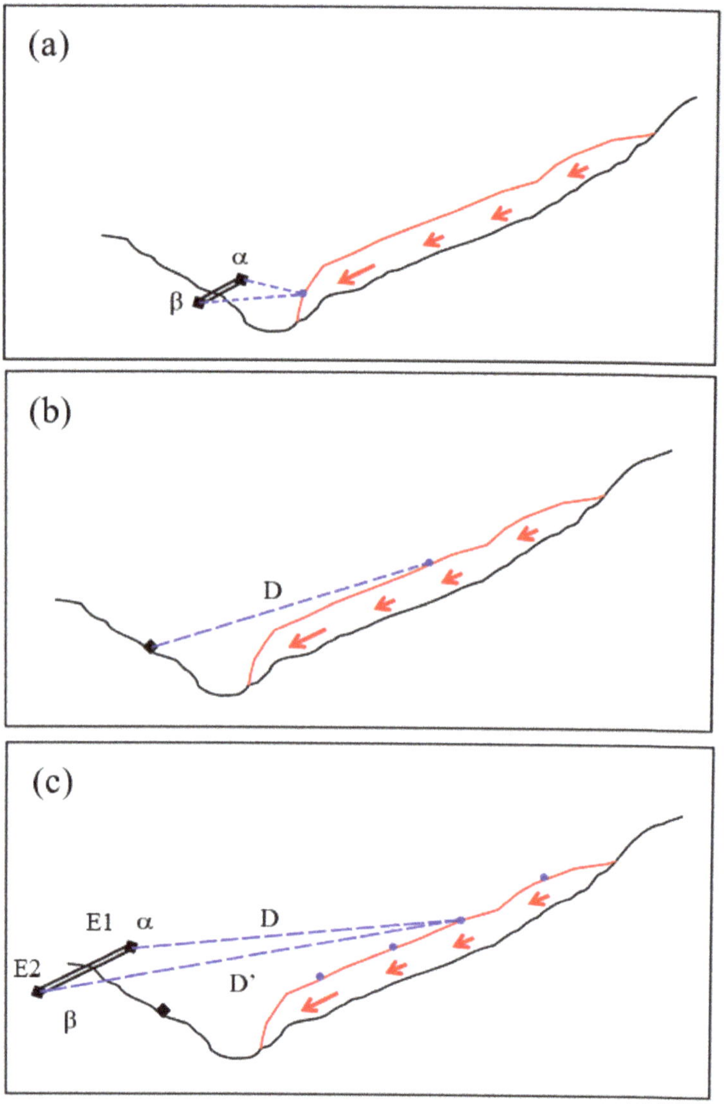

Figure 6. Schemes for the triangulation (**a**), single distance variation (**b**) and triangulateration (**c**) methods as used in the Vallcebre landslide from 1987 to 1995.

In terms of the precision of the observations, the EDM measurements proved to be more reliable (typically 1 cm) than angle measurements with theodolites (around 4 cm at typical distances), at least with the setup, equipment and sighting distances in use in Vallcebre. This is due to the fact that the precise determination of angles needed greater experience and better environmental conditions (no mist or smog in the line of sight, proper target illumination, lack of air vibration due to strong insolation and so on), often not completely available in mountain areas at the time when the measurements have to be taken.

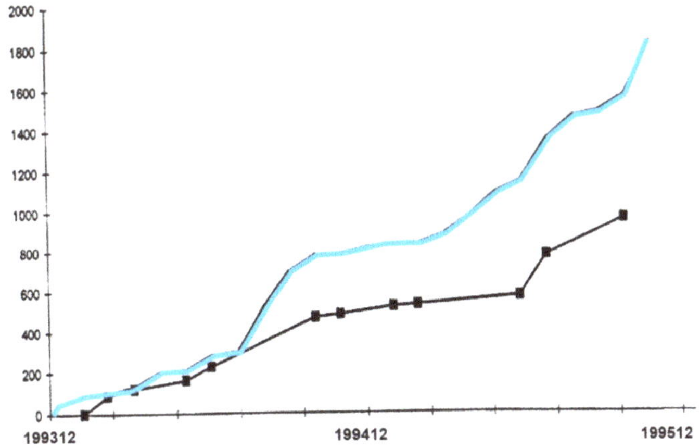

Figure 7. First tentative correlation between the displacement of a point in the lower unit (mm, black line) and the cumulative rainfall (mm, blue line), during 1994–1995.

2.3. *The GPS-GNSS Surveys*

Following the discussion on the precision of the observations, the expected errors for the Global Positioning System (1 to 2 cm, depending on the method in use [5]) were lower when compared to theodolite and EDM errors. Additionally, the GPS precision was more balanced in the three axes [5]. These facts led us to apply GPS techniques to perform systematic monitoring of the Vallcebre landslide. In December 1995, a complete double survey (EDM and GPS) was carried out in order to link the measurements taken with classical methods with the first GPS campaign. The equipment used then was a Trimble 4000 SSi model with two dual frequency receivers (Trimble, Sunnyvale, CA, USA, 1995, Figure 8). Currently, we are using two Topcon HiperPro dual constellation receivers (Topcon Positioning Systems, Livermore, CA, USA, 2005).

Figure 8. Examples of Vallcebre monitoring points with different antenna setup: telescopic pole (**a**); pole with mini-tripod (**b**); ground-plane antenna over a tripod, centered with an optical plummet over an inclinometer borehole (**c**).

Most of the old targets were recovered with minor modifications. As new points have been added since 1996, the monitoring network has now around 50 points (Figure 3), consisting on engravings in rock blocks outcropping in the hillside, steel rods, stakes, and the top of the casing of the inclinometric boreholes (Figure 8); the radar Corner Reflectors vertexes were incorporated to the GPS network as of 2006. There are seven additional points on the limestone around the sliding zone. These were the fixed points used to check the GPS accuracy. This network allowed both the measurement of displacements and the comparison with movements obtained with the borehole equipment (inclinometers and wire extensometers) and the Radar measurements.

Although the Fast-Static GPS method (more precise and robust) was partly applied in the first years, RTK (Real Time Kinematic) is currently in use for productivity reasons [5]. Currently, we can measure the entire network within a single day, travelling across the slope by car and on foot.

Fourteen GPS campaigns were carried out from December 1995 to February 1998, one survey every 2 months approximately (Figure 9a). Later on, the campaigns were continued on a yearly basis (Figure 9b), 36 GPS surveys until now. Average displacement rates derived from GPS campaigns show a sustained regime of velocities: the annual displacement depends on the precipitation of the year (total value and peak periods). More details about the GPS and special considerations for its application to landslide monitoring can be found in [5,18,19].

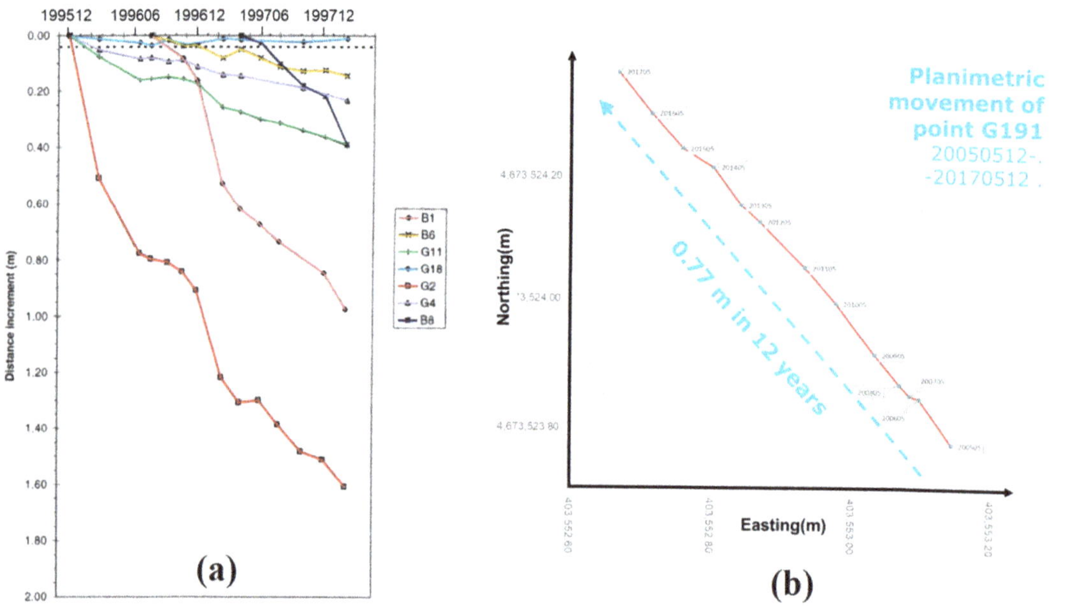

Figure 9. Examples of results of the GPS monitoring (**a**) planimetric displacements from the initial position in the period 1995–1998 (**b**) planimetric movement of one point during the yearly campaigns.

2.4. In-Hole (or Geotechnical) Instrumentation

Between July 1996 and March 1997, 14 boreholes were drilled in the slope for reconnaissance; 3 more were drilled in 2005 (Figure 10). Some of them were equipped with inclinometers, wire extensometers and open standpipe piezometers. Inclinometers and piezometers were standard devices. As an example, Figure 11 shows one inclinometer head and its typical record, where the slip surface is clearly marked. Measurements were made every 2–3 weeks until the casing deformation prevented the safe and proper sliding of the probe.

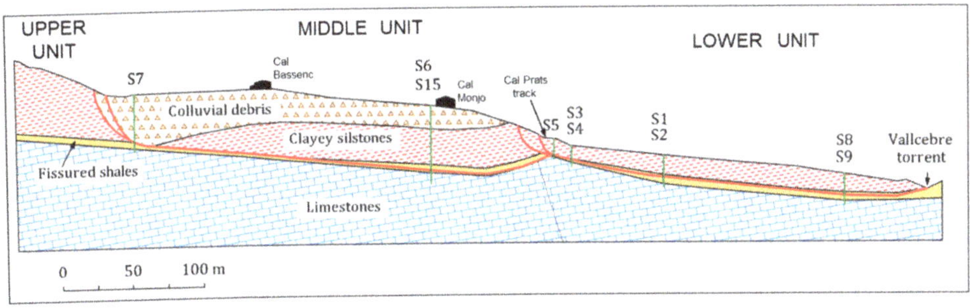

Figure 10. NE-SW geological cross-section of the Vallcebre landslide according to [1,2]. Some of the 17 boreholes drilled are shown, which were equipped with inclinometers, wire extensometers (for shear displacement) and vibrating wire piezometers.

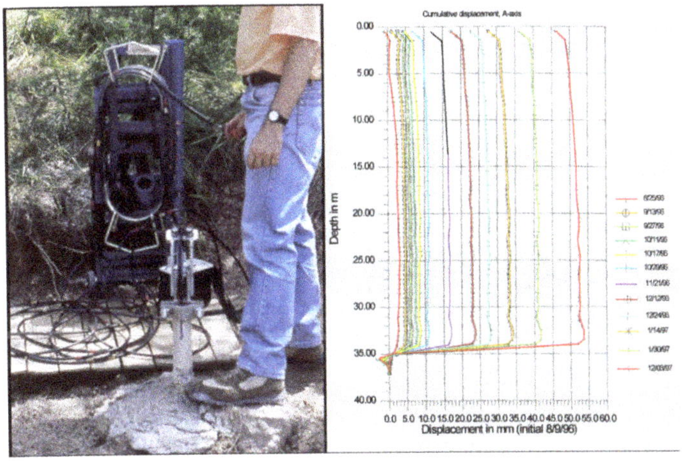

Figure 11. Left: view of the head of a borehole during the measurement with the inclinometer. **Right**: graph with the successive displacement profiles from August 1996 to December 1997. The main slip surface appeared to be 34 m deep. The pretty constant deformation profile corresponds with a translational slide.

On the other hand, the extensometers were wire-type, specially built following a design described in [20]. They consist of a protected steel wire anchored to the bedrock, below the slip surface, inside a piezometric pipe (Figure 12). After some computation, the wire displacement at the pulley can be related with the horizontal displacement of the landslide (Figure 13). We can quote two major advantages of this device. Firstly, with a potentiometer, it allows the continuous recording of the displacement, particularly necessary to collect information during the concentrated rainfall periods, characteristic of the Mediterranean climates (Figure 14). Secondly, the wire extensometer works properly with big landslide displacements, much larger than the 20–30 cm that would typically break the inclinometric pipe; in fact, in 2012, 16 years after their installation, 3 wire extensometers were still working, with a cumulated total wire displacement of up to 6 m. Full details of the wire extensometer can be found in [4].

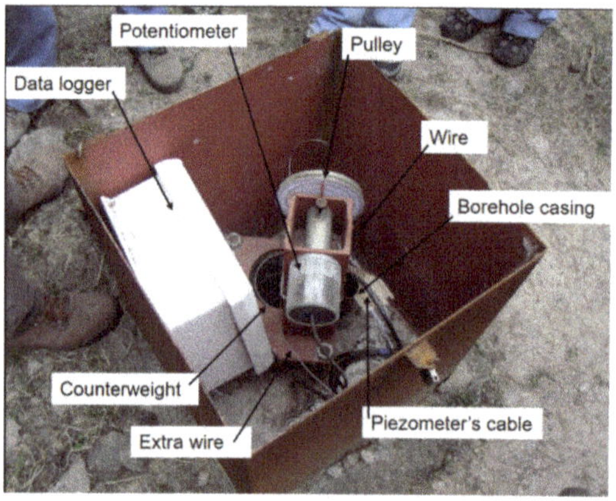

Figure 12. Head of one of the borehole wire extensometers installed at Vallcebre. The cable for the piezometer (black) can also be appreciated.

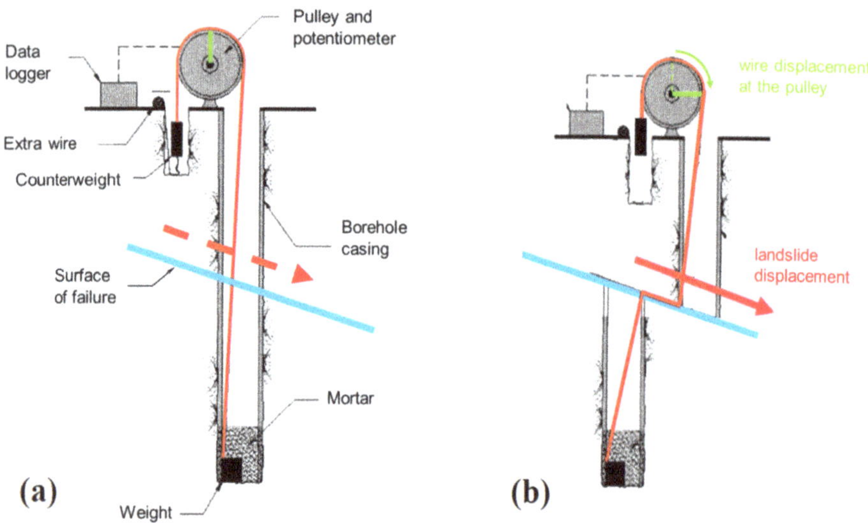

Figure 13. Sketch of the borehole wire extensometer. (**a**) just after installation; (**b**) after some shear displacement of the landslide (idealized). The relationship between the wire displacement at the pulley and the horizontal displacement of the landslide depends on the width of the shear zone and the borehole diameter [4]. The smaller the borehole diameter, the faster the wire response.

2.5. DInSAR Monitoring Using Corner Reflectors

The Space Borne SAR monitoring of landslides has been reported in [21–23] and elsewhere. When applying DInSAR to landslide monitoring in forested areas, a condition that is difficult to fulfil is that a sufficient number of targets within the area of interest remain coherent during the observation period. One way to overcome this limitation is the deployment of artificial corner reflectors (CRs) that ensure coherent, high-quality DInSAR estimates. Using CRs, however, demands additional resources; it also prevents historical deformation studies based on archive SAR imagery.

After a detailed discussion of the Vallcebre site suitability for DInSAR with CRs, [6] describe the installation, in 2006, of 7 metallic trihedral CRs (Figure 15) to be observed with the ENVISAT (C-band, with a 35-day revisiting period). The Vallcebre landslide movement is favorably oriented (towards the west) and has a gentle slope. Additionally, the deformation rates are moderate enough in order to prevent, the phase from "rolling over" (problem due to the ambiguous nature of the DInSAR). One corner (CR_2) was installed directly on the stable rock to be used as reference. The maximum distance between the corners was 300 m to keep the atmospheric bias negligible. The installation was done with utmost care, with the CR symmetry axis oriented towards the radar line of sight [6].

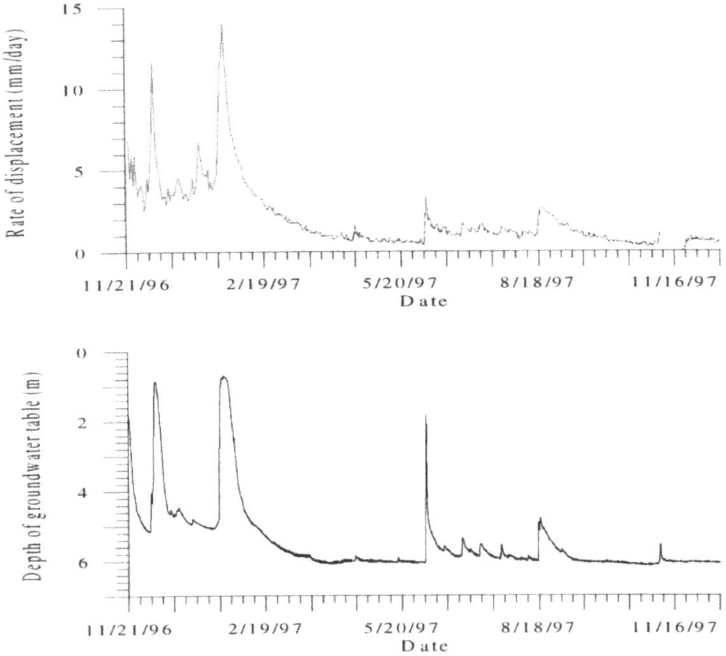

Figure 14. Measurements at borehole S2. Above: rate of displacement derived from the wire extensometer. Below: groundwater table fluctuation measured by the piezometer. A perfect synchronism between rate and water level can be appreciated. This correlation means that groundwater has a big influence in the balance of forces controlling the landslide dynamics.

Figure 15. Corner Reflectors (CRs) used for DInSAR monitoring at the Vallcebre landslide [6]. Examples of installation on a big rock block (**left**) or directly over the terrain (**right**).

In Figure 16a, the CRs are clearly distinguishable in an amplitude image of the landslide. Processing interferometrically four 2007 descending ENVISAT images, the map of the rate of movements was obtained (Figure 16b). As an example, Figure 17 displays the resulting displacements, which were small between the end of December 2006 and the beginning of March 2007, whereas for CR1 and CR4 there is a substantial increase in March and April associated with an abundant rainfall period.

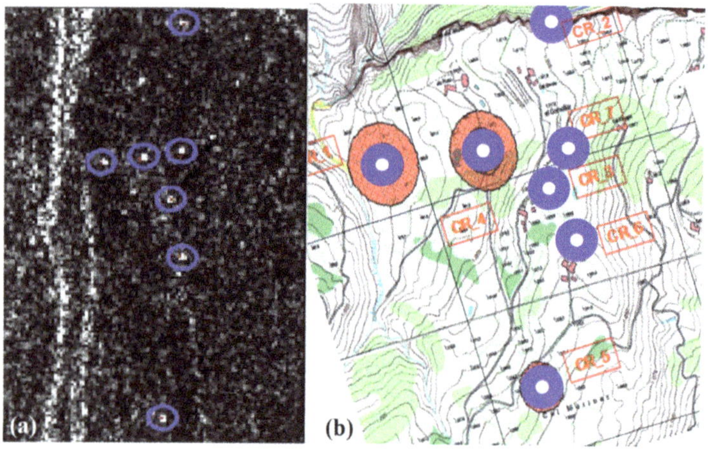

Figure 16. (**a**) the 7 Corner Reflectors appear as bright spots in a 2007 radar amplitude image; (**b**) the concentric circles on the map represent the rate of movement of the CRs during the first half of 2007.

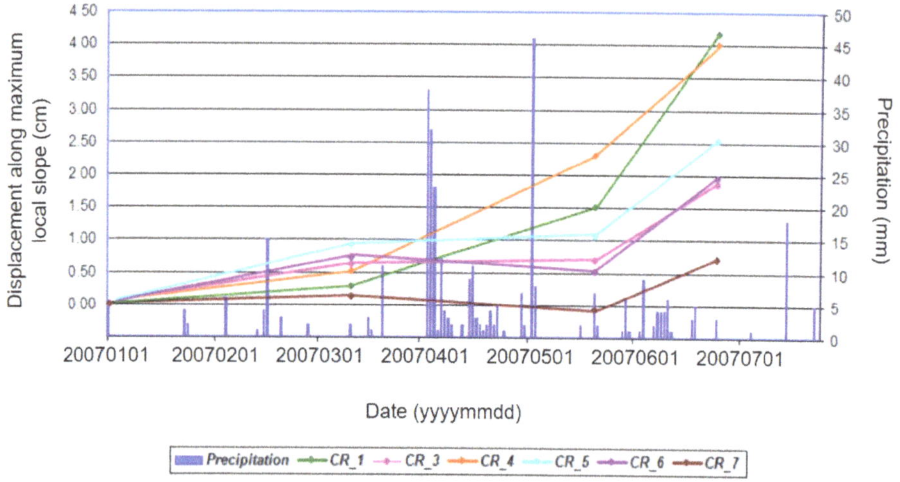

Figure 17. Evolution of the displacements (cm) of the CRs during the first half of 2007. The relative Line-Of-Sight (L.O.S.) displacement was corrected to the direction of the maximum local slope. Precipitation (mm) is also presented as vertical bars.

2.6. Noninterferometric GBSAR Monitoring

The authors of [24,25] presented the application of the Ground-Based SAR (GBSAR) to landslides. In 2010–2011, the lower part of the Vallcebre landslide was monitored using a GBSAR. In this case, we did not use interferometry based on the phase but a new procedure to process the amplitude component of the GBSAR data acquired in discontinuous mode.

This methodology intends to overcome several well-known drawbacks of the radar interferometry, especially in a landslide environment: loss of coherence; lapse of time between acquisitions; aliasing effect; atmosphere influence, and so on. The use of geometric features of the amplitude images combined with a matching technique is fully described in [7].

Between February 2010 and September 2011, this technique was applied to the lower unit of the Vallcebre landslide in order to evaluate the performance of the noninterferometric GBSAR approach. In this period, eight measurement campaigns were carried out using the IBIS-L Ku-band GBSAR (Ingegneria Dei Sistemi S.p.A., Pisa, Italy, 2010; Figure 18). In each campaign, 15 small CRs (Figure 19) were deployed in and around the target area (11 inside, 4 outside as reference).

Figure 18. Picture of the IDS GBSAR system. The synthetic aperture is obtained through the movement of the radar sensor (yellow box) along the rail.

Figure 19. One of the small Corner Reflectors deployed over the landslide lower unit in order to fix up the vegetation and snow problems. The operator is holding a surveying circular prism, which is observed from a reference point with a Total Station, for validation purposes.

In parallel to the GBSAR measurements, five Total Station surveying campaigns were carried out to validate the results obtained with the GBSAR. The analysis of the total measurements obtained during the 19 months permitted the validation of the technique. Displacements up to 80 cm were measured in the lower part of the lower unit (Figure 20); the movement of the points almost *en bloc* corresponds to a translational landslide.

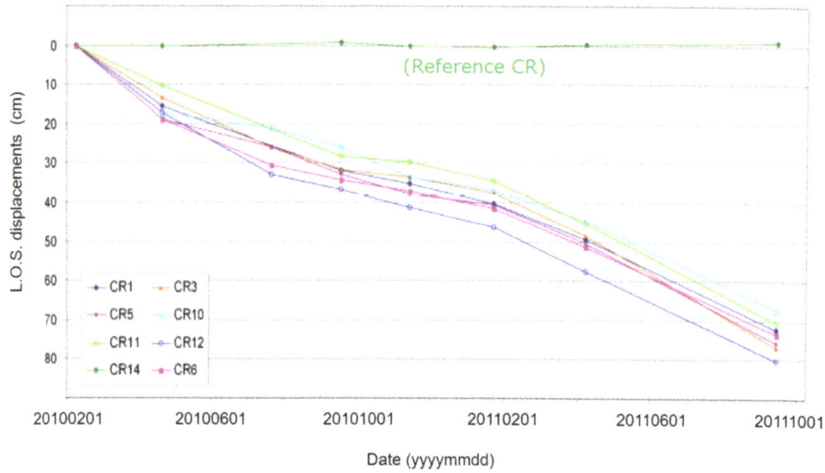

Figure 20. L.O.S. displacements measured with noninterferometric GBSAR between February 2010 and September 2011 at eight small CRs. The CR14, in a stable zone, acts as reference CR.

3. Discussion

As the different systems were applied to certain common points, the resulting displacements could be compared, and practical data, including advantages and drawbacks, could be derived.

For the superficial methods, we observed that the GPS measurements showed better trends and stability than the Total Station measurements, at least with the Vallcebre conditions and equipment. An example is given in Figure 21. Moreover, the GPS surveying is "all-weather" in practice, which is an additional advantage when working in mountainous areas. A theoretical composition of errors, along with independent checks of the GPS-RTK results obtained at fixed stable points, let us establish the following standard deviations [5]: 16 mm in the horizontal plane and 24 mm in elevation. The GPS network in Vallcebre is still in use.

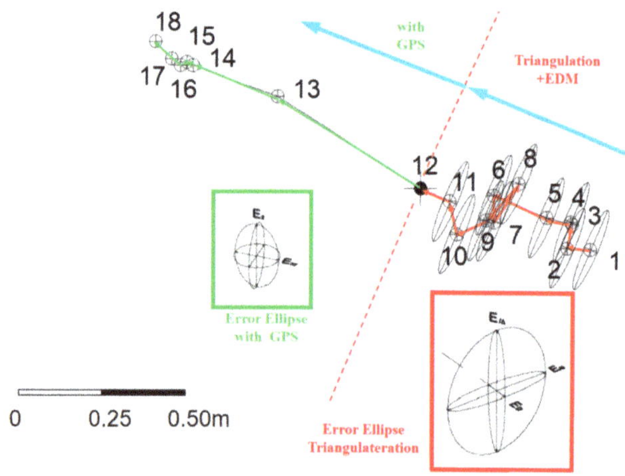

Figure 21. Apparent change in the behavior of point G2 when monitored with classical surveying (dots 1 to 12) or with GPS (dots 12 to 18). The insets with the error ellipses for both systems show a more balanced and better precision for the GPS measurements. Blue arrow points towards the Total Station.

The GPS measurements were also used to check inclinometer and wire extensometer displacements at several boreholes [4]. As an example, in Figure 22, a cross-check is presented. In Vallcebre, the inclinometric readings were possible only until 200 mm of total displacement for the top of the boreholes, due to the casing deformation in the failure zone. Nevertheless, the wire extensometers and GPS measurements have been continued without trouble beyond these figures.

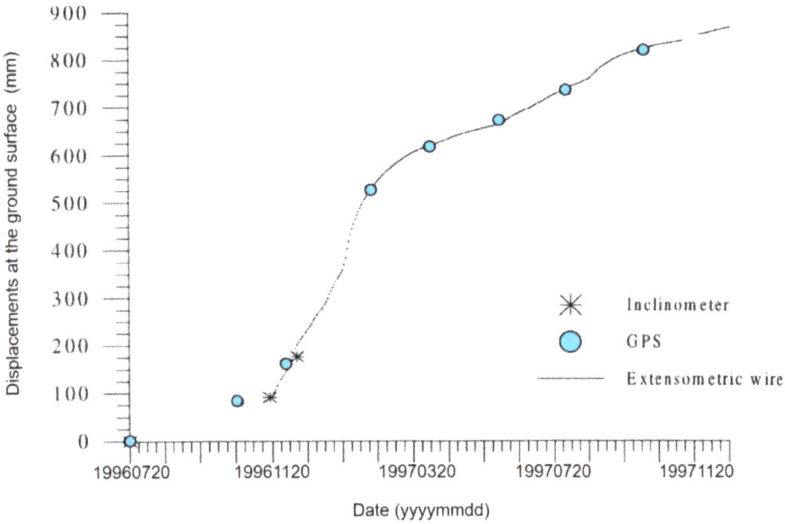

Figure 22. Comparison of the corrected wire extensometer displacement (D_h) against the GPS and the inclinometric measurements of the Vallcebre borehole S2.

Compared with the standard GPS (campaign by campaign), the extensometric technique has the advantage of being an automatic and continuous measuring system which can be correlated with rainfall and piezometric heads. On the other hand, the GPS yields the direction of the movement and the ΔZ as well as the ΔDistance. It is worth noting that the GPS can also be installed and operated in continuous (see [18] for instance).

Cross-checking was also useful for the radar measurements, validating or improving the remote sensing techniques. Every GPS campaign included the CR and the borehole heads (Figure 23). For instance, in Figure 24 a phase unwrapping problem is recovered. The figure shows two displacement time series of CR_4. The continuous blue line represents one phase unwrapping solution, which has an accumulated displacement of 6.9 cm and which shows an upward displacement of 1.1 cm between April and June 2008. Considering the kinematics of the landslide, this type of movement can be considered very unlikely. For this reason, a second solution was chosen, depicted with a dashed line and which shows an accumulated displacement of 11.9 cm in 11 months. The same figure shows the displacement measured by the wire extensometer S5 (green curve), which is located relatively close to CR_4. It is interesting to compare the time series of CR_4 (dashed line) and S5. They display quite a similar temporal evolution, which includes a strong deformation gradient between June and July, followed by a stationary period until September and a second, less strong deformation gradient. This similarity confirms the one phase shift applied to the CR_4 curve. The curves do not fully overlap because one point is 84 m from the other.

Figure 23. Cross-checking between techniques: every GPS campaign included the CRs (CR_2 in the picture) and the top of the boreholes where the wire extensometers and/or the inclinometers were installed.

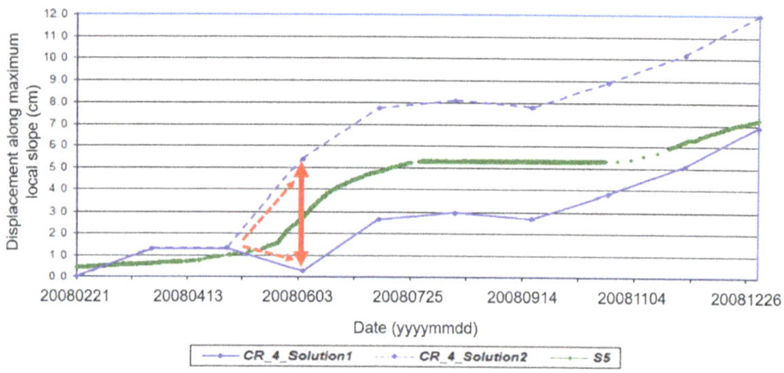

Figure 24. Displacement time series estimated with DInSAR over the CR_4 Corner Reflector, and measured by the wire-extensometer S5, located at a distance of 84 m. A DInSAR ambiguity is shown in red.

The GBSAR amplitude technique presented in Section 2.6 was also validated against the continuous monitoring given by the wire extensometer system. Prior to performing a proper comparison, the CR5 and CR7 Line-Of-Sight (L.O.S.) results (Figure 20) has to be projected onto the total displacement vector (estimated from GPS measurements in the area). The method can be found in [7] and is outlined in Figure 25.

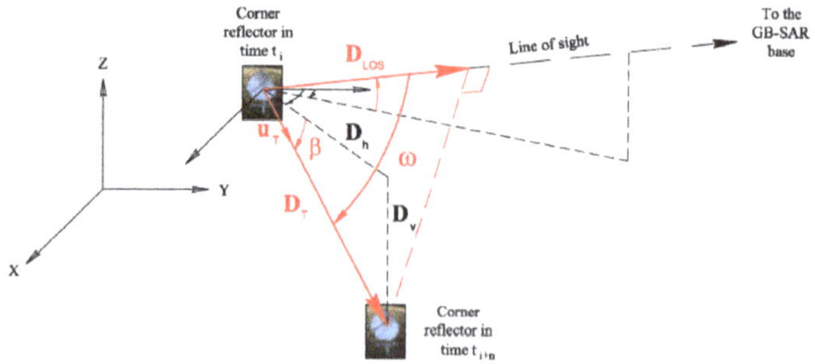

Figure 25. The GBSAR total displacement, D_T, is computed from D_{LOS} using the 3D average direction derived from GPS (unit vector U_T) [7].

After the projection, the CR5 and CR7 GBSAR displacements can be compared with the closer wire extensometer (S11 and S2 respectively, Figure 26). Dashed straight lines have a 1:1 slope, i.e., the perfect fit, which is almost reached.

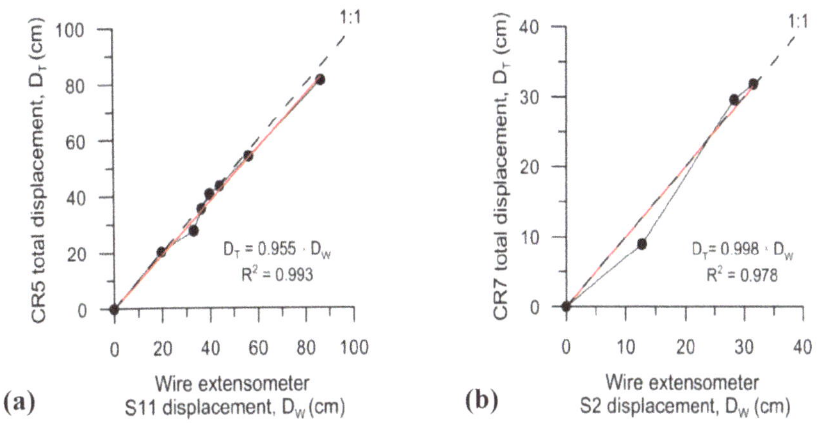

Figure 26. Charts showing GBSAR total displacement, D_T, against wire extensometer displacement, D_W: (**a**) for CR5 and S11, and (**b**) for CR7 and S2.

Finally, after all this precision considerations and system crosschecks, the SAR and/or cable displacements can be visualized as 3D vectors, the 3D direction is given by the GPS data (Figure 27). This overall presentation of the results for the landslide lower unit, with 15 displacement vectors, show a displacement pattern corresponding to a very orthodox translational landslide, with a fair general trend towards the NW, and a small gradient (increase in rate) when approaching the torrent.

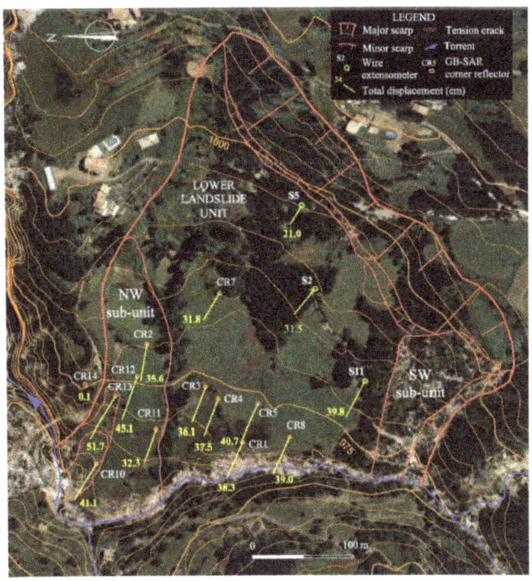

Figure 27. Displacement field of the lower unit of the Vallcebre landslide reconstructed using GBSAR CRs and borehole wire extensometers (S). The yellow figures, in cm, are the total displacements observed from February to November 2010 [7].

4. Summary and Conclusions

A general overview of the methods used in Vallcebre for measuring the displacements of the landslide has been given. More details of the systems presented here, along with the full formulation, validation and conclusions, can be found in [1,4–7]. To conclude, we present here some lessons learned during the progressive monitoring of the Vallcebre landslide since 1987.

A summary of the main characteristics of the methods is given in Table 1. This table is in accordance with [5,9], but the values are valid only for the specific constraints and equipment in Vallcebre. The characteristics of the methods are summarized along with the precision and main results achieved. The 'Precision' figures seek to characterize the 'real' or field performance of the systems in our landslide instead of the theoretical values that can be found in the instrument's technical specifications. The latter were determined in lab conditions and were found to be too optimistic, in general. The 'Complexity' column aims to quantify the refinement of the equipment and personnel involved, and the 'Cost' column includes the amount of instrumentation, installation and work needed to obtain the data.

Table 1. Overview of the methods used in Vallcebre landslide monitoring.

	Method	Results	Typical Range in Vallcebre	Typical Precision in Vallcebre	Complexity	Cost	Comments
Surface	Terrestrial Photogrammetry	Maps (points with $\Delta X,\Delta Y,\Delta Z$, contour lines)	30 to 45 m	10 mm for well- defined points	Medium	High	High precision equipment and skill is needed [b]
	EDM& Theodolite	Δdist. or $\Delta X,\Delta Y,\Delta Z$	15 to 1500 m	7 mm for Δdist, >25 mm for the rest	Medium	Low	Practical experience in surveying is needed [c]
	GPS [d]	$\Delta X,\Delta Y,\Delta Z$	800 to 2300 m	12–16 mm (horiz.) 18–24 mm (elev.)	Low	Low	Discontinuous over time [c]
	CR based DInSAR	Displacement L.O.S.	Non applicable	a few mm	High	High	Each 35 days with Envisat
	Noninterferometric GBSAR	Displacement L.O.S.	500–800 m	below 1 cm	High	High	Continuous. CRs must be installed each campaign
	Distributed Fiber Optics (in trench) [e]	Zones with strain changes	Non available	Non available	High	High	Not installed yet
In-hole	Inclinometer	$\Delta X,\Delta Y$	15 to 45 m depth	1 to 2 mm	Low	Medium	Discontinuous over time [c]
	Wire extensometer [d]	$\Delta D_{wire}\rightarrow\Delta D_{horiz}$, ΔD_{vert}	15 to 45 m depth	0.5 mm in the wire length [a]	Low	Medium	Continuous log, discontinu-ous/continuous download
	Piezometer [d]	$\Delta Pw \sim \Delta Hw$	15 to 45 m depth	30 mm water head	Low	Low-Medium	Continuous log, discontinu-ous/continuous download

[a]: Wire extensometer. After calibration of the wire equation, 2 mm in the surface displacement can be achieved. [b]: The application of Terrestrial or Low-aerial photogram. is becoming easier because of drones and VSfM based SW. [c]: The technique, with a slightly different setup, may become continuous (continuous GPS, Robotic Total St., In-Place-Incl. [d]: GPS, Wire extensometer and Piezometer are green because they are still operational. [e]: Distributed Fiber Optics is red because it has not yet been applied.

The methods in the table should not be considered as excluding alternatives, but rather, as complementary systems to be applied progressively or in different areas, as has been explained in Section 2. For instance, the photogrammetry was used in the toe of the slide to monitor a fairly small but very active area. EDM and triangulation (and later the GPS) were implemented to cover the whole hillside and capture mean movements as big as 1 m per year with minimum investment.

The surveying with EDM and GPS were carried out with a given frequency (i.e., campaigns each two months or each year), so the results are discontinuous over time. As stated in note 'c' in Table 1, although not implemented in Vallcebre, it is technically possible to automate the procedure for the continuous monitoring of the displacements.

On the other hand, the GPS results helped to calibrate the parameters of the wire extensometer equation [4]. The installation of these in-hole devices was possible only when additional funding was obtained in order to drill the boreholes and instrument them. The inclinometers and the in-hole wire extensometers complement each other very well. As the landslide was fairly active, the inclinometers had a short 'life', but produced high-quality information on landslide displacement profiles, velocities and the position of the shear surface quite immediately after their installation. In contrast, at the early stages of deformation, the wire extensometers may only record negative displacements, which are not easily related to the superficial ones. However, once the inclinometers were lost, the wire extensometers provided continuous recording, even for very large displacements. Since 1996, when they were first installed, the wire extensometers have provided a measurement readout each 20 min; at some points, the cumulative displacement has been as high as 6 m. The accelerations in the rate of displacement can be easily related to rainfall and groundwater rises (Figure 14), especially during critical rainy events, when other systems are not in operation. Most of the piezometers and 3 wire extensometers installed in Vallcebre were still operational in 2012. Around 2010 an automatic meteorological station was installed in the center of the landslide in order to get the significant precipitation, because the rainfall may vary sharply from one point to another.

Redundancy between methods is advisable in case of malfunctions in the devices or as a means to cover gaps in data. As shown in the previous sections, all the results are in fairly good accordance when adequate corrections are applied (Figures 22 and 23). Prior to the comparison of the different techniques, some error considerations must be made, in particular, the filtering of any systematic errors (Figures 24 and 25, for instance). In order to 'automatically' filter some systematic errors, during the monitoring operation (acquisition and data handling) it is convenient to be very 'systematic' (to follow exactly the same field and processing procedures).

Contrary to structural monitoring, when working with natural materials like soils and rock, we have to accept natural variability. To address this variability, it is better to apply continuous techniques, both in the time and spatial domains, i.e., the two axes presented in Figure 28. It must be highlighted that the tentative classification of the different techniques in the figure (time frequency/spatial resolution) corresponds to the Vallcebre landslide implementation as reflected in this paper; other setups could be adopted at Vallcebre and other sites. The actual installation depends not only on the site, but also on the level of risk and the available budget.

The systems that were operated continuously relies on automatic acquisition. We learned that 'automatic' does not means 'free of maintenance'. On the contrary, the continuous systems need continuous surveillance and maintenance [2]. Another recommendation for long-term monitoring networks is that, when feasible, it is better to avoid points or benchmarks that may exhibit their own movements (large boulders, light poles, trees, concrete pillars, old stone walls and so on). The discreet marking of points diminishes natural disturbance and vandalism. Finally yet importantly, it is mandatory to include within the network several fixed points in stable areas around the landslide for verification purposes.

Over the years, some improvements in the electrical supply (solar panels) and data transmission (remote download) have been introduced in the Vallcebre setup. Currently, wire extensometer and the piezometer readings can be downloaded from a remote location by means of the mobile phone network.

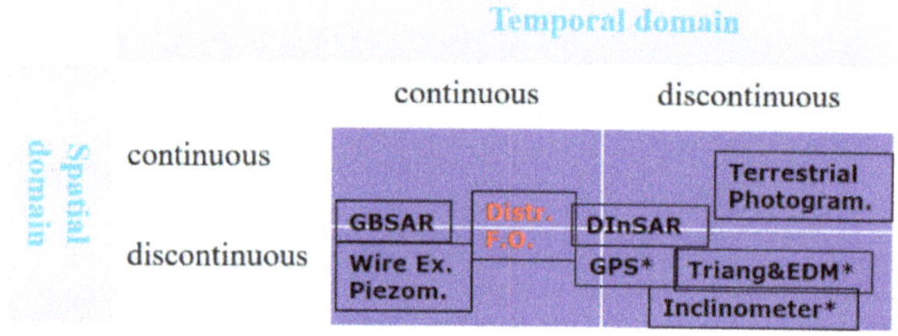

Figure 28. Tentative disposition of the different techniques as specifically used in Vallcebre: following [26], systems are classified according to their continuity in the Spatial and Temporal domain. Distributed Fiber Optics is in red because it has not yet been applied. The asterisks indicate that the technique, with a slightly different setup, may become continuous over space or time.

All the works described in this paper are capable of qualifying the Vallcebre landslide as a real-scale "in situ" lab or observatory. The relatively slow rate of motion and the sustained long-term mobility make it suitable for the testing and validation of several new systems within the frameworks of European and National Projects. In the near future, we plan to install an experimental array of distributed fiber optics [27] in order to assess its field performance. The projected array includes around 600 m of FO buried in a trench besides a country road, roughly along the 1000 m contour line shown in Figure 27. This area is subjected to shear displacements produced by the landslide lateral limits and the transition from the middle to the lower unit. The wire extensometer and the GPS control points present in the zone will permit the cross-calibration of the FO array results and performance.

Basic research on the performance of the different instruments and methods that have been applied in the Vallcebre landslide should help to choose a monitoring strategy in new sites. In making such choices, we must bear in mind that the monitoring must address some fundamental geotechnical questions [28,29]. Instruments and methods that help to answer these questions should be selected. As John Dunnicliff stated, "If there is no question, there should be no instrumentation".

Author Contributions: Geological investigation and methodology, J.A.G., J.M. and J.C.; Radar monitoring concept and processing, M.C. and O.M.; data analysis and validation, J.A.G. and J.M.; supervision and search for funding, J.C. and J.M.; writing, J.A.G. All authors have read and agreed to the published version of the manuscript.

Funding: The three decades work summarized here has been funded mainly by EC and Spanish Research Projects. In chronological order: TESLEC (*Temporal stability and activity of landslides in Europe with respect to climatic change*; EC; contract EV5V-CT94.0454); NEWTECH (*New technologies for landslide hazard assessment and management in Europe*; EC contract ENV4-CT96-0248 and Spanish CICYT contract AMB96-2480-CE); CICYT project (*Cartografía de la susceptibilidad a los deslizamientos del terreno en el área del Alto Llobregat ... Vallcebre mediante los SIG*; Spanish CICYT contract AMB97-1091-C06); MODEVALL (*Modelación hidrológico-mecánica del deslizamiento de Vallcebre*; Spanish Science and Education Ministry, MEC, contract CGL2005-05282); SAFELAND (*Living with landslide risk in Europe: Assessment, effects of global change, and risk management strategies*; EC, 7thFP, grant agreement 226479); PYRMOVE (*Research network* funded by AGAUR, Government of Catalonia; grant number 2014CTP00051); PYRMOVE (*Prevention and cross-border management of risk associated with landslides*; funded by Interreg V-A, POCTEFA, and by the European Regional Development Fund, ERDF; project EFA364/19). The last project is still active.

Institutional Review Board Statement: Not applicable.

Informed Consent Statement: Not applicable.

Data Availability Statement: No new data was created or analyzed in this study. Data presented here was gathered from previous published work [1–7]. Most of the Vallcebre monitoring data is not publicly and/or directly available due to the highly heterogeneous nature of the results, which in a significant proportion belong to the last century (paper-based), prior to the standardization of public datasets and digital repositories. However, in case of interest for some pieces of data, the authors are open to requests.

Acknowledgments: We would like to thank the individuals from the UPC and from the former Institute of Geomatics who helped us with the field work: J. Spagnolo (in memoriam), A. Lloret, A. Ledesma, T. Pérez-Revilla, J. Marín, D. Serral, A. Pérez-Carreras, E. Sanjuán, J. Rius, D.González, G. Luzi, E. Biescas and M. Cuevas. The authors wish to thank ICGC (Institut Cartogràfic i Geològic de Catalunya, formerly ICC) for the geodesic and photogrammetric work, ESA for the radar data, and Topcon, Trimble and Leica (formerly Wild) for their equipment and software assistance along the years. The comments and suggestions given by three anonymous reviewers during the editorial processing are also greatly acknowledged.

Conflicts of Interest: The authors declare no conflict of interest.

References

1. Corominas, J.; Moya, J.; Ledesma, A.; Lloret, A.; Gili, J.A. Prediction of ground displacements and velocities from groundwater level changes at the Vallcebre landslide (Eastern Pyrenees, Spain). *Landslides* **2005**, *2*, 83–96. [CrossRef]
2. Moya, J.; Corominas, J.; Gili, J.A.; Ledesma, A.; Lloret, A. Twenty years of continuous monitoring of the Vallcebre landslide (Eastern Pyrenees): Lessons learned. In Proceedings of the JTC1 Workshop on Advances in Landslide Understanding, Barcelona, Spain, 24–26 May 2017; Alonso, E., Pinyol, N., Eds.; CIMNE, Pub.: Barcelona, Spain, 2017; pp. 101–104, ISBN 978-84-946909-1-4.
3. Gili, J.A.; Moya, J.; Corominas, J.; Crosetto, M.; Monserrat, O.; Luzi, G. Twenty-five years of Vallcebre landslide monitoring: From theodolite to Radar. In Proceedings of the 1st IMEKO TC-4 International Workshop on Metrology for Geotechnics, Benevento, Italy, 17–18 March 2016; Daponte, P., Simonelli, A., Eds.; Athena Srl, Pub.: Benevento, Italy, 2016; pp. 261–268, ISBN 978-92-990075-0-1.
4. Corominas, J.; Moya, J.; Lloret, A.; Gili, J.A.; Angeli, M.G.; Pasuto, A.; Silvano, S. Measurement of landslide displacements using a wire extensometer. *Eng. Geol.* **2000**, *55*, 149–166. [CrossRef]
5. Gili, J.A.; Corominas, J.; Rius, J. Using global positioning system techniques in landslide monitoring. *Eng. Geol.* **2000**, *55*, 167–192. [CrossRef]
6. Crosetto, M.; Gili, J.A.; Monserrat, O.; Cuevas-González, M.; Corominas, J.; Serral, D. Interferometric SAR monitoring of the Vallcebre landslide (Spain) using corner reflectors. *Nat. Hazards Earth Syst. Sci.* **2013**, *13*, 923–933. [CrossRef]
7. Monserrat, O.; Moya, J.; Luzi, G.; Crosetto, M.; Gili, J.A.; Corominas, J. Non-interferometric GB-SAR measurement: Application to the Vallcebre landslide (eastern Pyrenees, Spain). *Nat. Hazards Earth Syst. Sci.* **2013**, *13*, 1873–1877. [CrossRef]
8. Ferrari, A.; Ledesma, A.; González, D.A.; Corominas, J. Effects of the foot evolution on the behaviour of slow-moving landslides. *Eng. Geol.* **2011**, *117*, 217–228. [CrossRef]
9. Mikkelsen, P.E. Field instrumentation. In *Landslides Investigation and Mitigation*; Chapter 11; Turner, A.K., Schuster, R.L., Eds.; T.R.B. Special Report 247; National Academy Press: Washington, DC, USA, 1996.
10. Dunnicliff, J. *Geotechnical Instrumentation for Monitoring Field Performance*; John Wiley & Sons: Hoboken, NJ, USA, 1993; p. 608.
11. Westoby, M.J.; Brasington, J.; Glasser, N.F.; Hambrey, M.J.; Reynolds, J.M. 'Structure-from-Motion' photogrammetry: A low-cost, effective tool for geoscience applications. *Geomorphology* **2012**, *179*, 300–314. [CrossRef]
12. Bemis, S.P.; Micklethwaite, S.; Turner, D.; James, M.R.; Akciz, S.; Thiele, S.T.; Bangash, H.A. Ground-based and UAV-based photogrammetry: A multi-scale, highresolution mapping tool for structural geology and paleoseismology. *J. Struct. Geol.* **2014**, *69*, 163–178. [CrossRef]
13. Abellán, A.; Jaboyedoff, M. Advances in rock slope characterization and monitoring using 3D/4D datasets (LiDAR, drones, Structure-from-Motion photogrammetry). In Proceedings of the ROCEXS 2017, 6th Interdisciplinary Workshop on Rockfall Protection, Barcelona, Spain, 22–24 May 2017; pp. 10–13. Available online: http://congress.cimne.com/rocexs2017/frontal/Doc/Ebook.pdf (accessed on 20 January 2020).
14. Buill, F.; Núñez-Andrés, M.A.; Lantada, N.; Prades, A. Comparison of photogrammetric techniques for rockfalls monitoring. In Proceedings of the World Multidisciplinary Earth Sciences Symposium (WMESS 2016), Prague, Czech Republic, 5–9 September 2016; Institute of Physics (IOP Publishing): Prague, Czech Republic, 2016; Volume 44, pp. 1–7. [CrossRef]
15. Fernández, T.; Pérez, J.L.; Cardenal, J.; Gómez, J.M.; Colomo, C.; Delgado, J. Analysis of landslide evolution affecting olive groves using UAV and photogrammetric techniques. *Remote Sens.* **2016**, *8*, 837. [CrossRef]
16. Rossi, G.; Tanteri, L.; Tofani, V.; Vannocci, P.; Moretti, S.; Casagli, N. Multitemporal UAV surveys for landslide mapping and characterization. *Landslides* **2018**, *15*, 1045–1052. [CrossRef]
17. Giordan, D.; Hayakawa, Y.; Nex, F.; Remondino, F.; Tarolli, P. The use of remotely piloted aircraft systems (RPASs) for natural hazards monitoring and management. *Nat. Hazards Earth Syst. Sci.* **2018**, *18*, 1079–1096. [CrossRef]
18. Malet, J.P.; Maquaire, O.; Calais, E. The use of Global Positioning System techniques for the continuous monitoring of landslides: Application to the Super-Sauze earthflow (Alpes-de-Haute-Provence, France). *Geomorphology* **2002**, *43*, 33–54. [CrossRef]
19. Coe, J.A.; Ellis, W.L.; Godt, J.W.; Savage, W.Z.; Savage, J.E.; Michael, J.A.; Kibler, J.D.; Powers, P.S.; Lidke, D.J.; Debray, S. Seasonal movement of the Slumgullion landslide determined from Global Positioning System surveys and field instrumentation, July 1998–March 2002. *Eng. Geol.* **2003**, *68*, 67–101. [CrossRef]

20. Angeli, M.G.; Gasparetto, P.; Silvano, S.; Tonetti, G. An automatic recording system to detect the critical stability of slopes. In Proceedings of the 5th International Symposium on Landslides (ISL1988), Lausanne, Switzerland, 10–15 July 1988; A.A. Balkema, Pub.: Rotterdam, The Netherlands, 1988; Volume 1, pp. 375–378.
21. Berardino, P.; Costantini, M.; Franceschetti, G.; Iodice, A.; Pietranera, L.; Rizzo, V. Use of differential SAR interferometry in monitoring and modelling large slope instability at Maratea (Basilicata, Italy). *Eng. Geol.* **2003**, *68*, 31–51. [CrossRef]
22. Colesanti, C.; Ferretti, A.; Prati, C.; Rocca, F. Monitoring landslides and tectonic motions with the permanent scatterers technique. *Eng. Geol.* **2003**, *68*, 3–14. [CrossRef]
23. Squarzoni, C.; Delacourt, C.; Allemand, P. Nine years of spatial and temporal evolution of the La Valette landslide observed by SAR interferometry. *Eng. Geol.* **2003**, *68*, 53–66. [CrossRef]
24. Tarchi, D.; Casagli, N.; Fanti, R.; Leva, D.D.; Luzi, G.; Pasuto, A. Landslide monitoring by using ground-based SAR interferometry: An example of application to the Tessina landslide in Italy. *Eng. Geol.* **2003**, *68*, 15–30. [CrossRef]
25. Atzeni, C.; Canuti, P.; Casagli, N.; Leva, D.; Luzi, G.; Moretti, S.; Pieraccini, M.; Sieber, A.; Tarchi, D. A portable device for landslide monitoring using Radar interferometry. *Landslide News* **2003**, *14/15*, 19–22.
26. Janeras, M.; Jara, J.; Royan, M.; Vilaplana, J.M.; Aguasca, A.; Fabregas, F.; Gili, J.; Buxó, P. Multi-technique approach to rockfall monitoring in the Montserrat massif (Catalonia, NE Spain). *Eng. Geol.* **2017**, *219*, 4–20. [CrossRef]
27. Lienhart, W. Case studies of high-sensitivity monitoring of natural and engineered slopes. *J. Rock Mech. Geotech. Eng.* **2015**, *7*, 379–384. [CrossRef]
28. Dunnicliff, J.; Burland, J.; Chapman, T.; Skinner, H.; Brown, M. Types of geotechnical instrumentation and their usage. Chapter 95. In *ICE Manual of Geotechnical Engineering: Volume II*; ICE Virtual Library: London, UK, 2012; pp. 1379–1403.
29. Mazzanti, P. Introduction. In Proceedings of the 6th International Course on Geotechnical and Structural Monitoring (6thICGSM), Rome, Italy, 27–31 May 2019; pp. 1–10.

applied sciences

MDPI

Article

New Landslide Disaster Monitoring System: Case Study of Pingding Village

Yao Min Fang [1], Tien Yin Chou [1,*], Thanh Van Hoang [1], Quang Thanh Bui [2], Duy Ba Nguyen [3] and Quoc Huy Nguyen [2,4]

1 Geographic Information Systems Research Center, Feng Chia University, Taichung 40724, Taiwan; frankfang@gis.tw (Y.M.F.); van@gis.tw (T.V.H.)
2 Faculty of Geography, VNU University of Sciences, 334 Nguyen Trai Str., Thanh Xuan, Hanoi 100000, Vietnam; qthanh.bui@gmail.com (Q.T.B.); st_huy@gis.tw (Q.H.N.)
3 Department of Photogrammetry and Remote Sensing, Hanoi University of Mining and Geology, No. 18 Vien Street—Duc Thang Ward—Bac Tu Liem District, Hanoi 100000, Vietnam; nguyenbaduy@humg.edu.vn
4 PhD program of Civil Engineering, Water Resources Engineering, and Infrastructure Engineering, College of Construction and Development, Feng Chia University, Taichung 40724, Taiwan
* Correspondence: jimmy@gis.tw; Tel.: +886-42451-6669

Received: 19 August 2020; Accepted: 18 September 2020; Published: 25 September 2020

Abstract: The Linbeiken area is located in the village of Pingding, Taiwan. Since the Mindulle and Aere Typhoons in 2004, and as a result of the landslide triggered by the continuous heavy rainfall on 9 June 2006, there has been a persistent collapse of side slopes in the area. This paper describes the equipment that was installed to collect on-site topographic and hydrological information in the Linbeiken area upstream of the Pingding River and to monitor changes in the landslide area, as well as the measurements that were collected during the 2008 Typhoon Sinlaku. A case study of a landslide in Pingding, Taiwan was used to monitor the accurate coordinate changes in the potential landslide areas during typhoons. The goal of this study was to establish warning indexes, and to strengthen the software and hardware at the local disaster response center in the hope of gaining a full idea of the surface movement in landslide areas in future flood seasons. This is important for boosting the preparedness to adapt to landslide hazards, for improving disaster warnings, and for reporting efficiently to better protect the lives and property of local residents. The results show that the landslide disaster monitoring and warning system in Taiwan, as applied during Typhoon Sinlaku in 2008, is both effective and comprehensive.

Keywords: Global Positioning System (GPS); landslide; disaster prevention monitoring; disaster mitigation

1. Introduction

Landslides cause significant damage to both people and property. Many studies have researched landslides, such as that of Chen et al. [1], which showed the application of the three-dimensional deterministic model to a landslide event in Taiwan. Hsu and Liu integrated the TRIGRS (Transient Rainfall Infiltration and Grid-Based Regional Slope-Stability) and DEBRIS-2D (debris flow-two dimensional) models for landslides in Taiwan [2], while Bunn et al. and Ramos-Bernal et al. [3,4] presented the LIDAR (Light Detection and Ranging) digital elevation model and ASTER (Advanced Spaceborne Thermal Emission and Reflection Radiometer) imagery research on landslides. Liu et al. [5] used a geographically weighted regression model for studying landslides in the QingChuan area of China. Liu et al. [6] used a variety indexes like the C-, X-, and L-band synthetic aperture radar (SAR) datasets for landslide analysis in China. Assilzadeh et al. applied GIS application for landside prevention Penang Island in the Straits of Malacca [7], while Long et al. [8] analyzed the mapping of rainfall-induced landslides in Vietnam. Chou et al. applied unmanned aerial vehicle technology to produce a digital elevation model-based dataset with a 5 m resolution for disaster monitoring and management operations [9]. In recent decades, many different techniques have been developed that focus on early warning disaster monitoring and management systems. The study by Anita et al. [10] was based on using sensor nodes to analyze the data for disaster monitoring. Kwak analyzed multiple data sources of various satellite for disaster risk reduction [11]. McCarthy et al. created a GIS (Geographic Information System) expert system framework for the detection and monitoring of hazards [12]. Arattano et al. combined analytical sensors for the warning and monitoring of debris flow [13]. Xuan et al. provided vivid examples of and suggestions for improving monitoring systems [14]. Lucas et al. integrated characterization and monitoring [15]. Trocone et al. analyzed slope response by using the material point method [16].

With the Taiwan Strait on the left, connecting the island and Eurasia, and the Pacific Ocean on the right, Taiwan is located in one of the areas around the world that experiences the most monsoons. Prevailing winds change significantly as seasons alter; the rainy season and typhoons bring abundant rainfall and often cause landslides and landslips. Landslides are referred to as heterotaxy and occur instantly, while landslips are slow and are known as long-term heterotaxy. The Global Positioning System (GPS), which has been widely used in many academic subjects and fields, has proved to be a powerful tool for monitoring artificial and natural structure deformations, as well as slope displacements. GPS has a great number of advantages compared with traditional measurement technologies. Concisely, GPS is more precise, effective, and highly automatic, with low labor intensity required.

Wu analyzed a landslide event in Taiwan (during the 2009 Typhoon Morakot) by combining landslide susceptibility data and drawing a map [17]. In recent years, many different types of technologies have been developed, such as in case studies of Korea and Germany [18–20], which were focused on several technologies for disaster mitigation. Estrela et al. [21] used cameras to observe ground movement and reduce associated risks, while Zhao et al. researched SAR images for landslide monitoring [22]. An operational web-based GIS for the early warnings of landslides in Bangladesh, created by Ahmed et al. [23], is capable of providing alerts five days in advance. Other research works have resulted in the flooding model [24], the flooding forecasting system [25], and dynamic modeling for reservoir watersheds [26], which are all widely applied in disaster management.

On 25 June 2003, the 42–45 K section (Wuwanzi) of A Li Mountain Highway was affected by the most serious landslide in its history due to continuous torrential rain. With ceaseless rain-wash, the slumped section enlarged, and an area of nearly 120 m slipped off of the valley. As a result, the A Li Mountain Public Works Section commissioned Feng Chia University to install various transmission devices and equipment in the 31 K section of the Taiwan 18th Line Highway. As an outcome of this installation, it was suggested that reference values for rainfall levels, water levels of barrier lakes, drilled underground water levels of deposition areas, and overflow water levels be established and then used as alarms when exceeding such values.

Abnormal phenomena such as up-warp and crack-openings once again affected the dipped-slope site of the collapse zone in November 2002. To understand whether the Jiufenershan Collapse Zone was continuously sliding, the Soil and Water Conservation Bureau immediately organized the "Jiufenershan Collapse Zone Observation Project" to gain a primitive understanding of land surfaces. Observations of underground changes via geological drilling, land surface changes observation, the interior inclination of drilling holes, and underground observations can be used as references for emergency responses during typhoons or torrential rain [27].

The village of Pingding in the county of Yunlin, Taiwan was chosen as a study site. This village is located in a valley surrounded by steep mountains. The dynamic sediment process and the arrangement of the sensors are supposed to provide alerts six hours in advance of any potential landslides so as to protect this village. The advantage of this study is its demonstration of a real-time monitoring station in Pingding that provides the residents with a landslide warning and management system. The aim of this research was: (1) To receive real-time data and to monitor the entire study area (high landslide potential area); (2) to integrate software and hardware for preparedness to adapt to landslide hazards; (3) to send warnings to residents to avoid serious landslides. The research procedure is indicated in Figure 1.

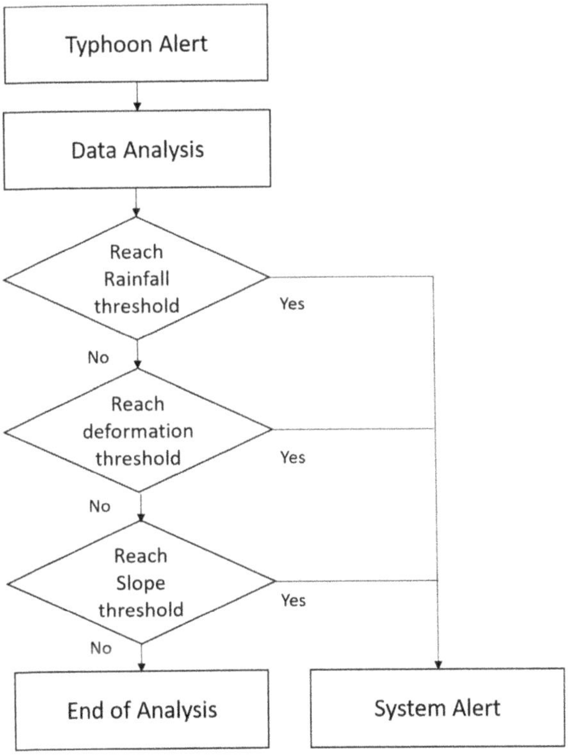

Figure 1. Flowchart of the research procedure.

2. Material and Methods

2.1. Study Area: The Village of Pingding

Pingding is situated in the township of Linnei in the county of Yunlin, Taiwan, and its geographical coordinates are 23°45′36″ N and 120°36′38″ E (Figure 1).

Linnei is divided into mountainous areas and plains by the Taiwan 3rd Line Highway. With mountains on its southeastern side and plains in the northwest, Linnei's altitude is 70–320 m above sea level, and its topography is a slope extending roughly from the southeast to the northwest. Pingding is situated on a mesa-type hill, with an altitude of 325 m above sea level. With a tropical and humid climate, the village has a yearly rainfall of 1500–2000 mm, which is concentrated in the summer. The exposed geological formation is the Toukoshan Formation Huoyanshan Conglomerate Section; the gravel components of the conglomerate are of less than 20 cm in size and are unevenly distributed. The sandstone and mudstone, which are sometimes mixed in the conglomerate, are mainly composed of sandstone and quartz sandstone. The exposed parts of the conglomerate often form cliffs or precipices and gorges if there are any valleys. The Pingding Collapse Zone is located within Pingding in the mountains east of Linnei. Linnei, which is located in the northeast of Yunlin county (one of the counties in Taiwan), adjoins Pingding, Pingding Mountain, and the township of Zhushan, Nantou.

In terms of its geographical location, Pingding is located between Shexi and Gaoxikou in the township of Gaoshu (Figure 1). The study site is flat, tilting north and west, and the west is similar to a valley (Figure 2).

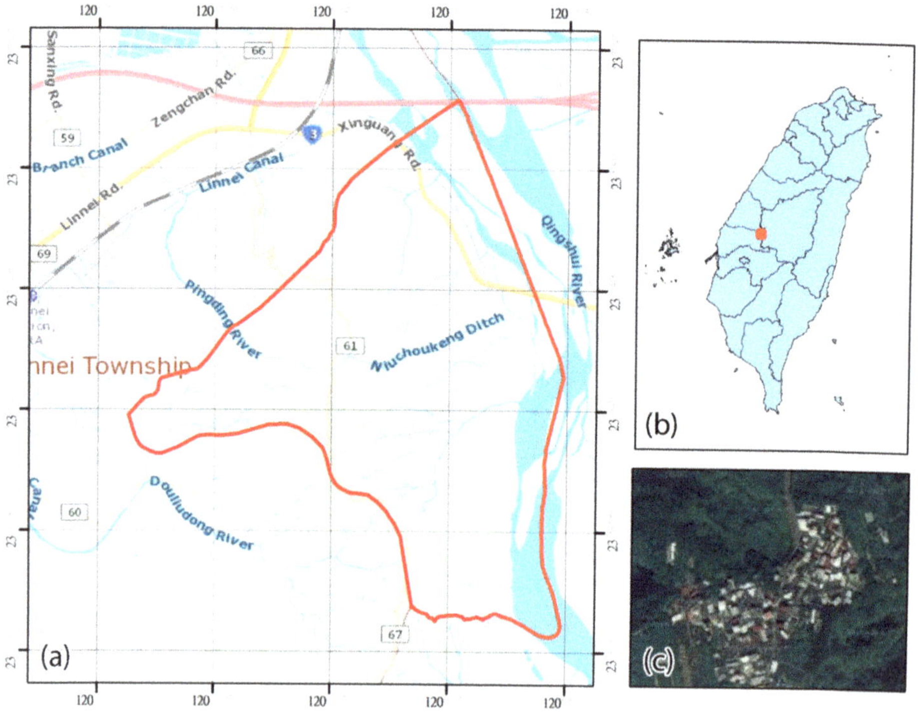

Figure 2. Pingding in Taiwan: (**a**) Pingding in the township of Linnei; (**b**) Pingding in Taiwan; (c) satellite image of Pingding.

According to the "Central Taiwan Geological Map—From Zhushan to Jiayi" document from the fourth issue of the Central Geological Survey Article Collection and Annual Report of Taiwan Council of Agriculture by the Soil and Water Conservation Bureau, the exposed strata near the Pingding Collapse Zone include the Huben Stratum and the Sanxing Stratum of the Pleistocene Age, as well as the Accretion of the Holocene. Judging from the positions and patterns of these strata, the Pingding Collapse Zone is located in the range of the Huben Stratum and the Sanxing Stratum, while both of the shores of the Pingding sub-stream are in accretion. According to the on-site geological surveys and the six (BH-1–BH-6) (geographical surveying site from BH-1 to BH-6) 12–30 m deep geological drilling holes, the strata can be further divided into reddish brown silty clay, brown sandy silty soil in the colluvial rocks, brown silty soil in the gravel stratum, and brown or grey siltstone.

2.2. Analysis of the Causes of Collapse in Pingding

The southwestern hills of Pingding belong to the Toukoshan Formation; the poor cementing of the hills' conglomerate and sandstone tissue raises the possibility of collapse due to washouts. As a result of the frequent headward erosion in the Linbeiken area (Pingding River), the platform has shrunk in the southeastern direction, leaving only the Pingding area behind. Enveloped in serious erosion, the entire mountain area collapses easily when rainfall occurs. The proposed cause of this quick downward flow of the seepage water in the above soil stratum is because of the excellent permeability of the red soil and gravel strata on the surface. The seepage water, however, gets stuck here when confronted with gray sandy mud strata, which are poor in permeability and move horizontally. The underground water finally causes a collapse due to the decompression effect following seepage through the soil from the side slope surface. As the water content in the mud stratum increases, the mud stratum's ability decreases its shear resistance strength. These are key factors that contribute to the damage caused to the side of slopes, and can be judged from the locations and types of previous collapses. Generally, large-scale types of mass collapse are the main factors causing damage, while collapse locations are influenced by the load of slope tops, the water content of local soil (rock) strata, and the gradient.

2.3. Historical Disasters

The 921 Earthquake in 1999: The serious collapse areas that developed during the 921 Earthquake included the northeast slumping side slopes and the northwest Linbeiken Collapse Zone. The Linbeiken Collapse Pit experienced collapses of different scales, and despite many restoration works being done, the Linbeiken Collapse Pit has doubled in size in the past decade. The Yunlin County Government completed construction of the reinforced soil retaining wall in April 2003.

Typhoons Mindulle and Aere in 2004: Landslides once again struck the south bank of Linbeiken in September 2004 (Figure 3). The areas with reinforced soil retaining walls and part of the north-side tea garden collapsed into the Linbeiken Collapse Pit. The collapse created a stage terrain with a 30 m height difference. The collapse area of the first landslide covered approximately 1 ha. Afterward, it was followed by several collapses and slumps of different scales. The damage caused by this collapse included the slump of a 2 m^2 tea garden, but there were no casualties. Residents from four dwellings moved for safety purposes. Regarding public facilities, a diversion ditch of approximately 23 m in length and three areas of PVC (polyvinyl chloride) piping collapsed and were pulled apart.

Continuous torrential rain in June 2006: The flood on 9 June 2006 caused the collapse of the gabion slope protection on the north side of the Linbeiken source bank and the wash-away of part of the Yunlin 61st Line Highway roadbed.

22 May 2007: At 6 a.m., with no influence of rainfall, the original collapse area of the Yunlin 61st Line Highway suffered from another topple, damaging two-thirds of the main access road of the nearby villagers, making it impassable. Emergency work such as fence establishment and warning sign installations were carried out. The main reason for this collapse was the gravity effect influenced by saturation of the soft soil caused by the cracking of tap water pipes of the area, which led to the seepage of tap water and drizzle on previous days.

2.4. Data Collection

This section shows the organization of the research methods used. Figure 4 provides a flowchart of the research methods adopted to create a disaster monitoring and management system. First, for simulating the landslide disaster monitoring system, we visited the site to carry out surveys. Second, based on the results of the survey, we decided on and installed the appropriate equipment for the study site. Third, the thresholds for the sensors were defined. Fourth, based on the results of the analysis, we identified the accuracy of the measurements and validated the threshold value. Finally, a landslide monitoring displaying system in real time was constructed.

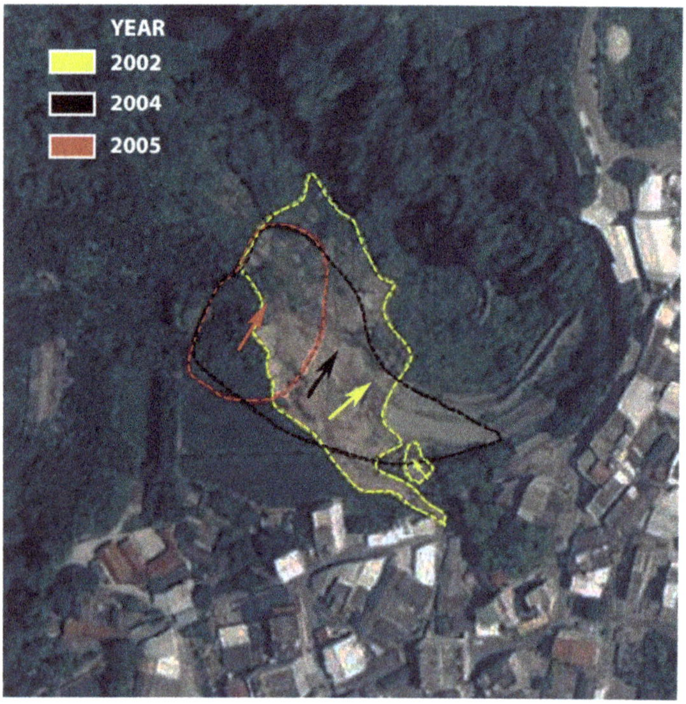

Figure 3. Collapse areas of Pingding River upstream, Linbeiken, and Pingding across different periods of time [28]. Note: The yellow parts refer to the collapse on June 2002 (caused by Jiji Earthquake and Typhoon Toraji); the black parts are the collapses on 9 September 2004, while the parts in red are the collapses on 11 January 2005. The arrows indicate the sliding directions of each collapse.

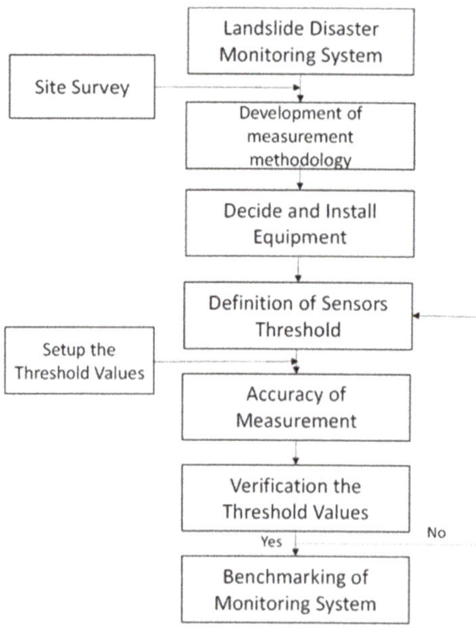

Figure 4. Research methodology flowchart.

2.4.1. Equipment for the Collapse Monitoring System

The Linbeiken area in Pingding, upstream of the Pingding River, has faced constant side slope collapses during torrential rain seasons since the 921 Earthquake in 1999. To assess the condition, this study monitored the collapse changes of the Pingding area and established a side slope collapse monitoring system to understand the real-time activities of the collapse zone during flooding (Figure 3). The following work was planned: The establishment of relevant monitoring equipment, including camcorders and devices and equipment related to rainfall monitoring and land surface deformation monitoring, as well as recording, storage, transmission, and display systems for monitoring the above equipment. In light of the simulation analysis results of Kun-yi Chen [29], the completed work included an inbuilt GPS-4 and GPS-5, a retractable cable meter (group 3), a retractable cable meter (group 4) on the sliding surfaces, an inbuilt retractable magnetic induction meter (groups 1), and a retractable meter (groups 1 and 2) in two open spaces that contained obvious cracks and may have suffered from possible continuous displacements. Finally, GPS-1, GPS-2, GPS-3, and slope circles were placed on the top of the fixed soil retaining wall. The details of the monitoring device numbers are shown in Table 1; the monitoring station is equipped with the following sensors: (1) Two sets of rain gauges; (2) three sets of CCD (A charge-coupled device) camcorders; (3) two sets of high-pressure illuminators; (4) one set of remote camcorders; (5) four sets of retractable meters; (6) one set of slop circles; (7) one sets of GPSs. The device arrangement is displayed in Table 1, Figure 5, which highlights the slope deformation monitoring system contained in the GPS receiver base station and five GPS receiver mobile stations. The precision of the real-time coordination solution was ±3 cm, while the solution speed was over one piece of information (including one) every 3 s.

Table 1. Quantities and descriptions of the observation facilities [30].

Devices	Numbers	Brand	Device Functions
Rain gauge	Two sets	Taketa Keiki Industry Co., Ltd. Itabashi City, Tokyo 173-0024, Japan	Observing rainfall as a reference for real-time rainfall vigilance values
CCD (A charge-coupled device) camcorder	Three sets	AXIS Lund, Sweden	Observing local slopes and transmitting back real-time local images
High-pressure illuminator	Two sets	Micro Balance Wuqi District, Taichung City, Taiwan	Assisting the night-time observation of CCD camcorders
Remote camcorder	One set	AXIS Lund, Sweden	Helping to make the controls on the way to the Command Post remote to aid in the understanding of the on-site situation
Retractable meter (retractable displacement meter)	Four sets (10 m)	Novotechnik Ostfildern, Germany	Monitoring the changes in land surface cracks as a reference for emergency evacuation
(Dual direction) Slope circles	One set	Micronor Inc. Camarillo, CA 93012, United States	Measuring the changes in buildings' trace angles of inclination
GPS slope deformation monitoring system	One set (six GPS)	Leica Wetzlar, Germany	Monitoring the slope spatial 3D change condition through GPS as a reference for emergency evacuation

GPS, Global Positioning System; 3D, three-dimensional.

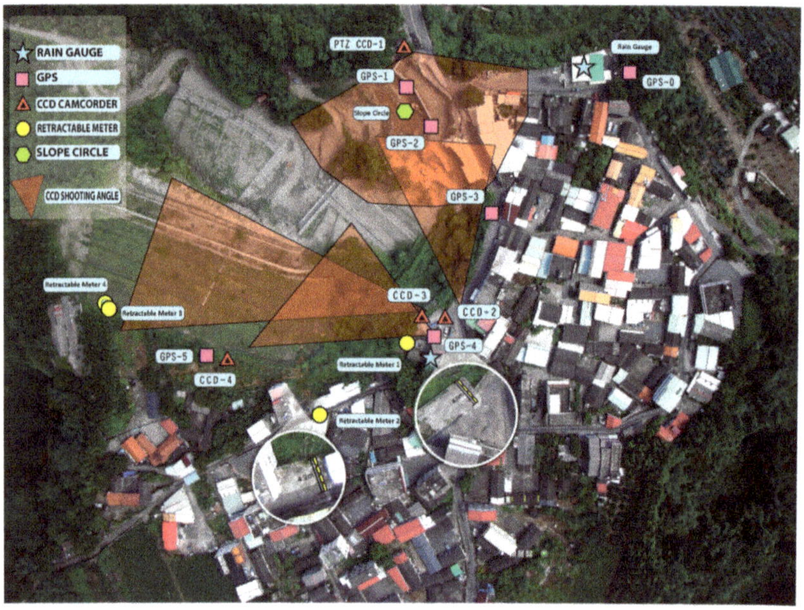

Figure 5. Arrangement of the devices presented in Pingding. The monitoring station is equipped with a rain gauge, a CCD camcorder, a high-pressure tiltmeter, a remote camcorder, a geophone, and slope circles (dual direction).

2.4.2. Establishment of Management Values

The fact that side slope failures often take place after rainstorms has led to the belief in a certain relationship between side slope failures and rainfall. Indeed, this relationship between side slope disasters and rainfall has been noticed and proven by researchers such as Brand [31], Lumb [32], and Slosson and Larson [33]. Nonetheless, some other prior hydrological conditions are essential for the triggering of side slope failures. Pre-storm rainfall injects water into side slope surfaces and allows water to flow freely in the slope. In other words, the soil surface needs to be saturated to trigger the following rainstorm's side slope failure mechanism. The pre-storm rainfall required by side slope failures depends on the soil surface cover, the water conduction capacity of the soil, the seepage rate, evapotranspiration, and the hydrological condition of the side slope.

Knowing that the influences of pre-storm rainfall on side slope stability have been studied for years, Lumb [32] discovered the impact that pre-storm rainfall has on side slope failures. In particular, he found out that higher pre-storm rainfall leads to worse side slope failures. He categorized side slope failures caused by rainfall into the following four groups:

- Severe events that cause over 50 side slope failures per day;
- Serious events that cause 10–50 side slope failures per day;
- Minor events that cause less than 10 side slope failures per day;
- Independent events that cause only one side slope failure per day.

3. Results

3.1. Monitoring Warning Systems

The data of different devices can be displayed in the following ways: Image data displays, cover real-time image displays, and historical image displays, as demonstrated by Figure 6.

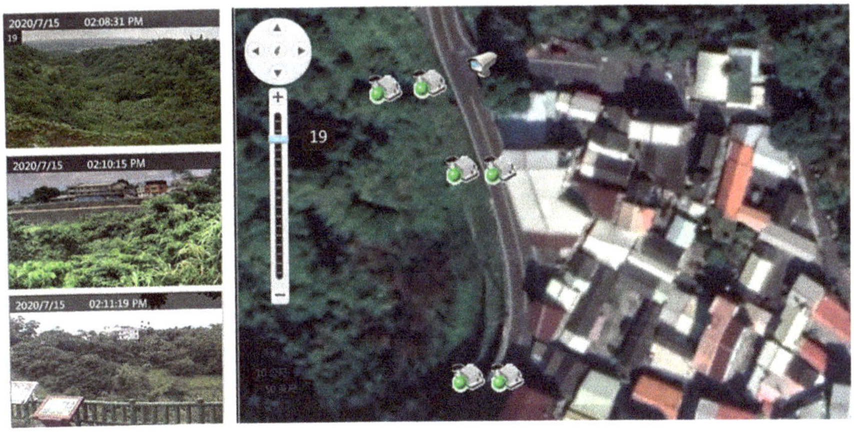

Figure 6. Real-time image displays of the status of the study site (green dots are normal status).

Two types of real-time image displays are available: Single-station displays and loop displays of data from all of the observation stations. Historical image displays allow users to browse the CCD images of certain chosen time periods through the single-station display mode. The slope displacement condition is shown through the displays and analyses regarding the space transformation condition of the retractable meters (tensile strain meters) and the GPS slope land deformation monitoring systems. Users are allowed to access the displacement condition through direct clicks on the devices shown in the device arrangement figure, which is a vigilance data reference provided for involved staff, as displayed in Figure 7. There are four groups of retractable meters, while the data displays show the conditions of the devices using red, yellow, and green lights, as per the display mode of the device set up procedures. When the displacement quantities of the retractable meters exceed the set threshold of the vigilance values, the system will switch the light to red or yellow to remind the responsible personnel of the meters' current condition. The GPS slope land deformation monitoring systems also display the devices' condition through red, yellow, and green lights, and the lights are switched to red or yellow as a reminder for personnel responsible for the current GPS conditions, as shown in Figure 8. In both situations, users are also permitted to click on the red or yellow light icons to inquire about the detailed displacement condition of a particular piece of equipment. The slop circle data displays analyze and display data, focusing on the condition of the angle changes of the slop circle. Users are allowed to access the displacement condition through direct clicks on the devices shown in the device arrangement figure, which is a vigilance data reference provided for relevant staff, as displayed in Figure 9. The real-time monitoring data of all of the devices and the arrangement figures of the relevant devices are shown in Figure 10.

3.2. Monitoring Data Analysis

During Typhoon Sinlaku, the emergency response team of the Soil and Water Conservation Bureau was active from 08:40 on 11 September 2008 to 18:32 on 19 September 2008. It is a hyetograph of the total cumulative rainfall of the entire island, while the cumulative rainfall of the Yulin area, which reached over 300 mm, is demonstrated in Figure 11.

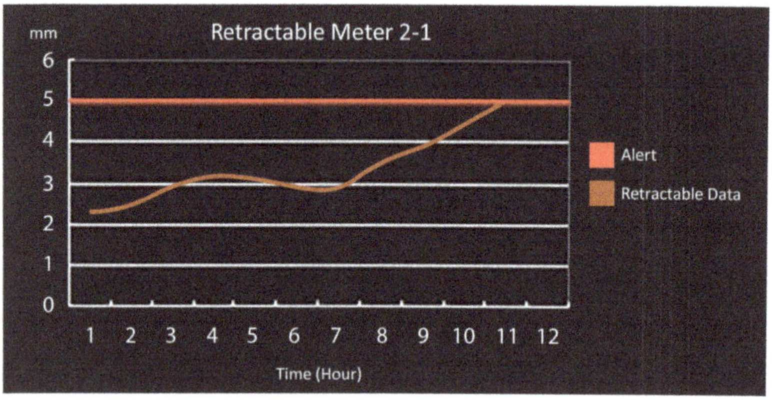

Figure 7. A detailed data display of the retractable meter.

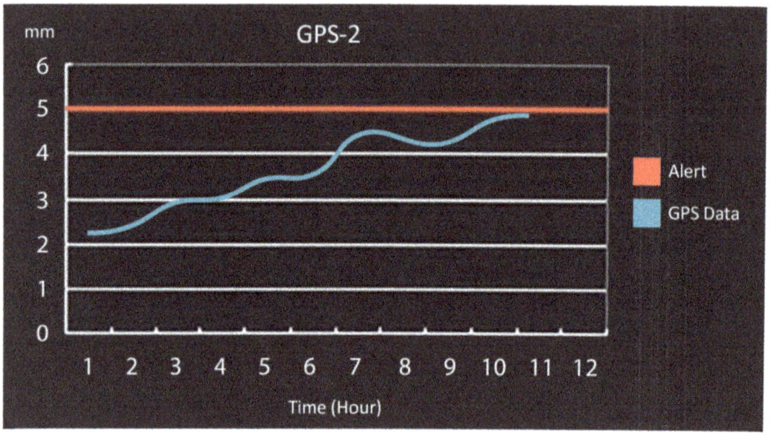

Figure 8. A detailed data display of the GPS.

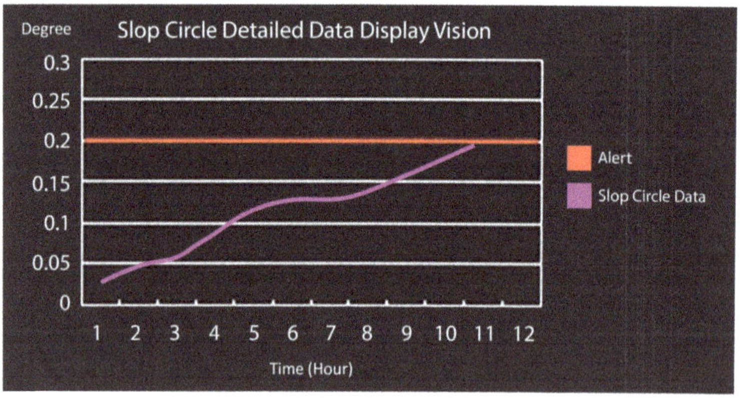

Figure 9. A detailed data display of a slop circle.

Figure 10. Integrated monitoring data display.

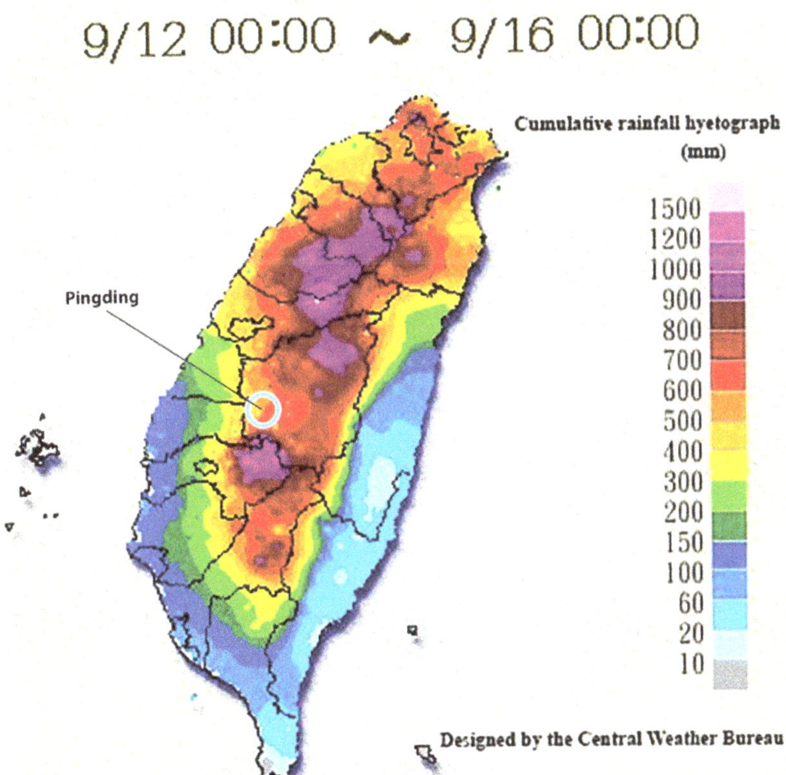

Figure 11. The cumulative rainfall from 12 to 16 September during Typhoon Sinlaku in 2008.

3.2.1. Rainfall Data

Influenced by Typhoon Sinlaku, the total cumulative rainfall from 11 September to 16 September at the observation stations was 349 mm, with the greatest rainfall intensity being 40.5 mm/h. A hyetograph of the cumulative rainfall during the time when the emergency response team was active, showing the rainfall intensity and the cumulative rainfall conditions, is provided in Figure 12. The yellow line in the figure represents the vigilance values, while the red line refers to the action values. The hourly rainfall reached 12 mm, and the daily cumulative rainfall was 86.5 mm at 6 a.m. on 14 September, which reached the threshold of the yellow level of vigilance. The hourly rainfall reached 20.5 mm, and the daily cumulative rainfall was 140.5 mm at 2 p.m. on 14 September, which reached the threshold of the red level of vigilance. No abnormal phenomena were observed in the comparison between the monitoring results of the GPS and retractable meters (Figure 13). The local village heads, the Yunlin County Government, and the related soil and water conservation departments were informed through the disaster report process of this system, as shown in Figure 14. It is assumed that the topographic features of the Pingding area, which increase the possibility of torrential rainfall within short periods of time, lead to higher rainfall intensity and heavier cumulative rainfall. Therefore, more data concerning local typhoons and torrential rainfalls will be collected in the future to determine whether the vigilance and action values of the rainfall in this area should be raised.

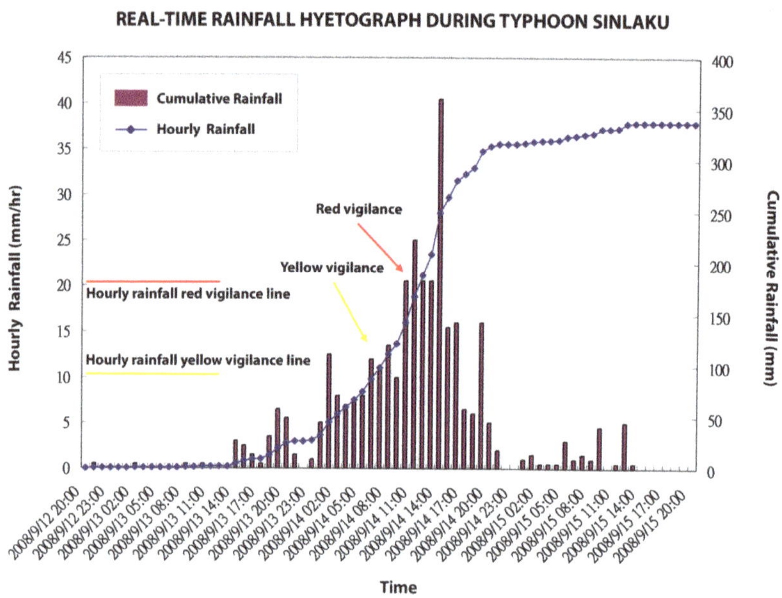

Figure 12. Real-time rainfall hyetograph during Typhoon Sinlaku.

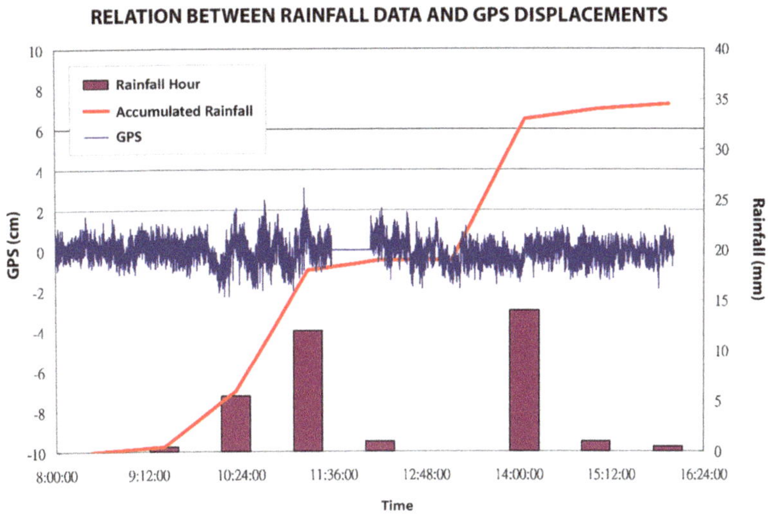

Figure 13. Relationship between Hourly Rainfall and GPS displacements.

3.2.2. GPS Monitoring Data

Figure 15 shows the GPS monitoring data in normal time without typhoons, as well as a comparison between the data and management values. As shown in the figure, all of the GPS points were in a stable condition, and the vibration of the displacement quantity was within 2 cm, indicating that its precision was approximately 2 cm, thus not exceeding the set threshold of the vigilance values. The GPS displacement of all of the action stations did not exceed the set threshold of vigilance values during Typhoon Sinlaku.

3.2.3. Monitoring Data of the Retractable Meters

The monitoring data from the magnetic induction and cable retractable meters during Typhoon Sinlaku, which are shown in Figure 16, were compared with the management values. Except for the micro-displacements caused by the indium steel wires of the cable retractable meters, all of the retractable meters remained in a stable condition and the set threshold of the vigilance values was not exceeded.

3.2.4. Slope Circle Monitoring Data

The slop circle monitoring data from the Typhoon Sinlaku period are displayed in Figure 17. The figure shows two bigger inclinations, which indicate soil retaining wall inclinations caused by the increase in soil pressure resulting from an increase in the water of the soil after rainfall. The data return to the initial position after the water in the soil evaporates or runs off. The data were compared with those of the management values, and all of the slopes were found to remain in a stable condition and the set threshold of the vigilance values was not exceeded. Figure 18 shows a detailed data display during Typhoon Megi—International Number 1013 is the 13th tropical cyclone named in the Pacific typhoon season in 2010, and was also the strongest storm of the year.

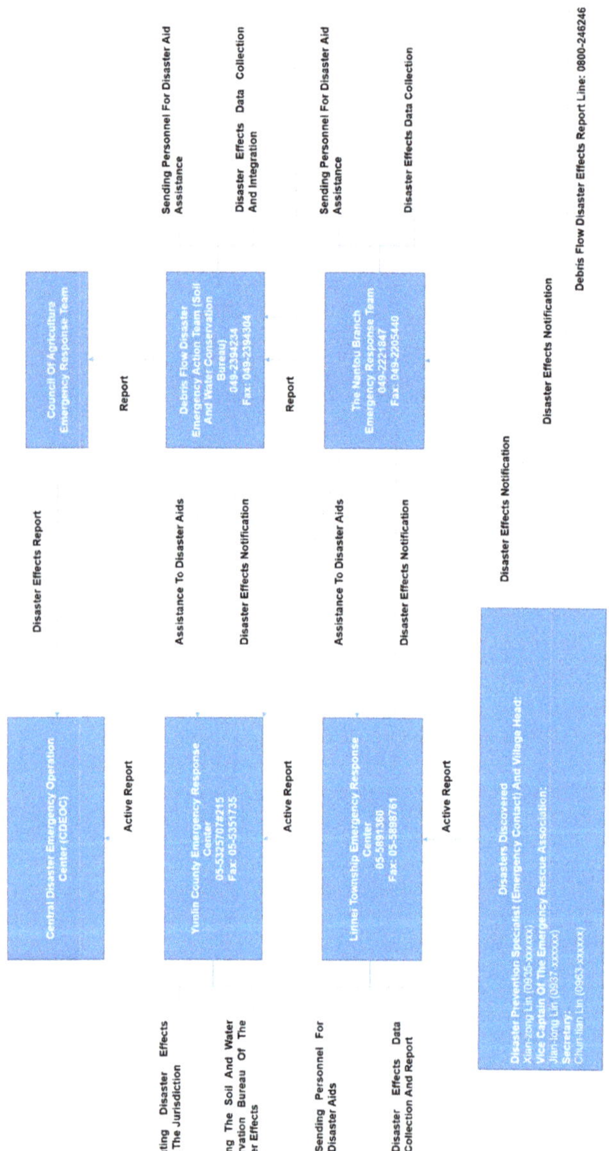

Figure 14. The reporting process for slope land collapse disasters.

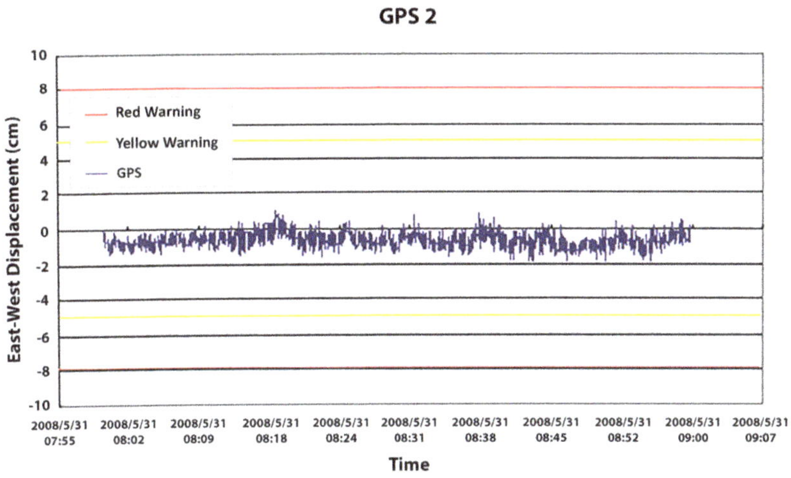

Figure 15. GPS-2 horizontal displacement quantities.

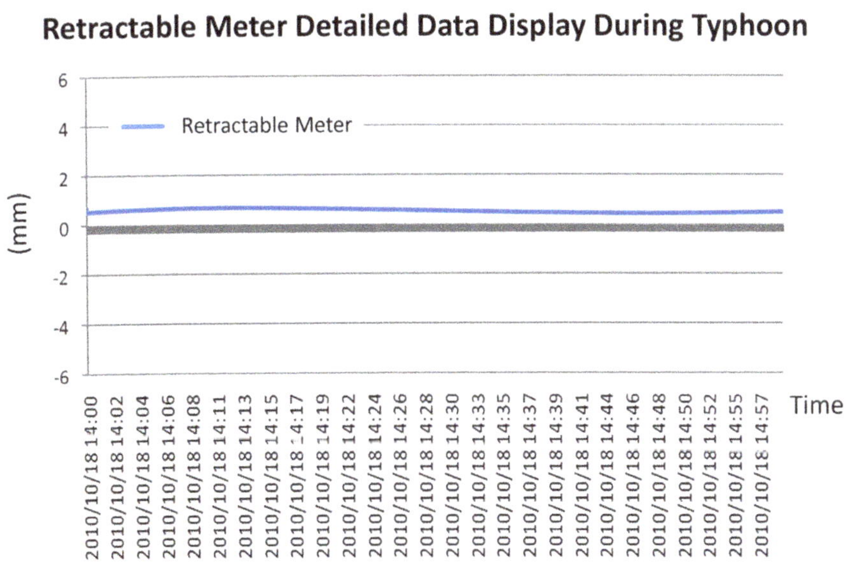

Figure 16. Detailed data display of the retractable meter during Typhoon Sinlaku (17–18 October 2010).

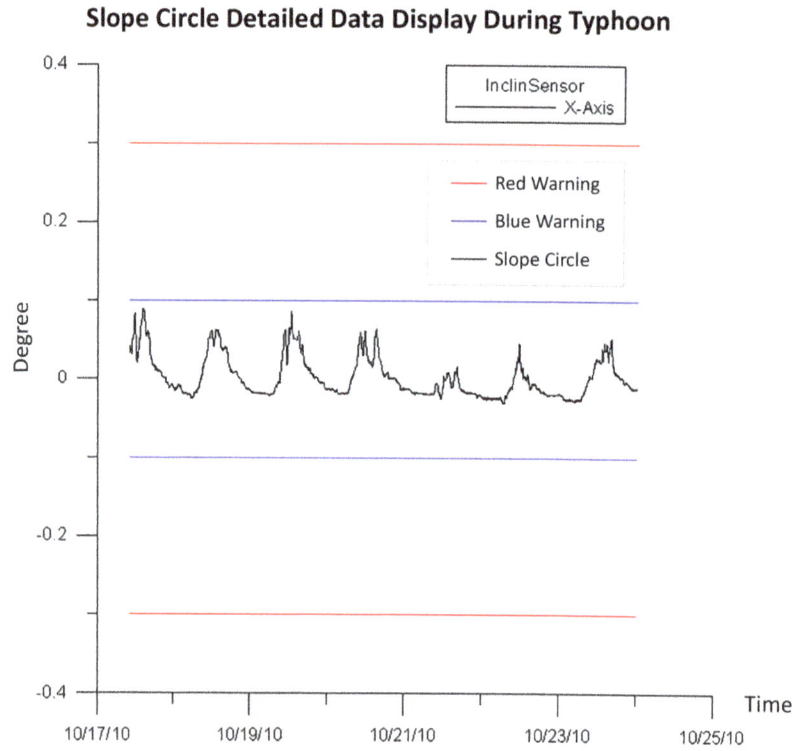

Figure 17. Detailed data display of the slop circle during Typhoon Megi (17–25 October 2010).

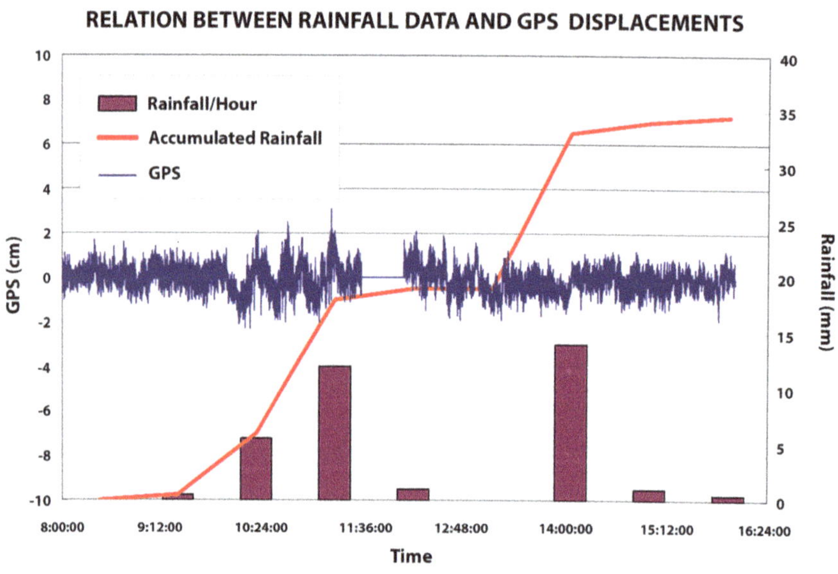

Figure 18. Relationship between rainfall and GPS displacement data.

3.2.5. Comparison between the Monitoring Devices and Rainfall

A synthetic comparison between the monitoring results is advised if the monitoring data surpass the threshold of the management values to determine whether a disaster has struck or whether an abnormal phenomenon of a single device has happened. As Figure 18 shows, a comparison between rainfall and GPS displacement involves determining the relationship between cumulative rainfall and displacement, as well as the relationship between rainfall intensity and displacement speed. Figure 19 shows a comparison between the CCD images of the time periods with the greatest rainfall intensity, aiming to ensure the rainfall monitoring is mistake-free.

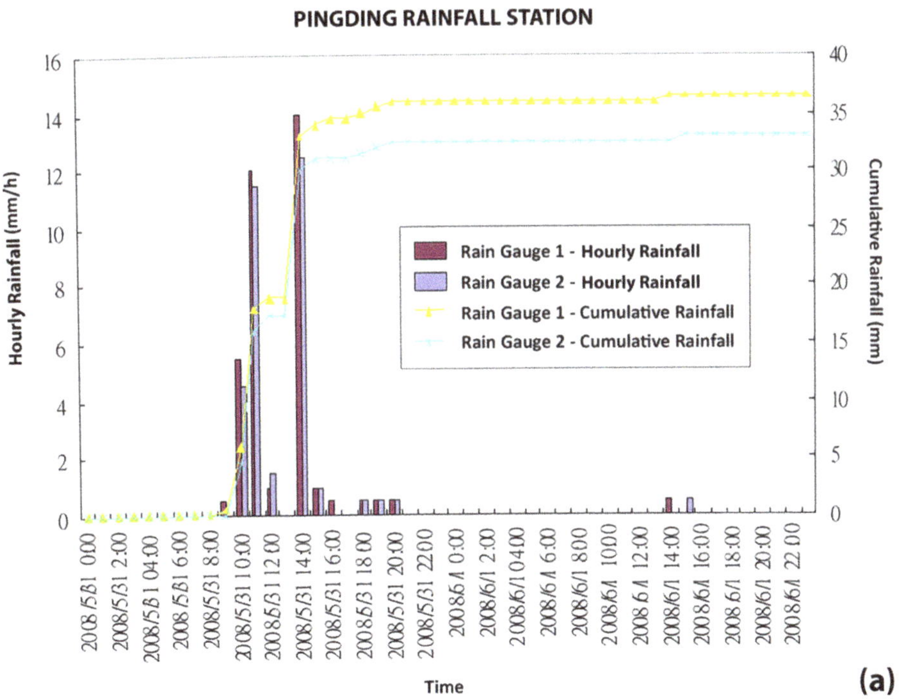

Figure 19. *Cont.*

Figure 19. Relationship between rainfall and CCD data: (**a**) Cumulative rainfall of Pingding; (**b**) CCD camera 1; (**c**) CCD camera 2; (**d**) CCD camera 3.

During a typhoon and heavy rain, we can observe this system to determine the real-time accumulated rainfall intensity. This monitoring process demonstrated that there is no displacement of the ground or the slope of the retaining wall, which proves that our system is workable.

4. Discussion

According to the obtained rainfall data, Lumb established the scope of events of different gradations and provided an explanation based on 15-day pre-storm rainfall and 24 h rainstorms. Severe events brought over 100 mm of rainfall in 24 h and over 350 mm of pre-storm rainfall in 15 days, while serious events generated more than 100 mm of rainfall in 24 h and 200 mm of pre-storm rainfall. This proof of a relationship between side slope failures and rainfall encouraged many studies focusing on the threshold of rainfall for causing side slope failures. For example, it was determined that the mechanism of side slope failures is successfully triggered when the rainfall in Southern California reaches 140% of its normal rainfall (an average value of the recorded data of over 100 years) [34–36], suggesting that the threshold for Los Angeles is 125% (1993), which was also confirmed by Wieczorek [37]. Auer and Shakoor studied the collapse distribution of Nelson County, central U.S., after Hurricane Camille in 1969. According to their research, the hurricane, which moved from the west to the east, caused a more serious collapse of the side slopes in the west, northwest, and southwest directions. This indicates that there might be a connection between the directions of side slopes in collapse zones and hurricane routes [38]. Hong Kong has no records of collapse generated by earthquakes; however, the Geotechnical Engineering Office compared the risk of earthquakes and rainfall in terms of their generation of artificial slope collapse [39]. The study results showed that earthquakes bring a lower risk of causing artificial slope collapse than rainstorms.

Brand, Permchitt, and Phillipson [40] examined the side slope failures of Hong Kong through the following three steps:

- The data provided by Lumb [32] were referred to and the rainfall data collected through Hong Kong's 46 automatic rainfall recorders were recorded;
- The considered rain delay was 1 h and 24 h of rainstorms, which was matched with the 30-day pre-storm rainfall;
- The timing of side slope failures was documented based on the reports of the National Fire Agency.

According to the above research, based on the relationship between disaster numbers and rain intensity, the authors [40] concluded that over half of the studied side slope failures in Hong Kong were caused by local brief rain showers, and that the threshold for triggering side slope failures is crossed when the rain intensity reaches 70 mm/h.

Shi-Jie Jian [41], who discussed the influences of different factors on the mechanism of side slope slips through on-site measurement data, focused on the slopes (28 K + 900 to 31 K + 500) along the roads near Wuwanzi in Gongtian village, which is in the township of Fanlu in Jiayi county. As part of the old landslip region, this area has suffered from landslips for a long time—since the opening of the road. Several measurement devices have been installed for two-year periods since the beginning of 2000, carrying out consistent observations of local strata, land surface deformation, the underground water level, and rainfall. Relevant discussions have also been carried out in coordination with basic theory and indoor experiments. According to the observation results, the condition of this particular area is extremely unstable. It has also been discovered that the slip behaviors of this area are highly related to rainfall. The monitoring data since 2000 indicate that the cumulative rainfall needed for accelerated side slope slips is 80–270 mm, which equals a cumulative rainfall of 3–5 days.

Xing-Fu Ye [42] discussed the influence of rainfall seepage recharge on slope land collapse through the concept of watershed-based water balance using the following steps: (1) Estimating the base flow volume through the river flow volume qualification line method and the base flow estimation mode, and related to base flow as the seepage recharge of underground water; (2) taking into consideration the correlation system of rainfall, seepage, run-off, evapotranspiration, and underground water recharge through the unsaturated soil water balance method. These two modes were both shown to have similar results. The next step was to carry out a sensitivity analysis through STEDWIN (STEDwin is the smart editor to simplify working with Purdue or PennDOT* STABL programs) and to discuss its influence on the side slope stability. This revealed that the internal friction angle variability affects the side slope stability most significantly, followed by sloped variability and the rise of the underground water level, while the coherence and unit weight come next. From the figure of the area's rainfall and safety factor relationship, it can be seen that as the steepness of the slope reaches 35 degrees and the rainfall reaches 400–500 mm, according to the analysis of Janbu's simplified method, the safety factor reaches 1.0, forming an imminent risk of collapse. Regarding the discussion of practical examples, a correlation was found between the locations in the Cingshuei River basin that have suffered from collapse and the estimated rainfall recharge factors.

Yi-Feng Qiu [43] discussed the effects of the relevant factors that have affected the mechanism of side slope slips through on-site measurements and theoretical analyses. Long-term monitoring work regarding information such as land surface and stratum heterotaxies, rainfall, the underground water level, and the underground water flow volume of the surrounding areas of Wuwanzi along the Taiwan 18th Line Highway (29 K + 900 ~ 31 K + 500) has been carried out since the beginning of 2000. Serious failures took place on 26 June 2003, leading to the depletion of 150 m of the road base of the 31 K + 340 section and a 1.5-month highway disruption. The causes of this event's side slope failures were analyzed, discussed, and compared with the on-site monitoring results. With the theoretical analyses and the estimation of the stability of falling residual slopes, a final possible restoration solution was proposed. Ming-ren Gao [44] studied the slope lands near Wuwanzi along the Taiwan 18th Line Highway. An unexpected slip failure struck the 31 K + 340 section on 26 June 2003, causing disruption of the entire highway. A simulation of dry seasons and rainy seasons based on the underground water level was conducted using SLOPES/W software (SLOPE/W is the leading slope stability software for soil and rock slopes, GeoStudio company, head office in Calgary, Alberta, Canada). The load input was increased during rainy seasons to achieve the side slope's failure conditions. The failure mode was then integrated, analyzed, and compared with the observation statistics of the tiltmeter.

The Japanese Society for Landslip Solution Technologies [45] suggested a classification of the displacement speed effect on unstable side slope activities, as shown in Table 2. Such a classification is the displacement speed obtained through dividing the obtained stratum displacement by the observation time. The basis of classification depends on the level of speed and the trend of displacement. The table shows that unstable side slope activities can be roughly classified into four categories. The first three categories are side slopes with (1) emergency movements, (2) definite movements, and (3) semi-definite movements, which are moving slopes with confirmed problems that require immediate effective solutions. The fourth type of movement suffers from potential hazards in terms of the stability of side slopes if there is a certain trend in the displacement direction, because the displacement speed of this type of potentially moving side slope is extremely low—i.e., 0.5–2.0 mm every month, thus producing a displacement quantity of 6–24 mm every year. Further measurements are needed to confirm the stability of side slopes with displacement speeds of over 0.5 mm/month but without certain cumulative displacement directions (forward displacements and backward displacements have both been observed). The stability of such side slopes can be affected by measurement errors of devices or by the non-compact surrounding stuffing of inclination observation tubes when installed. The relevant precaution solutions and management principles provided by the Japanese Society for Expressways [46] are displayed in Table 3.

Table 2. Table of displacement speed and side slope stability judgment and suggestions [45].

Movement Types	Daily Movement (mm)	Monthly Movement (mm)	Accumulative Inclination of Certain Directions	Activity Judgement	Abstract
Emergency movements	Over 20 mm	Over 500 mm	Significantly apparent	Rapid mass wasting	Mass wasting movements and mudslides
Definite movements (movement values)	Over 1	Over 10	Apparent	Moving actively	Colluvial soil slips and deep slips
Semi-definite movements (vigilance values)	Over 0.1	Over 2.0	Slightly apparent	Moving slowly	Clay slips and backfill slips
Potential movements (alarm values)	Over 0.02	Over 0.5	Small	Further observation is needed	Clay slips and deep slips

Table 3. Monitoring management benchmarks of stratum slips [46].

Monitoring Methods	Management Classification		
	Attention	Vigilance	Evacuation Required
Land surface retractable meters	0.5–10 mm/day	>10 mm/day	>50 mm/day
Rainfall intensity		10–20 mm/h	>20 mm/h
Cumulative rainfall		50 mm	100 mm

To date, no domestic regulations have been set for relevant management values; in addition, no long-term monitoring results have been gained. Consequently, referring to Japan's recommended values and the device-measurable precision of side slope stability, judgement is inevitable, since Japan already has a certain amount of related experience. The retractable meters' action values are 5–10 times that of the vigilance values. The tentative monitoring management values are shown in Table 4 based on the premise that all of the monitoring devices' action values are five times that of their vigilance values. In the future, the data obtained by the monitoring of the devices will be analyzed, interpreted, and judged, and interactive comparisons will be made between different monitoring devices.

Table 4. Management values of the monitoring devices [30].

Monitoring Devices	Rain Gauge, Rainfall Intensity, and Cumulative Rainfall	Retractable Meters (Magnetic Induction)	Retractable Meters (Cable)	GPS (Horizontal)	Slope Circles
Action values (red light)	>20 mm/h and 100 mm/day	>±50 mm/day or ±5 mm/min	>±250 mm/day or ±25 mm/min	±80 mm/min	±0.3°/min
Vigilance values (yellow light)	10–20 mm/h and 50 mm/day	>±10 mm/day or ±1 mm/min	>±50 mm/day or ±5 mm/min	±50 mm/min	±0.1°/min
Precision	0.5 mm	0.1 mm	1 mm	30 mm	0.01°

In this study, by using these sensors, we integrated software and hardware to monitor landslides in Pingding. We aimed to review the benchmarking management values and to further investigate the management values. In addition, improvement of the real-time monitoring system was also expected after amending the management values based on different viewpoints to better determine the effects and efficiency of monitoring and vigilance.

In Pingding, Taiwan, particularly in summer (i.e., the rainy season), rainfall is the main contributing factor to the occurrence of landslides, rockfalls, and mudflows. Rainfall information is mainly used to monitor whether there is rain alert on the spot. Station rain gauges, automatically adopt the rain station's operation method, and provide a self-sufficient method. Such stations play a role in recording data, storage, transmission, and display systems for monitoring landslides. The results of this research not only monitor the landslide in real-time, but also manage the space transformation conditions when disasters occur.

The advantages of such a station are: (1) The station is equipped with advanced sensors and technologies; (2) provision of early warning information in real-time for residents who live in the area, as well as for managers who are in charge of this issue, by sending an SMS message to a mobile device; (3) a successful landslide monitoring station can support the monitoring network in Taiwan.

The feasibility of monitoring collapses and displacements through GPS was proven by this study. In addition, with the precision of different monitoring devices, the judgment of the management benchmark values of collapse monitoring established by domestic and overseas documentations was found to be reasonable, according to the monitoring results. Further data concerning typhoons and torrential rainfall will be collected in the future to confirm or amend the management benchmark values.

Finally, this study plays a vital role in the early warning system of landslides in Pingding, as highly effective monitoring can provide information on the current status that is not only applicable for weather forecasts, but also for disaster mitigation policy.

5. Conclusions

This study established the Pingding Monitoring System in the surrounding area of the Pingding Collapse Zone. On-site observation equipment was installed, and topographic and hydrological information was also collected as a reference for local disaster prevention and response.

Centered on emergency response, this study mainly focused on the displacement measurement of land surfaces. In light of the analysis results of side slope stability, the main cause of collapse and slips appears to be the water content of the soil. The previous collapse disaster confirmation suggests that inbuilt meters that can detect the soil moisture in mudstones should be used to serve as a reference for the necessity of advanced vigilance or as an alert when the water content in the soil is on the rise.

Regarding modern disaster prevention management thinking, the strategy for tackling disasters is a process that involves the application of the following four steps: (1) Readiness—establishing emergency response measures and a management system to enable a quick response once a disaster strikes; (2) response—taking immediate actions before, during, and after disasters to reduce the number of casualties and the property loss and to accelerate recovery; (3) restoration, including the restoration of basic vital resource supply systems within short periods of time and the long-term responsibility of restoring people's normal lives; (4) disaster reduction—advocating for certain policies and applying particular measures to alleviate the impacts brought by future disasters. The above thinking suggests that solving disaster management problems through single modes is not sufficient; instead, in addition to traditional engineering treatments, it is essential to consider multilateral aspects such as multilateral and cross-field discussions involving the following issues: Climatology, economics, engineering, geography, geology, law, meteorology, planning, psychology, public policy, and sociology.

In the future, we would like to apply this research to further applications of the system and its subsequent monitoring in order to establish an early-warning system (social-economic goal) and to conduct post-event analysis when a landslide has actually occurred (scientific goal).

Author Contributions: Conceptualization, T.Y.C. and T.V.H.; data curation, Y.M.F.; formal analysis, Y.M.F. and T.V.H.; funding acquisition, T.Y.C. and Y.M.F.; investigation, Y.M.F. and Q.T.B.; methodology, Y.M.F. and Q.H.N.; project administration, T.Y.C. and Y.M.F.; software, Y.M.F. and Q.H.N.; supervision, T.Y.C. and D.B.N.; validation, T.Y.C. and D.B.N.; visualization, Y.M.F. and T.V.H.; writing—Original draft, T.-Y.C. and T.V.H.; writing—Review and editing, T.Y.C. and T.V.H. All authors have read and agreed to the published version of the manuscript.

Funding: This article is the result of a state-level project financed by the Soil and Water Conservation Bureau and Geographic Information Systems Research Center, Feng Chia University, Taiwan (grant number MOST20051110).

Acknowledgments: The authors thanks the Soil and Water Conservation Bureau and Geographic Information Systems research Center, Feng Chia University, Taiwan for their support.

Conflicts of Interest: The authors declare no conflict of interest.

References

1. Chen, L.-K.; Liu, C.-H.; Ho, J.-Y. Application of a Three-Dimensional Deterministic Model to Assess Potential Landslides, a Case Study: Antong Hot Spring Area in Hualien, Taiwan. *Water* **2020**, *12*, 480. [CrossRef]
2. Hsu, Y.-C.; Loi, K.-F. Combining TRIGRS and DEBRIS-2D Models for the Simulation of a Rainfall Infiltration Induced Shallow Landslide and Subsequent Debris Flow. *Water* **2019**, *11*, 890. [CrossRef]
3. Bunn, M.D.; Leshchinsky, B.A.; Olsen, M.J.; Booth, A. A Simplified, Object-Based Framework for Efficient Landslide Inventorying Using LIDAR Digital Elevation Model Derivatives. *Remote Sens.* **2019**, *11*, 303. [CrossRef]
4. Ramos-Bernal, R.N.; Vazquez-Jimenez, R.; Remero-Calcerrada, R.; Arrogante-Funes, P.; Novillo, C.L. Evaluation of Unsupervised Change Detection Methods Applied to Landslide Inventory Mapping Using ASTER Imagery. *Remote Sens.* **2018**, *10*, 1987. [CrossRef]
5. Li, Y.; Liu, X.; Han, Z.; Dou, J. Spatial Proximity-Based Geogrpahically Weighted Regression Model for Landslide Susceptibility Assessment: A Case Study of QingChuan Are, China. *Appl. Sci.* **2020**, *10*, 1107. [CrossRef]
6. Liu, X.; Zhao, C.; Zhang, Q.; Peng, J.; Zhu, W.; Lu, Z. Multi-Temporal Loess Landslide Inventory Mapping with C-, X- and L-Band SAR Datasets—A Case Study of Heifangtai Loess Landslides, China. *Remote Sens.* **2018**, *10*, 1765. [CrossRef]
7. Assilzadeh, H.; Levy, J.K.; Wang, X. Landslide Catastrophes and Disaster Risk Reduction; A GIS Framework for Landslide Prevention and Management. *Remote Sens.* **2010**, *2*, 2259–2273. [CrossRef]
8. Long, N.T.; De Smedt, F. Analysis and Mapping of Rainfall-Induced Landslide Susceptibility in A Luoi District, Thua Thien Hue Province, Vietnam. *Water* **2019**, *11*, 51. [CrossRef]
9. Chou, T.Y.; Yeh, M.L.; Chen, Y.C.; Chen, Y.H. Disaster Monitoring and Management by The Unmaned Aerial Vehicle Technology. In Proceedings of the Conference ISPRS TC WII Symposium—100 Years ISPRS, Vienna, Austria, 5–7 July 2010. IAPRS, Vol. XXXWIII.
10. Anita, R.S.; Sushabhan, C.; Bhupendra, S. Wireless Disaster Monitoring and Management System for Dams. *Proc. Comput. Sci.* **2015**, *48*, 381–386. [CrossRef]
11. Kwak, Y.J. Nationwide Flood Monitoring for Disaster Risk Reduction Using Multiple Satellite Data. *ISPRS Int. J. Geoinform.* **2017**, *6*, 203. [CrossRef]
12. McCarthy, J.D.; Graniero, P.A.; Rozic, S.M. An Integrated GIS-expert System Framework for Live Hazard Monitoring and Detection. *Sensors* **2008**, *8*, 830–846. [CrossRef] [PubMed]
13. Arattano, M.; Marchi, L. Systems and Sensors for Debris-flow Monitoring and Warning. *Sensors* **2008**, *8*, 2436–2452. [CrossRef]

14. Xuan, W.; Chen, X.; Zhao, G. Early Warning Monitoring and Management of Disaster. In Proceedings of the 2007 IEEE International Geoscience and Remote Sensing Symposium, Barcelona, Spain, 23–27 July 2007; pp. 2996–2999.
15. Lucas, D.; Frankhauser, H.; McArdell, B.; Grob, R.; Herzog, R.; Bleiker, E.; Springman, S.M. Slope Stability of a Scree Slope Based on Integrated Characterisation and Monitoring. *Water* **2020**, *12*, 447. [CrossRef]
16. Troncone, A.; Conte, E.; Pugliese, L. Analysis of the Slope Response to an Increase in Pore Water Pressure Using the Material Point Method. *Water* **2019**, *11*, 1446. [CrossRef]
17. Wu, C. Landslide Susceptibility Based on Extrem Rainfall-Induced Landslide Inventories and the Following Landslide Evolution. *Water* **2019**, *11*, 2069. [CrossRef]
18. Park, S.-J.; Lee, C.-W.; Lee, M.-J. Landslide Susceptibility Mapping and Comparison Using Decision Tree Models: A Case Study of Jumunjin Area, Korea. *Remote Sens.* **2018**, *10*, 1545. [CrossRef]
19. Heo, B.Y.; Park, J.H.; Heo, W.-H. Sustainable Disaster and Safety Management of Government: Integrated Disaster and Safety Budget System in Korea. *Sustainability* **2018**, *10*, 4267. [CrossRef]
20. Leader, A.; Gaustad, G.; Tomaszewski, B.; Babbitt, C.W. The Consequences of Electronic Waste Post-Disaster: A Case Study of Flooding in Bonn, Germany. *Sustainability* **2018**, *10*, 4193. [CrossRef]
21. Estrela, V.V.; Saotome, O.; Loschi, H.J.; Hemanth, J.; Farfan, W.S.; Aroma, J.; Saravanan, C.; Grata, E.G.H. Emergency Response Cyber-Physical Framework for Landslide Avoidance with Sustainable Electronics †. *Technologies* **2018**, *6*, 42. [CrossRef]
22. Zhao, F.; Mallorqui, J.J.; Iglesias, R.; Gili, J.A.; Corominas, J. Landslide Monitoring Using Multi-Temporal SAR Interferometry with Advanced Persistent Scatterers Identification Methods and Super High-Spatial Resolution TerraSAR-X Images. *Remote Sens.* **2018**, *10*, 921. [CrossRef]
23. Ahmed, B.; Rahman, M.S.; Islam, R.; Sammonds, P.; Zhou, C.; Uddin, K.; Al-Hussaini, T.M. Developing a Dynamic Web-GIS Based Landslide Early Warning System for the Chittagong Metropolitan Area, Bangladesh. *ISPRS Int. J. Geoinform.* **2018**, *7*, 485. [CrossRef]
24. Myo Lin, N.; Rutten, M.; Tian, X. Flood Mitigation through Optimal Operation of a Multi-Reservoir System by Using Model Predictive Control: A Case Study in Myanmar. *Water* **2018**, *10*, 1371. [CrossRef]
25. Chang, C.-H.; Chung, M.-K.; Yang, S.-Y.; Hsu, C.-T.; Wu, S.-J. A Case Study for the Application of an Operational Two-Dimensional Real-Time Flooding Forecasting System and Smart Water Level Gauges on Roads in Tainan City, Taiwan. *Water* **2018**, *10*, 574. [CrossRef]
26. Chen, Y.-C.; Wu, Y.-H.; Shen, C.-W.; Chiu, Y.-J. Dynamic Modeling of Sediment Budget in Shihmen Reservoir Watershed in Taiwan. *Water* **2018**, *10*, 1808. [CrossRef]
27. Soil and Water Conservation Bureau, Council of Agriculture, Executive Yuan. *Jiufenershan Collapse Zone Observation Project (2nd Year)—Results Report*; Industrial Technology Research Institute: Hsinchu City, Taiwan, 2004.
28. The 3rd Engineering Office, Soil and Water Conservation Bureau, Council of Agriculture, Executive Yuan. *Pingding Village Pingding Collapse Zone Inspection Project Engineering Mandated Technical Services—Final Report*; Cheng-Wen Engineering Consultants Limited: Taichung City, Taiwan, 2005.
29. Chen, K.-Y. The Influence of Heterogeneous Soil Water Content on Side Slope Stability—A Case Study of the Analysis Method of the Soil Side Slope Stability Value of Pingding Collapse Zone, Linnei. Master's Thesis, Yulin University of Science and Technology, Yulin County, Taiwan, 2007. Unpunlished work.
30. Soil and Water Conservation Bureau, Council of Agriculture, Executive Yuan. *Pingding River Upstream Slope Land Monitoring System Establishment Project—Results Report*; Feng Chia University: Taichung City, Taiwan, 2008.
31. Brand, E.W. Analysis and design in residual soil. In Proceedings of the ASCE Geotechnical Engineering Division Specialty Conference-Engineering and Construction in Tropical and Residual Soils, Geotechnical Engineering—ASCE, Honolulu, HI, USA, 11–15 January 1982; pp. 89–141.
32. Lumb, P. Slope failures in Hong Kong. *Q. J. Eng. Geol.* **1975**, *8*, 31–65. [CrossRef]
33. Slosson, J.E.; Larson, R.A. Slope failures in southern California: Rainfall threshold, prediction, and human causes. *Environ. Eng. GeoSci.* **1995**, *1*, 393–401. [CrossRef]
34. Slosson, J.E. The role of engineering geology in urban planning. In *The Governor's Conference on Environmental Geology*; Colorado Geological Survey: Denver, CO, USA, 1969; pp. 8–15.

35. Slosson, J.E.; Krohn, J.P. Southern California landslides of 1978 and 1980. In *Storms, Floods, and Debris-Flows in Southern California and Arizona, 1978 and 1980: Committee on Natural Disasters (National Research Council, Report CSS-CND-019)*; Brooks, N.H., Ed.; National Academy Press: Washington, DC, USA, 1982; pp. 291–319.
36. Wilson, R.C.; Mark, R.K.; Barbato, G. Operation of a real-time warning system for debris flows in the San Francisco bay area, California. In *Hydraulic Engineering 1993: Proceedings of the 1993 Conference*; Shen, H.W., Su, S.T., Wen, F., Eds.; Hydraulics Division, American Society of Civil Engineers: Reston, VA, USA, 1993; pp. 1908–1913.
37. Wieczorek, G.F. Effect of rainfall intensity and duration on debris flows in central Santa Cruz mountains, California. In *Reviews in Engineering Geology*; Costa, J.E., Wieczorek, G.F., Eds.; The Geological Society of America: Boulder, CO, USA, 1987; pp. 93–104.
38. Auer, K.; Shakoor, A. Geotechnical characterization of drainage basin stability with respect to debris topples in center Virginia. *Bull. Assoc. Eng. Geol.* **1989**, *26*, 387–395.
39. Wong, H.N.; Ho, K.K.S. *Preliminary Risk Assessment of Earthquake-Induced Landslides at Man-Made Slopes in Hong Kong (GEO Report No. 98)*; Geotechnical Engineering Office: Kowloon, Hong Kong, 2000.
40. Brand, E.W.; Permchitt, J.; Phillipson, H.B. Relationship between rainfall and landslides in Hong Kong. In *Proceedings of the 4th International Symposium on Landslides, Toronto, Canada, 16–21 September 1984*; Canadian Geotechnical Society: Toronto, ON, Canada, 1984; pp. 377–384.
41. Jian, S.-J. Discussion about the Mechanism and Stability of the A Li Mountain Wuwanzi Slope Slip Event on 26 Jun. Master's Thesis, Yulin University of Science and Technology, Yulin County, Taiwan, 2002. Unpunlished work.
42. Ye, X.-F. Study on the Influences of Rainfall Seepage on Slope Stability—A Case Study of the Cingshuei River Basin. Master's Thesis, National Cheng Kung University, Tainan City, Taiwan, 2004. Unpunlished work.
43. Qiu, Y.-F. Discussion about the Mechanism of the A Li Mountain Wuwanzi Slope Slip Event on 26 Jun. Master's Thesis, Yulin University of Science and Technology, Yulin County, Taiwan, 2004. Unpunlished work.
44. Gao, M.-R. Study on Side Slope Slips—A Case Study of the Wuwanzi Collapse Zone. Master's Thesis, National Cheng Kung University, Tainan City, Taiwan, 2006. Unpunlished work.
45. Japan Landslide Society. *Japanese Society for Landslip Solution Technologies: Tips Essential for Landslip Solution Technology Design and Practice*, 2nd ed.; Landslip Inspection Construction; Japan Landslide Society: Tokyo, Japan, 1978.
46. Japanese Society for Expressways. *Dynamic Observation and Construction Research Report on Landslide Dangerous Area*; Japanese Society for Expressways: Tokyo, Japan, 1988; Volume 3.

applied sciences

MDPI

Article

Separating Landslide Source and Runout Signatures with Topographic Attributes and Data Mining to Increase the Quality of Landslide Inventory

Jhe-Syuan Lai

Department of Civil Engineering, Feng Chia University, Taichung 40724, Taiwan; jslai@fcu.edu.tw; Tel.: +886-4-2451725 (ext. 3118)

Received: 4 September 2020; Accepted: 21 September 2020; Published: 23 September 2020

Abstract: Landslide sources and runout features of typical natural terrain landslides can be observed from a geotechnical perspective. Landslide sources are the major area of occurrences, whereas runout signatures reveal the subsequent phenomena caused by unstable gravity. Remotely sensed landslide detection generally includes runout areas, unless these results have been excluded manually through detailed comparison with stereo aerial photos and other auxiliary data. Areas detected using remotely sensed landslide detection can be referred to as "landslide-affected" areas. The runout areas should be separated from landslide-affected areas when upgrading landslide detections into a landslide inventory to avoid unreliable results caused by impure samples. A supervised data mining procedure was developed to separate landslide sources and runout areas based on four topographic attributes derived from a 10–m digital elevation model with a random forest algorithm and cost-sensitive analysis. This approach was compared with commonly used methods, namely support vector machine (SVM) and logistic regression (LR). The Typhoon Morakot event in the Laonong River watershed, southern Taiwan, was modeled. The developed models constructed using the limited training data sets could separate landslide source and runout signatures verified using the polygon and area constraint-based datasets. Furthermore, the performance of developed models outperformed SVM and LR algorithms, achieving over 80% overall accuracy, area under the curve of the receiver operating characteristic, user's accuracy, and producer's accuracy in most cases. The agreement of quantitative evaluations between the area sizes of inventory polygons for training and the predicted targets was also observed when applying the supervised modeling strategy.

Keywords: data mining; landslide detection; landslide inventory; Typhoon Morakot

1. Introduction

Natural hazards occur frequently in Taiwan. Among them, landslides can be triggered by heavy rainfall events, especially in mountainous areas. The precipitous terrain, complicated geology, and dense population may increase vulnerability, causing a serious loss of life, property damage, and economic loss. Furthermore, these events can affect water resource supply and cause other livelihood problems. Therefore, landslide analysis and assessment have become critical in natural hazard and disaster mitigation, prevention, and reconstruction in Taiwan. A growing number of studies have investigated landslide-related topics on different scales, such as slope stability analysis for a specific slope site [1,2]; regional landslide detection and mapping [3–7]; characterization of the relationship between landslides and environmental or triggering factors [8–13]; susceptibility, hazard, risk assessment, and management [14–19]; and modeling or estimating physical, environmental, and rainfall-based parameters [20–24]. Landslide risk assessment and management are crucial, systematic, and extensive frameworks in the related works. Dai et al. [14] divided this topic into several parts. In this framework, generating landslide inventory (inventory map and database are used synonymously in this study) is essential to connect the following assignments.

A basic landslide inventory should record the landslide's location, date (event-based) or period (multitemporal), and types of movement [25]. The definition, assumptions, requirements, production processes, and statistical properties of landslide inventories have been discussed in detail by Guzzetti et al. [25], Harp et al. [26], Malamud et al. [27], and Shao et al. [28]. Furthermore, Galli et al. [29] and Mondini et al. [11] have assessed the quality and completeness of landslide inventory maps by comparing two maps of the same area. Numerous studies have indicated that the continuing improvements in remote sensing and geographic information systems (GISs) have led to cooperation with data mining and machine learning algorithms to produce a regional landslide inventory. In particular, integrating GIS-based models with geo-spatial data [18,30] and generating event-based landslide inventory [11,25,31] have garnered interest.

Three common features of typical natural terrain landslides have been observed from a geotechnical perspective. These features were the landside source, trail, and deposition fan [32]. The surface of the rupture, comprising the main scarp and the scarp floor, is defined as the source area. A landslide trail may also occur predominantly as a result of the transport of the landslide mass. The majority of the landslide mass is deposited (i.e., deposition fan). The term runout is generally used to indicate the landslide trail and deposition fan [33]. Landslide detection performed using automatic and semiautomatic methods with remotely sensed images usually includes runout areas unless these results have been excluded manually through a detailed comparison with stereo aerial photos and other auxiliary data. The areas detected using remotely sensed landslide detection can be referred to as "landslide affected" areas. Mixing landslide source areas with runout regions may reduce the reliability of a landslide analysis [34], such as susceptibility and hazard assessments. These runout areas should be separate from landslide source areas because they have different mechanisms. More precisely, producing landslide inventory requires further processing after the remote sensing of landslide detections.

Data mining and machine learning-based algorithms have been used increasingly for landslide modeling, especially in landslide susceptibility assessments, such as decision trees, deep learning systems, evolution-based algorithms, fuzzy theory, neural networks, random forests (RF), and support vector machines (SVM) [35–40]. Related studies have also considered various composites and compared the effectiveness of strategies based on established methods in achieving a specific purpose [41–43]. However, few studies have employed these approaches for detecting landslides and producing inventories that consider the separation of runout areas from landslide affected regions. The RF method [44] consists of data mining and machine learning algorithms that have displayed excellent performance in the analysis of numerous remote sensing and landslide topics [45,46]. The RF algorithm is based on the ensembles of various decision tree results and exhibit desirable properties, such as high accuracy, robustness against overfitting the training data, and integrated measures of variable importance [47]. Furthermore, Lai and Tsai [48] and Lai et al. [34] have demonstrated that combining the RF algorithm with a cost-sensitive analysis [49] in landslide susceptibility assessments can reduce extreme omission (missing) or commission (false alarm) predictions because it adjusts the decision boundary.

The main purpose of this study was to explore the feasibility of separating landslide source and runout areas based on landslide affected extents extracted from remotely sensed images in order to upgrade landslide detections into a landslide inventory. The significance of topographic data in relation to the flow velocity, geomorphology, runoff rate, soil water content, and differences between landslide source and runout areas was demonstrated [34,50]. The term "signature" used in this study represents the patterns of landslide source and runout areas in a feature space. Feature space means that the dimension is the used factors with the samples based on a scatter plot form. Therefore, the developed RF-based data mining models and topographic attributes, such as aspect, curvature, elevation, and slope, constructed for the Typhoon Morakot event in 2009 were combined to compare the results with those obtained using SVM and logistic regression (LR) algorithms. SVM and LR algorithms are commonly applied in the related domains. The performance of cost-sensitive analysis was also evaluated to improve the constructed models by adjusting the decision boundary.

2. Materials and Methods

2.1. Study Site and Data

The study area was located in the Laonong River watershed in southern Taiwan and covered approximately 117 km^2, as illustrated in Figure 1a. The elevation in the study site ranged from 258 to 1666 m above sea level, measured using the digital elevation model (DEM) modified by Chiang et al. [51]. The average elevation was 716 m. The slope range, average slope, and standard deviation were 0–71.2°, 25.84°, and 11.98°, respectively, which indicated that the terrain of the Laonong River watershed is steep. Three geological formations and four soil types were identified in the study area. The geological formations were Lushan, Sanhsia, and Toukoshan. The four soil types were alluvium, colluviums, lithosol, and loam. Lai et al. [34] reported detailed information regarding geological formations and soil types within the study site. The Laonong River watershed is located in the tropical monsoon region, and the annual precipitation is approximately 3400 mm. Therefore, this area is frequently struck by typhoons. For example, Typhoon Morakot caused extensive rainfall in 2009, which resulted in numerous landslides and debris flow in southern Taiwan. An official report [52] indicated that 769 casualties or missing people were directly or indirectly caused by these landslides. Furthermore, Typhoon Morakot caused losses of approximately $526 million because of the damage to agriculture, forestry, and fishery. In particular, a riverside village called Xiaolin (sometimes spelled Shiaolin, Hsiaolin) was destroyed by the landslides and debris flows from a devastating landslide nearby, which caused approximately 500 fatalities.

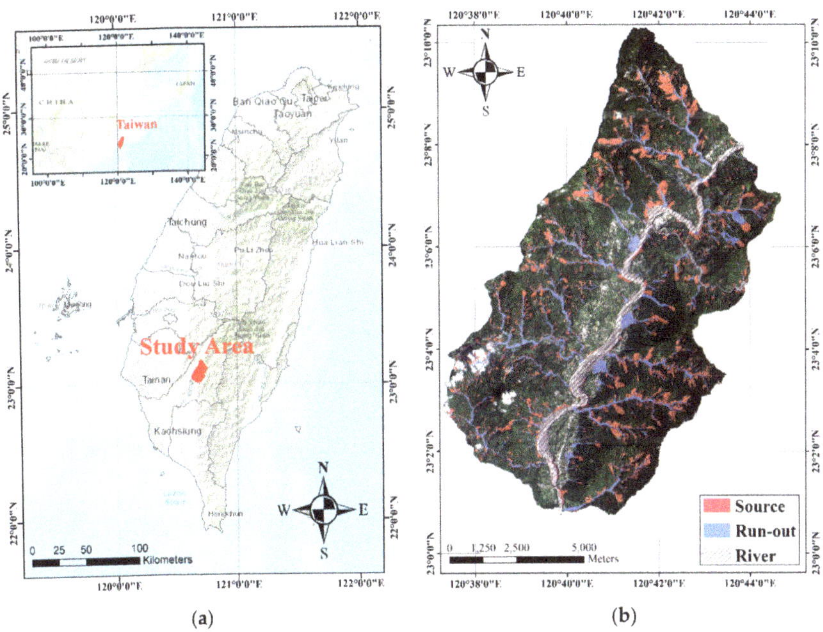

Figure 1. Location of the study site (**a**) and landslide inventory map (**b**).

A landslide inventory map of Typhoon Morakot revealed deep-seated landslides, as illustrated in Figure 1b. The landslide source, runout, and channel classes were interpreted manually [34] based on stereo aerial photos and auxiliary data. This study assumed that the used landslide inventory comprised the results of the detection of affected landslide areas extracted from the remotely sensed images. The landslide inventory was also used for quantitative verifications. The 10–m DEM was then analyzed to derive topographic attributes, including the aspect, curvature, elevation, and slope, as illustrated in Figure 2. Furthermore, the landslide inventory map was converted into a grid format of 10 m^2 to match the topographic factors for constructing data mining-based models.

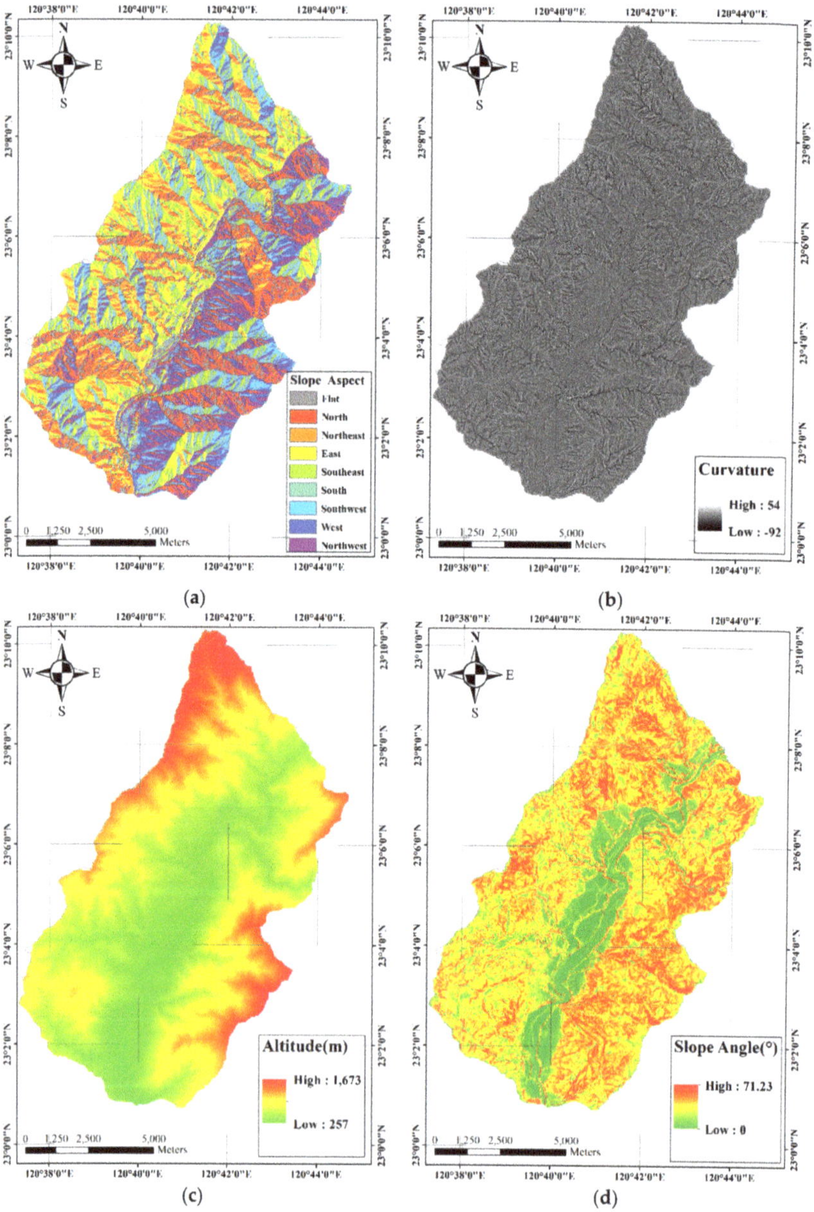

Figure 2. Topographic layers used in this study. (**a**) Aspect, (**b**) curvature, (**c**) elevation, and (**d**) slope.

2.2. Procedure and Data Mining Algortithms

Figure 3 illustrates the conceptual procedure implemented in this study. The topographic attributes were derived from the 10–m DEM, which connected the samples extracted from the landslide inventory for modeling (Sections 2.2.1 and 2.2.2) and verification (Section 2.3). Two experiments were designed for the demonstration of separating landslide source and runout signatures. First, the limited training datasets were selected (the five largest area sizes of landslide source and runout classes in the landslide inventory polygons) for polygon by polygon modeling. Second, each model was verified using other four polygon-based samples. Third, these models were used to predict the polygon samples with an area of ≥100,000 and 50,000–100,000 m^2 to identify the optimal model. Previous results are presented in Sections 3.2 and 3.3. Detailed information regarding the samples extracted from the landslide inventory is presented in Table 1. Finally, the spatial patterns of landslide source and runout areas were produced instance by instance based on the output labels of the optimal model, as demonstrated in Section 3.4.

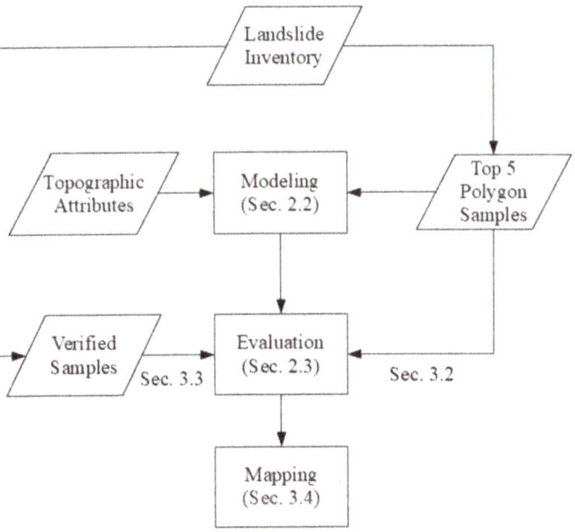

Figure 3. The conceptual procedure used in this study to model and verify the separation between landslide source and runout signatures.

Table 1. The training and verification samples extracted from the landslide inventory polygons.

Dataset	Landslide Source		Runout	
	Area (m^2)	No. of Samples	Area (m^2)	No. of Samples
No. 1	250,493	2508	385,381	3839
No. 2	154,709	1545	239,199	2386
No. 3	141,574	1412	206,055	2060
No. 4	132,005	1313	197,162	1980
No. 5	126,636	1265	190,338	1905
≥10,000 m^2 *	155,333 (avg)	9318	162,768 (avg)	27,640
50,000—100,000 m^2	68,670 (avg)	15,798	63,792 (avg)	8937

*Areas and number of samples in the table include the training polygon samples, but these samples were eliminated during the verified process.

2.2.1. Developed Models

The random forests (RF) algorithm [44] is a data mining method used for constructing a classification-based model that employs a supervised strategy. RF is an extension of the tree-based algorithms. The RF employs the information gain (IG) measure or Gini index to determine the degree of impurity of factors or variables, applying the bootstrap and vote operators to improve the results derived from the tree-based rules. The bootstrap method "randomly" selects subtraining data to generate several trees (termed "forests") to avoid the overfitting results. The vote component is used to determine the optimal results according to the developed trees, thus improving the RF classifier. A topographic factor with a larger IG or Gini value had to be selected because higher values indicate a higher priority for constructing a tree node, which should be ignored in the next computation. A sequence of tree-based rules was constructed after several iterations. The rules were then used to classify other instances. Nominal (or discrete) and numeric (or continuous) formats are commonly used in the field. Entropy theory was applied to calculate the IG in the nominal case, as displayed in Equations (1)–(3). In these equations, E(A) represents the entropy of all training data; S and R represent the numbers of landslide sources and runout samples, respectively; E′(a) and v are the entropy and number of subsets of a specific topographic factor, respectively; E(aj) is the entropy of the subset in a specific topographic factor, calculated using Equation (1); IG(a) represents the IG of a specific topographic factor. The Gini index was applied to determine the IG in the numeric cases, as displayed in Equations (4) and (5), where C represents a segmented point for a specific topographic factor used to divide continuous data into two parts, and N_1 and N_2 represent the numbers of $a \leq C$ and $a > C$, respectively. The steps were detailed by Guo et al. [53].

$$E(A) = -\frac{S}{S+R}\log_2\frac{S}{S+R} - \frac{R}{S+R}\log_2\frac{R}{S+R} \quad (1)$$

$$E'(a) = \sum_{i=1}^{v}\frac{S_i+R_i}{S+R}E(a_i) \quad (2)$$

$$IG(a) = E(A) - E'(a) \quad (3)$$

$$Gini(a \leq C \text{ or } a > C) = 1 - \sum_{i=1}^{m}\frac{n_i}{N} \quad (4)$$

$$IG(a, C) = \frac{N_1}{N}Gini(a \leq C) + \frac{N_2}{N}Gini(a > C) \quad (5)$$

The strategy of adjusting the decision boundary used in the study is termed as the cost-sensitive analysis. The cost-sensitive analysis is a postclassification method based on the cost matrix that reclassifies the instances and balances the accuracies of certain classes when missing or false alarm errors are unreliable [49,54,55]. The dimension of the cost matrix is equal to that of the confusion matrix. The confusion matrix used in this study is presented in Table 2. True negative (TN), false negative (FN), false positive (FP), and true positive (TP) were used to represent an agreement between the classified and reference labels in counts tabulated in a confusion matrix. The cost in this study indicated a weighting without the unit. The diagonal costs in the table represent correct results for the TN and TP, and the remaining costs indicate misclassification costs between the FN and FP. The diagonal costs and other elements are usually set to 0 and 1, indicating unadjusted and adjusted conditions of the decision boundary. A large amount of missing or false alarm errors are severe FP or FN errors, respectively. Adjusting the decision boundary by increasing the cost (weighting) of the FP or FN leads to the inclusion of more samples, thereby balancing the classification result of a certain class. Equation (6) displays the optimal prediction (R) of sample x in class i, where P(j|x) is the likelihood of estimating a classification of a sample into all classes (j), and C represents the costs. In binary cases, the optimal prediction is the landslide source label (class 1 or positive) if the expected cost of this prediction is less than or equal to the expected cost of predicting the runout label (class 0 or negative), as displayed in Equation (7). Given $p = P(1|x)$ and $C_{TN} = C_{TP} = 0$, Equation (7) can be simplified, as displayed in Equation (8). An adjusted threshold (p*) of the decision boundary, presented in Equation (10), can be derived from Equation (9) to classify a sample x as the landslide source label when P(1|x) is larger than or equal to the threshold [56]:

$$R(i|x) = \sum_{j} P(j|x)C_{ji} \tag{6}$$

$$P(0|x)C_{FP} + P(1|x)C_{TP} \leq P(0|x)C_{TN} + P(1|x)C_{FN} \tag{7}$$

$$(1 - p)C_{FP} \leq p\, C_{FN} \tag{8}$$

$$(1 - p^{*})C_{FP} \leq p^{*}C_{FN} \tag{9}$$

$$p^{*} = \frac{C_{FP}}{C_{FP} + C_{FN}} \tag{10}$$

Table 2. The confusion matrix used in this study.

Confusion Matrix		Ground Truth	
		Runout	**Landslide Source**
Prediction	Run-out	True Negative (TN)	False Negative (FN)
	Landslide source	False Positive (FP)	True Positive (TP)

2.2.2. Algorithms for Comparison

Two commonly used data mining methods were used for comparison with the developed models. The support vector machines (SVM) method has been wieldy applied in remote sensing studies [57] to classify land cover or use targets, extract biophysical parameters in different spatial resolutions, and perform landslide analyses [40,41]. The core role of the SVM classifier is to determine the largest margin of the classifier based on factor transformations with linear and nonlinear functions. These transformations are known as the kernel function. Equation (11) describes a concept of the SVM-based classification with a "sign" operator for the binary classification, where l, y, and k(xi,xj) indicate the number of used factors (support vectors), the outputted label, and a kernel function, respectively, and a and b represent the coefficients of the margin. The kernel functions used in this study were linear, polynomial, and radial basis functions.

$$f(x) = \text{sign}\left[\sum_{i=1}^{l} a_i y_i k(x_i, x_j) + b\right] \tag{11}$$

Logistic regression (LR) is a typical and traditional statistical method used to model landslide events [20,58]. The computation describes the relationship between the dependent and independent variables as defined in Equation (12), where p refers to the likelihood or probability, a is a constant, b is the regression coefficient, and x represents independent variables or used factors:

$$p = \frac{\exp(a + b_1x_1 + b_2x_2 + \ldots + b_nx_n)}{1 + \exp(a + b_1x_1 + b_2x_2 + \ldots + b_nx_n)} \tag{12}$$

2.3. *Accuracy Assessment*

The quantitative results of comparing the output and ground truth labels, identified based on the threshold, were derived from the confusion matrix to verify the constructed models. An instance was categorized into the landslide (positive) class in most cases when the likelihood of occurrence was ≥0.5; otherwise, the instance was assigned the nonoccurrence (negative or non-landslide) label. Similar to previous classifications, the positive and negative labels in this study were changed from landslide and non-landslide to landslide source and runout classes, respectively. The threshold of 0.5 for classification could also be adjusted by the cost-sensitive analysis (p*).

Four commonly used quantitative indices derived from a confusion matrix, namely overall accuracy (OA), user's accuracy (UA), producer's accuracy (PA), and area under the receiver operating characteristic (ROC) curve (AUC), were used to quantitatively evaluate the developed models. OA represents the ratio of samples correctly classified as TN and TP, as illustrated in Equation (13). UA and PA reflect the errors for each class (landslide source and runout) as presented in Equations (14)–(17), where R and S represent the runout and landslide source classes. The false alarm (commission errors) and missing (omission errors) assignments can be calculated as 1–UA and 1–PA, respectively. The AUC is commonly used to assess data mining-based models. The binary classification threshold is used for reclassifying instances, calculating the rates of FPs and TPs to draw the ROC curve. In general, the AUC ranges from 0.5–1, and results larger than 0.7 are acceptable, and larger than 0.8 are excellent [59].

$$OA = \frac{TN + TP}{TN + FN + FP + TP} \tag{13}$$

$$UA(R) = \frac{TN}{TN + FN} \tag{14}$$

$$UA(S) = \frac{TP}{FP + TP} \tag{15}$$

$$PA(R) = \frac{TN}{TN + FP}. \tag{16}$$

$$PA(S) = \frac{TP}{FN + TP} \tag{17}$$

3. Results

The data mining modeling and accuracy assessments used in this study were developed using the WEKA environment (http://www.cs.waikato.ac.nz/mL/weka/), which is a free and open-source platform. A four-stage step was designed to explore the separability between landslide sources and runout signatures. The first stage, as reported in Section 3.1, compared the topographic attributes between the landslide source and runout samples extracted from different sizes of landslide inventory polygons that had an area of ≥100,000 or 50,000–100,000 m^2. In the second stage, the RF and cost-sensitive analysis-based data mining algorithms were used to construct the models polygon by polygon, with the limited training datasets (the five largest area sizes in the landslide inventory polygons were considered), as reported in Section 3.2. The major parameter in the RF-based computation was the number of trees (Ntree). Du et al. [60] determined that 10–200 Ntree did not have any effect on the results, but an increase in Ntree increased the computation loading. Therefore, 100 trees were adopted in this study, as suggested by WEKA, to develop the RF-based models. In the third stage (Section 3.3), the constructed models were applied to predict other landslide source and runout samples extracted from the polygons with an area of ≥100,000 and 50,000–100,000 m^2, respectively. The developed models were compared with the commonly used methods, namely SVM and LR. The fourth stage consisted of visualizing the spatial distributions of predicting the landslide source and runout areas to assess the performance of the constructed model, as reported in Section 3.4.

3.1. Signatures of Topographic Attributes

The constraints on the area size of the inventory polygons were compared in the pairs of the topographic attributes to preliminarily examine differences between landslide source and runout signatures. The samples were extracted based on area sizes of ≥100,000 and 50,000–100,000 m^2. Furthermore, the value domain of the aspect, curvature, elevation, and slope layers was normalized into the range of 0–1 for the visualization of data distribution and separation. The patterns obtained by comparing the topographic datasets and area constraints are illustrated in Figures 4 and 5, respectively.

A larger area size enabled a larger separation of landslide source and runout features. The disagreement was not observed in the smaller landslide inventory polygons. This trend is similar to findings reported by Lai et al. [34]. This trend may reflect that the complete mechanisms of landslide source and runout areas appear in the larger landslide sizes. Moreover, the elevation is a critical factor in this study. These data distributions reveal the feasibility of distinguishing the landslide source and runout signatures when the sizes of landslide-affected areas were sufficient to reveal their mechanisms.

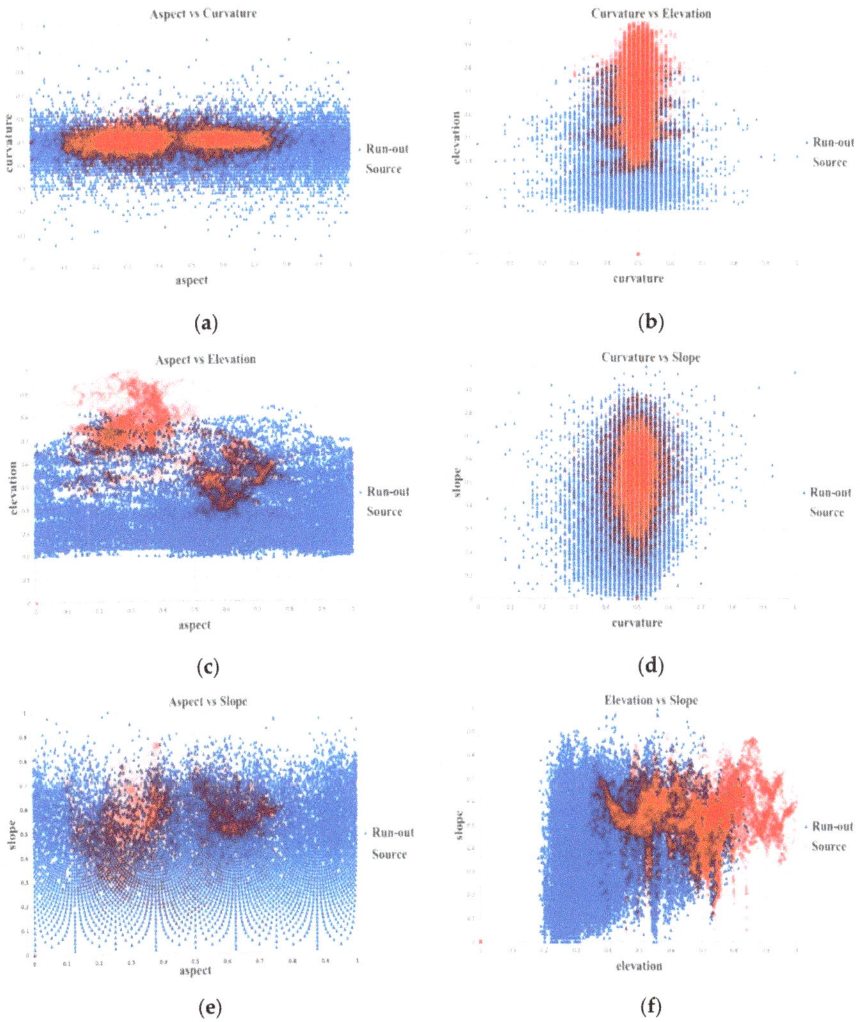

Figure 4. Patterns for comparing aspect, curvature, elevation, and slope layers with the samples extracted from the inventory polygon with an area of ≥100,000 m^2, where red and blue dots represent the landslide source and runout classes, respectively. (**a**): Aspect vs. curvature.; (**b**): curvature vs. elevation; (**c**): aspect vs. elevation; (**d**): curvature vs. slope; (**e**): aspect vs. slope; (**f**): elevation vs. slope.

3.2. Construction of Random Forests Based Data Mining Models

The RF-based data mining models were constructed using the top five polygons of area sizes in the landslide inventory polygons to further explore the feasibility of separating the landslide source and runout areas. Each polygon was transformed into a grid format of 10 m^2. These samples could be used to extract the corresponding topographic attributes and produce the training and verification datasets. The quantitative evaluations of OA, AUC, PA, and UA are presented in Figure 6. The accuracies exceeded 80% in most cases. Model No. 3 provided the most accurate predictions. However, the results of predicting (a) the No. 3 and No. 5 polygon samples by using the No. 1 and No. 2 models, (b) the No. 5 polygon samples by using the No. 3 and No. 4 models, and (c) the No. 1 and No. 2 polygon samples by using the No. 5 model revealed a disagreement between the training and verification datasets. Possible reasons for this disagreement include (a) imbalanced sample ratios between landslide source and runout classes, (b) dissimilar data distributions between different inventory polygons, (c) lack of other critical factors, and (d) limited training data.

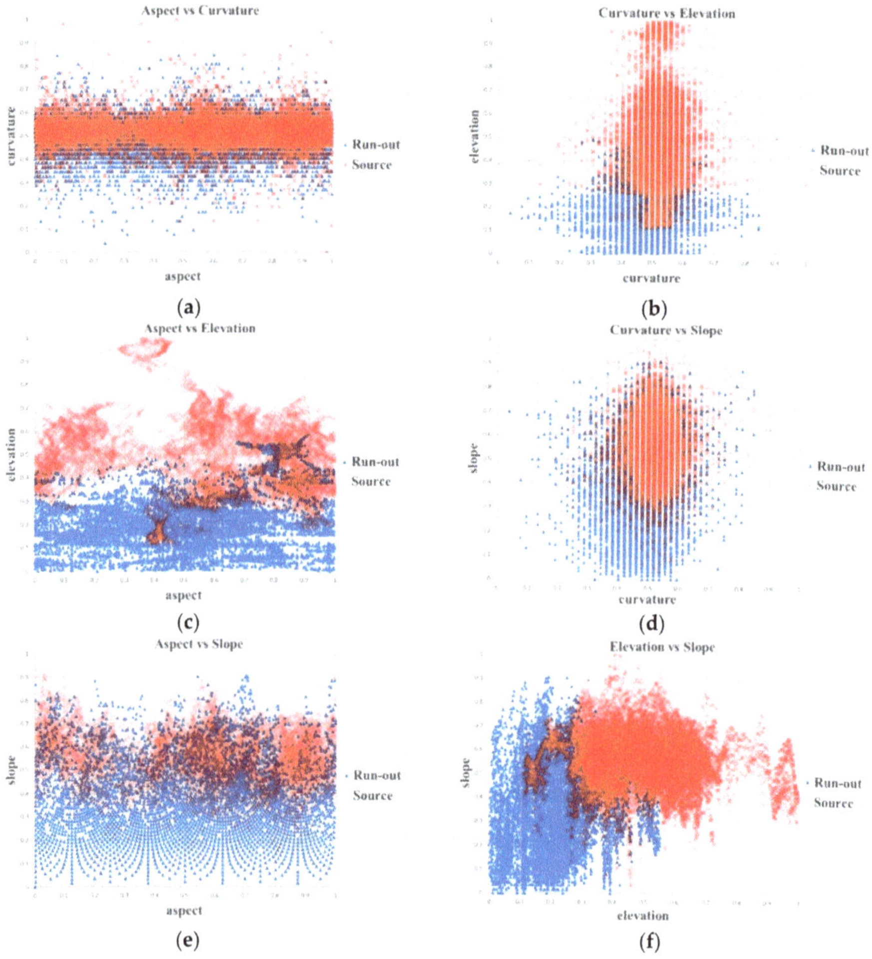

Figure 5. Patterns for comparing aspect, curvature, elevation, and slope layers with the samples extracted from the inventory polygon with an area of 50,000–100,000 m^2, where red and blue dots represent the landslide source and runout classes, respectively. (**a**): Aspect vs. curvature.; (**b**): curvature vs. elevation; (**c**): aspect vs. elevation; (**d**): curvature vs. slope; (**e**): aspect vs. slope; (**f**): elevation vs. slope.

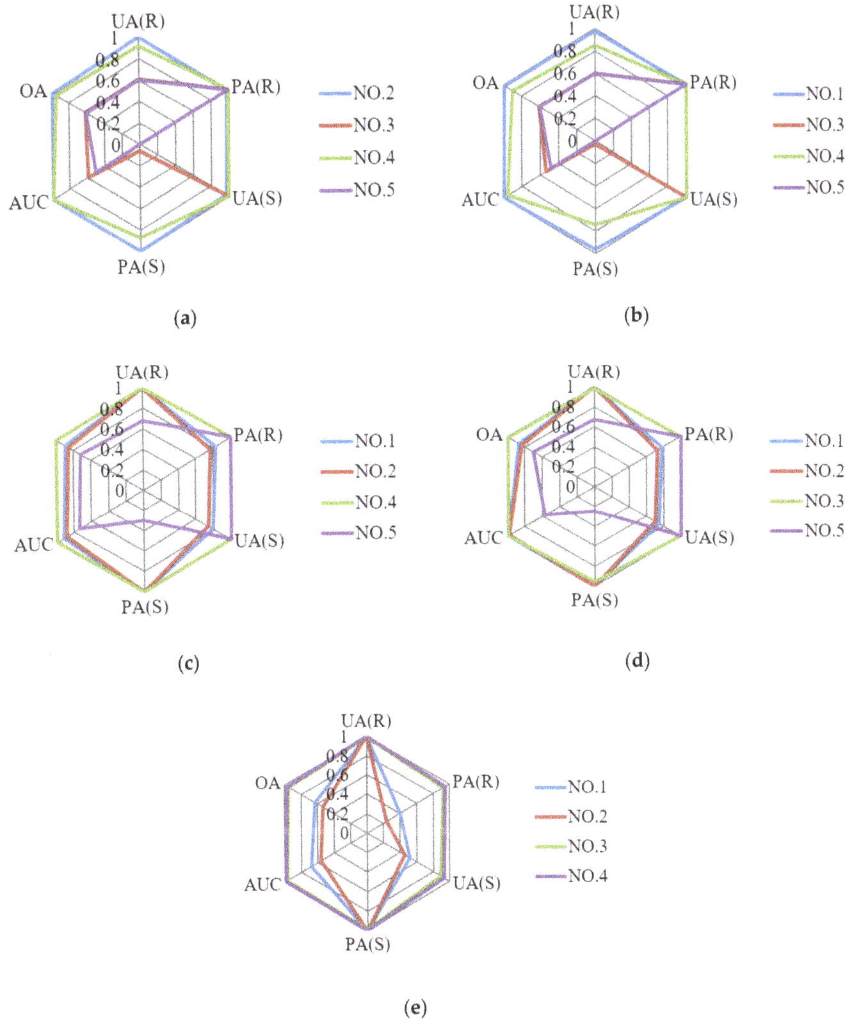

Figure 6. Accuracies of the random forest (RF)-based models using the top five polygon samples to predict each other, where OA, AUC, PA, UA, S, and R represent overall accuracy, the area under the ROC curve, producer's accuracy, user's accuracy, landslide source, and runout, respectively. (**a**) No. 1, (**b**) No. 2, (**c**) No. 3, (**d**) No. 4, and (**e**) No. 5 polygon samples for modeling.

To address this problem, cost-sensitive analysis procedures, similar to the procedure presented by Lai and Tsai [48], were employed to adjust the decision boundary during RF-based modeling. The results of the cost-sensitive analysis for different costs are illustrated in Figure 7. The developed model without the cost-sensitive analysis (i.e., the cost is equal to 1) had a high missing error for detecting landslide source regions, as displayed in Figure 7a–f, and runout areas, as displayed in Figure 7g,h, because of the assignment of higher FN and FP, respectively. The variations and improvements of accuracies could be observed after increasing the costs of FN and FP for the cases presented in Figures 7a–f and 7g,h, respectively. Based on Figures 6 and 7, Table 3 lists the representative models applying the RF algorithm and cost-sensitive analysis with different inventory polygon samples for further verification, as presented in Section 3.3.

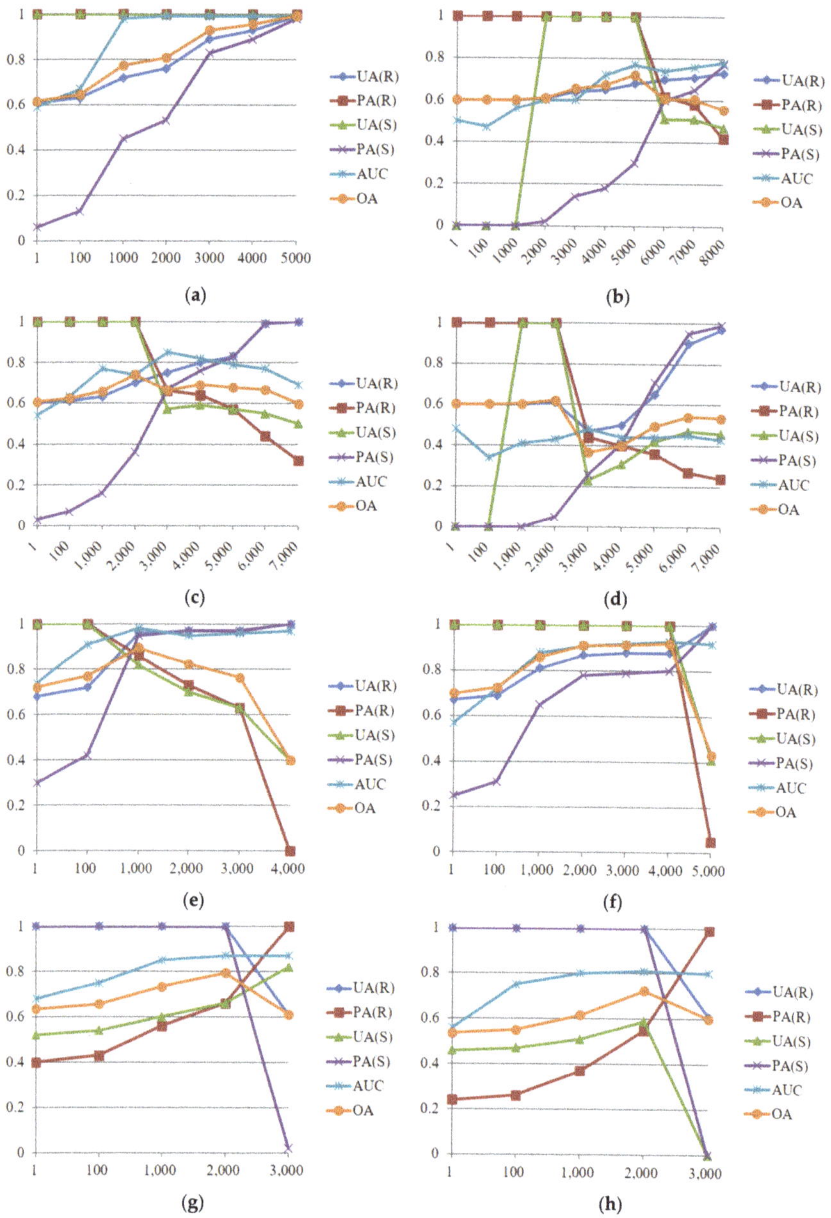

Figure 7. Accuracies (*Y*-axis) of the RF-based models with different costs (*X*-axis) and polygon samples for predictions, where OA, AUC, PA, UA, S, and R represent the overall accuracy, area under the ROC curve, producer's accuracy, user's accuracy, landslide source, and runout, respectively. (**a**) No. 1 model predicting No. 3 polygon samples, (**b**) No. 1 model predicting No. 5 polygon samples, (**c**) No. 2 model predicting No. 3 polygon samples, (**d**) No. 2 model predicting No. 5 polygon samples, (**e**) No. 3 model predicting No. 5 polygon samples, (**f**) No. 4 model predicting No. 5 polygon samples, (**g**) No. 5 model predicting No. 1 polygon samples, and (**h**) No. 5 model predicting No. 2 polygon samples.

Table 3. The representative models using the random forests (RF) algorithm with the costs for verifications based on pervious comparisons. For example, RF_3000 indicates that the RF algorithm with a cost of 3000 was used.

Model	No. 1	No. 2	No. 3	No. 4	No. 5
Algorithm_Cost	RF RF_3000 RF_4000 RF_5000 RF_6000 RF_7000	RF RF_3000 RF_4000 RF_5000	RF RF_1000	RF RF_2000 RF_3000 RF_4000	RF RF_2000

3.3. Model Performances and Comparisons

Based on settings presented in Table 3, the constructed models were used to predict the samples extracted from other inventory polygons with an area of ≥100,000 or 50,000–100,000 m^2, namely the larger and smaller inventory polygons, presented in Sections 3.3.1 and 3.3.2, respectively. Furthermore, the performances between the developed models and the approaches of SVM and LR were compared.

3.3.1. The Case of Larger Inventory Polygons

The prediction results of No. 1 models obtained using the RF and cost-sensitive analysis are presented in Figure 8a. The examination of this figure reveals that a cost of 5000 provided more favorable results, with the accuracy reaching over 80% in most indices. However, the evaluations of SVM-based approaches in consideration of linear, the first- to third-order polynomial, and radial basis functions are displayed in Figure 8b. Overall, the second-order polynomial function outperformed others. Compared with previous LR results, Figure 8c indicates that the developed model in Section 3.2 obtained the highest accuracies, although the UA of the landslide source was slightly lower.

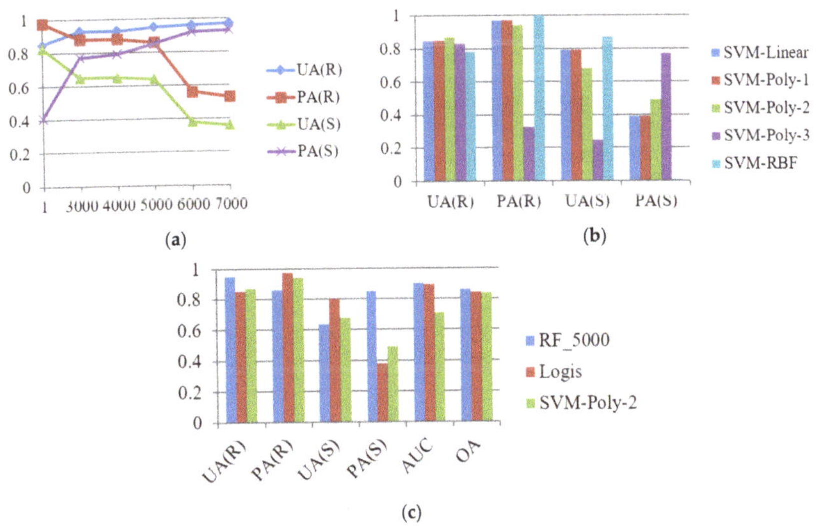

Figure 8. Performances (Y-axis) of No. 1 model, where OA, AUC, PA, UA, S, and R represent the overall accuracy, area under the ROC curve, producer's accuracy, user's accuracy, landslide source, and runout, respectively. (**a**) Using the algorithm with the costs; (**b**) using the support vector machine (SVM) with linear, the first- to third-order polynomial, and radial basis (RBF) functions; (**c**) comparing the optimal results of (**a**) and (**b**) by using logistic (Logis) regression.

Similar procedures to those used to verify the No. 2 model were applied to assess the algorithms of RF with the cost-sensitive analysis, SVM, and LR. Figure 9a further compares the RF with costs, as displayed in the third column in Table 3. Based on this figure, the RF algorithm appeared to outperform the other models without increasing costs. As illustrated in Figure 9b, the accuracies of all the functions of SVM modeling were similar, with higher missing errors for detecting landslide sources. The RF and the second-order polynomial function of SVM were compared with LR, as illustrated in Figure 9c, which revealed that the AUC of the LR was the highest. The accuracies of the three other approaches were similar.

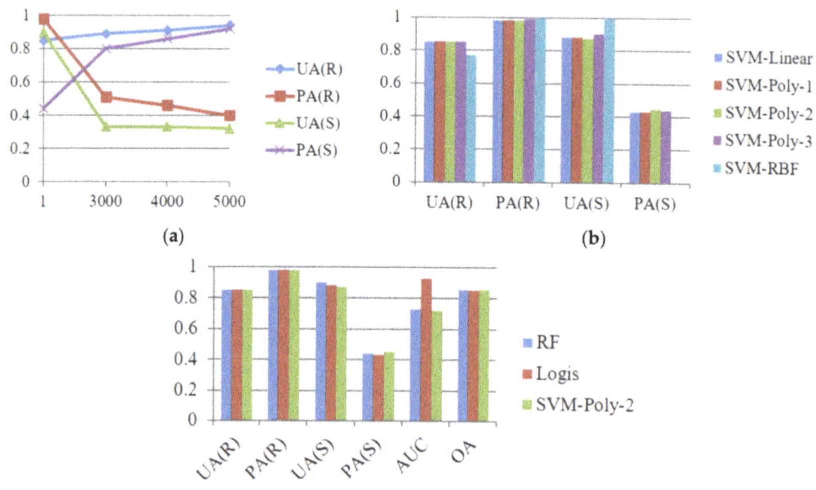

Figure 9. Performance (*Y*-axis) of the No. 2 model, where OA, AUC, PA, UA, S, and R represent the overall accuracy, area under the ROC curve, producer's accuracy, user's accuracy, landslide source, and runout, respectively. (**a**) Using the RF algorithm with the costs; (**b**) using SVM with linear, the first- to third-order polynomial, and radial basis (RBF) functions; (**c**) comparing the optimal results of (**a**) and (**b**) by using logistic (Logis) regression.

For the No. 3 model, Table 4 presents the comparison of the RF algorithm and cost-sensitive analysis. The results indicate that the original RF algorithms provided higher accuracy, whereas the RF method with a cost of 1000 had lower PA and UA for the runout and landslide source classes, respectively. In the SVM case, Figure 10a reveals that the RBF obtained the least accurate results, especially PA and UA for the runout and landslide source signatures. The performances of linear and the first-order polynomial functions were similar. A further comparison of these models, illustrated in Figure 10b, revealed that the results produced by the RF method were more favorable in most cases, especially UA for the landslide source class.

Table 4. The performances of the No. 3 model by using the RF method with the cost of 1000, where OA, AUC, PA, UA, S, and R represent overall accuracy, the area under ROC curve, producer's accuracy, user's accuracy, landslide source, and runout, respectively.

Method	UA(R)	PA(R)	UA(S)	PA(S)	AUC	OA
RF	0.96	0.86	0.66	0.87	0.89	86.42%
RF_1000	1	0.56	0.41	0.99	0.92	66.39%

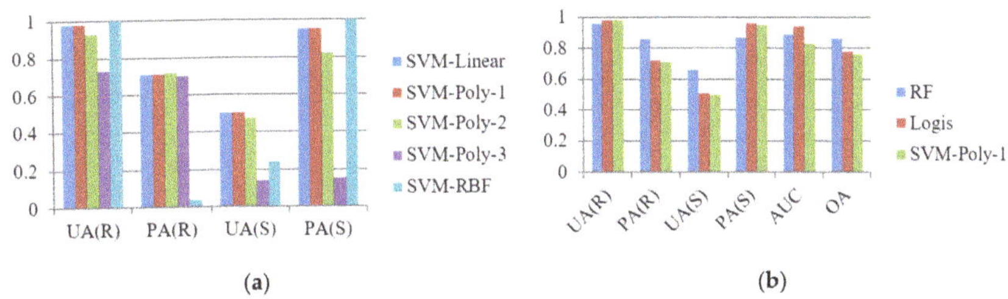

Figure 10. Performances (*Y*-axis) of the No. 3 model, where OA, AUC, PA, UA, S, and R represent the overall accuracy, area under the ROC curve, producer's accuracy, user's accuracy, landslide source, and runout, respectively. (**a**) Using SVM with linear, the first- to third-order polynomial, and radial basis (RBF) functions; (**b**) comparing the optimal results of Table 4 and (**a**) by using logistic (Logis) regression.

The No. 4 models were also examined as illustrated in Figure 11. The RF method without costs also provided the most accurate results, as presented in Figure 11a. Similarly, the performances of the linear and first-order polynomial functions outperformed the RBF method in SVM-based computation, as illustrated in Figure 11b. The comparison based on the aforementioned methods is illustrated in Figure 11c. The findings indicated that the AUC result of SVM was slightly higher than the other indices of the RF, SVM, and LR, which were similar.

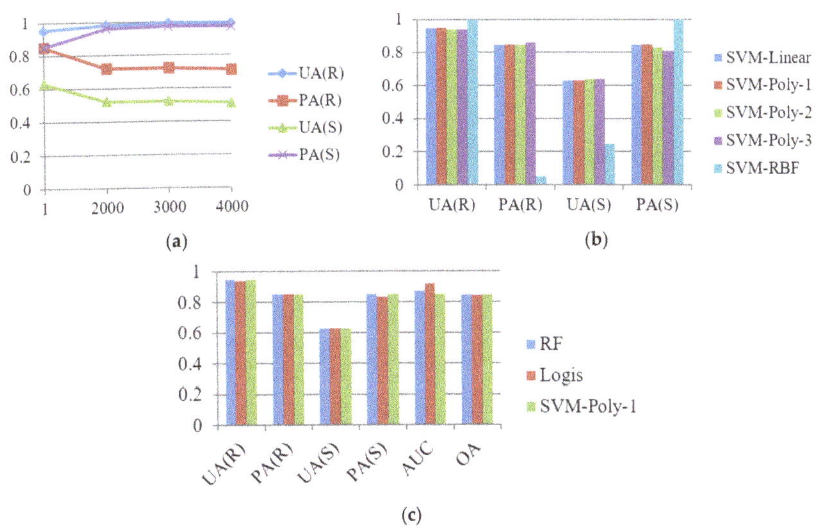

Figure 11. Performance (*Y*-axis) of the No. 4 model, where OA, AUC, PA, UA, S, and R represent overall accuracy, area under the ROC curve, producer's accuracy, user's accuracy, landslide source, and runout, respectively. (**a**) Using the RF algorithm with the costs; (**b**) using SVM with linear, the first- to third-order polynomial, and radial basis (RBF) functions; (**c**) comparing the optimal results of (**a**) and (**b**) with logistic (Logis) regression.

In the No. 5 model, the results produced by the RF algorithm with a cost of 2000 displayed high accuracy, as illustrated in Table 5. For SVM-based modeling, Figure 12a indicated that the RBF method had unbalanced predictions with a significant missing error for the landslide source and the highest PA for the runout. However, the results of the linear and polynomial functions were similar. The second-order polynomial SVM and LRs, illustrated in Figure 12b, displayed accuracies similar to the RF_2000.

Table 5. Performance of the No. 5 model using the RF method with a cost of 2000, where OA, AUC, PA, UA, S, and R represent the overall accuracy, area under the ROC curve, producer's accuracy, user's accuracy, landslide source, and runout, respectively.

Method	UA(R)	PA(R)	UA(S)	PA(S)	AUC	OA
RF	1	0.49	0.38	1	0.76	61.44%
RF_2000	1	0.76	0.56	1	0.91	81.46%

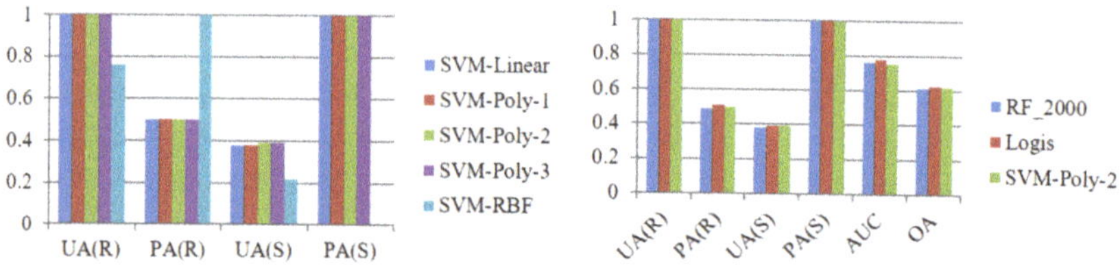

(a) (b)

Figure 12. Performances (Y- axis) of the No. 5 model, where OA, AUC, PA, UA, S, and R represent the overall accuracy, area under ROC curve, producer's accuracy, user's accuracy, landslide source, and runout-out, respectively. (**a**) Using SVM with linear, the first- to third-order polynomial, and radial basis (RBF) functions, (**b**) comparing the optimal best results of Table 5 and (**a**) with logistic (Logis) regression.

Significant improvements were observed in the No. 1 and No. 3 models. Moreover, the RF algorithm and cost-sensitive analysis provided favorable results in most cases. To identify the optimal model in which the samples extracted from different size areas of the inventory polygons, Figure 13a compares the RF method for the No. 2, 3, and 4 models, with costs of 5000 and 2000 for No. 1 and 5 models, respectively. Figure 13 indicates that the accuracies produced by the No. 1, 3, and 4 models were similar, whereas the No. 2 and 5 models displayed less favorable results for the detection of a landslide source class. Figure 13b indicated that the AUC of the No. 1 and 3 models were the most favorable. These evaluations indicated that the No. 1 and 3 models provided more reliable prediction results.

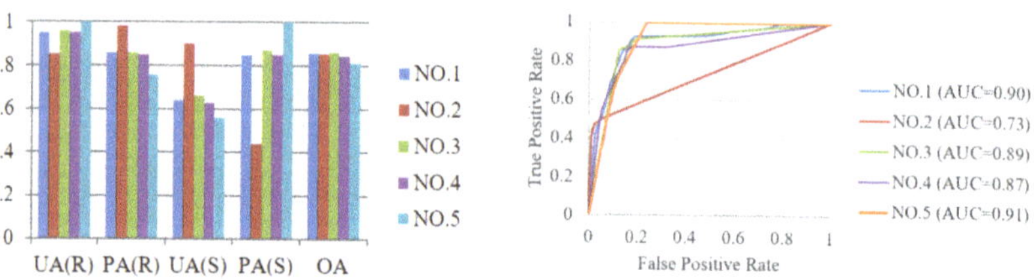

(a) (b)

Figure 13. Comparison of the representative results using the RF method for the No. 2, 3, and 4 models, with costs of 5000 and 2000 for No. 1 and 5 models, respectively. (**a**) User's accuracy (UA), producer's accuracy (PA), and overall accuracy (OA); (**b**) area under the ROC curve (AUC) for landslide source (S) and runout (R) detections.

3.3.2. The Case of Smaller Inventory Polygons

The performances of the No. 1 model based on the RF method with a cost of 5000 and the No. 3 model with the RF algorithm were assessed, as presented in Table 6, to predict the samples extracted from the polygon areas equal to or larger than 50,000 and smaller than 100,000 m^2. The accuracies were lower than those displayed in Figure 13a, especially for the runout UA and PA of the No. 1 model. Furthermore, the No. 3 model with the RF method outperformed the No. 1 model, indicating accurate results. The topographic roughness index (TRI) is commonly used for assessing landslide susceptibilities [61]. TRI was further used to improve this case as also listed in Table 6. However, the significant topographic attributes [62–64] and various indices [65] for separating landslide source and runout signatures need to be further explored. These results reveal the uncertainty between the training data and prediction targets. More precisely, the area size of the training inventory polygon affected the capability and applicability of the constructed models. In practice, manually interpreting the larger targets of landslide source and runout from remotely sensed images to produce the training data was easier than selecting smaller objects, which was a clear limitation of using the supervised strategy with four topographic variables in this study. Addressing this problem requires further consideration regarding the tradeoff between the cost of producing training data and the applicability of the developed models. The developed models were suggested for different area sizes of the inventory polygons.

Table 6. Performances of the No. 1 model using the RF method with a cost of 5000 and No. 3 model with the RF algorithm, where OA, AUC, PA, UA, S, and R represent the overall accuracy, area under the ROC curve, producer's accuracy, user's accuracy, landslide source, and runout, respectively.

Model	Method	UA(R)	PA(R)	UA(S)	PA(S)	AUC	OA
No. 1	RF_5000	0	0	0.64	1	0.67	63.87%
No. 3	RF	0.59	0.93	0.94	0.63	0.85	73.94%
No. 3*	RF	0.63	0.91	0.93	0.70	0.85	77.54%

* With the topographic roughness index.

3.4. Visualization of Landslide Detection

To visualize the separated results of the case in Section 3.3.1, the predicted patterns of the No. 3 models for distinguishing landslide source and runout targets were generated, as illustrated in Figure 14. The black and red and blue colors in this figure represent the ground truth and prediction results, respectively. As illustrated in Figure 14a, the developed models displayed suitable detection of the landslide sources (high PA(S)) with some false alarm errors (low UA(S)). Figure 14b also illustrated that the model provided reliable results for detecting runout areas with higher UA(R) and PA(R). These patterns accorded with the results of the No. 3 model in Figure 10b.

The detected distributions with the topographic situation in a three-dimensional (3D) space were further examined. Figure 15 displays the patterns of the No. 3 model for predicting the No. 1 and No. 2 landslide source and runout inventory polygons. In this figure, red and yellow represent the correct and incorrect predictions, respectively. Figure 15a,b displays accurate detections for the landslide sources. The results presented in Figure 15c indicate that the lower part of a runout area can be detected, whereas the upper region may be miscategorized as a landslide source. This finding reveals that elevation is a critical factor for the separation of landslide source and runout areas. This finding also accords with the result of the factor analysis presented in Section 3.1. Future studies can determine the elevation-based threshold to improve the capability of the developed models.

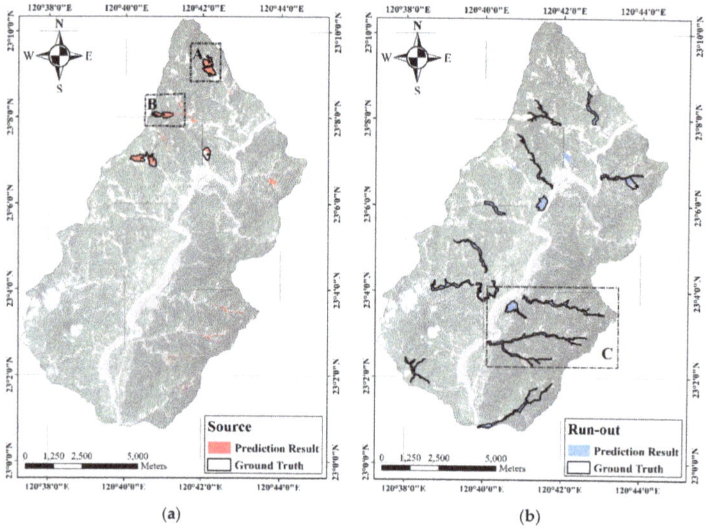

Figure 14. Prediction results of the No. 3 model using the RF algorithm to separate landslide source and runout signatures with the areas of inventory polygons of ≥100,000 m^2. Cases of (**a**) landslide source and (**b**) runout polygons.

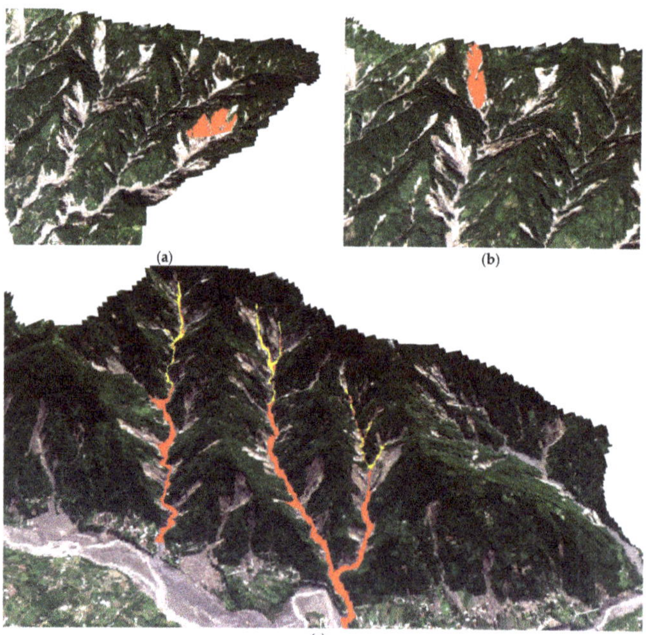

Figure 15. Three-dimensional (3D) view of the No. 3 model for predictions of separating landslide source and runout areas. The landslide source areas of the (**a**) No. 1 and (**b**) No. 2 polygons in the landslide inventory (located region A and B in Figure 14), and (**c**) the runout regions of the No. 1 and No. 2 inventory polygons (located region C in Figure 14). Red and yellow represent correct and incorrect results, respectively.

4. Discussion

The major finding in this study was that detecting landslides is not fully equal to inventory generation. The detected targets require further examination. Lai et al. [34] suggested that the runout area and topological relationship could be considered independent classes in landslide records when producing a landslide inventory. Therefore, this study focused on the separation of landslide source and runout classes caused by different mechanisms. The feasibility of distinguishing between landslide source and runout signatures to improve landslide detections was assessed. The limited training samples extracted from the top five polygons in the landslide inventory map were used to construct supervised data mining models in order to classify other instances. The top five polygon samples were used for training because manually recognizing the larger targets from a regional scale-based remote sensing image is relatively easier than interpreting smaller objects. Based on this scenario, the developed models with RFs and cost analysis and mapping results demonstrated the stability of the proposed procedure. The hierarchical analysis was designed in this study. Section 3.2 explored the feasible models as listed in Table 3. Section 3.3 demonstrated the performances of developed models verified by the samples of larger and smaller inventory polygons, comparing them with the SVM and LR methods as shown in Figures 8–13 and Tables 4–6. The quantitative evaluations of the optimal results as shown in Figure 13 were determined to be higher than 80% in most cases. The results further demonstrated that the developed models outperform the SVM and LR algorithms in most accuracies.

The primary limitation of this study was that the developed models were more favorable for larger area sizes but may be difficult to apply to smaller areas. The capability of the models based on the supervised strategy was related to the characteristics of the training datasets. The tradeoff between the cost of producing the training data and the area sizes of targets for predictions should be considered. Perhaps a multi-scale model trained by different area sizes of the inventory polygons is a possible solution. This method could also be improved by including more topographic attributes, as described by Sallem et al. [61].

Producing a high-quality landslide inventory is time consuming and expensive. Landslide inventories are crucial for landslide analyses, such as for performing tasks in the framework of landslide risk assessment and management [14]. The potential for reducing the degree of manual interpretation is demonstrated in mapping results. Based on the spatial patterns of predictions, the misclassification for detecting runout areas appeared in the top region. Two possible strategies may solve this problem. First, a threshold based on the elevation attribute could be designed. Second, studies could investigate only extracted landslide sources from the affected areas, and the remainder are assigned the runout label. These runout patterns may be adequate to further assess the related parameters, such as runout distance and damage corridor width [14].

5. Conclusions

A novel data analysis based on the scenario of separating landslide source and runout areas with the topographic attributes was presented in this paper to improve landslide detections. To explore the feasibility of the proposed approach, four hierarchical experiments were designed. First, elevation was a critical variable according to the factor analysis. Second, the candidate models were constructed using the RF and cost-sensitive analysis and the limited training samples extracted from the top five largest areas of the inventory polygons. The polygon-based cross-validation was also used for the verification. Third, the candidate models verified using different polygon-based samples indicated that the No. 1 model using the RF method with a cost of 5000 and the No. 3 model based on the RF were the most favorable for predicting the larger polygon (area $\geq$ 100,000 m^2) samples. For the smaller polygon (area of 50,000–100,000 m^2) samples, the No. 3 model provided acceptable results. The accuracies of developed models were above 0.8 in most cases, outperforming the performances of the SVM and LR. Finally, mapping the predicted distributions revealed that the detection of landslide sources provided accurate results with lower missing rates. The extraction for the runout areas revealed excellent user's and producer's accuracies.

The gap in recognizing landslides has existed in terms of remote sensing and geotechnical engineering. More precisely, the landslide detections extracted from remotely sensed images without additional interpretations to eliminate the runout areas should be termed the landslide affected area and should not be directly treated as a landslide inventory or database for further analyses. This study provided a systematic procedure following a supervised data mining–based approach to separate landslide source and runout signatures. The tradeoff between the size areas of the training data and the predicted range of the developed models was emphasized in the results. More research is warranted regarding the effect of the elevation-based threshold and the binary classification based on the detections of landslide sources only to produce fully automatic landslide inventory productions.

Funding: This study was supported, in part, by the Ministry of Science and Technology of Taiwan under project No. 109-2121-M-035-001 and 108-2119-M-035-003.

Acknowledgments: The author would like to thank S.-H. Chiang in Center for Space and Remote Sensing Research, National Central University, Taiwan, for providing the landslide inventory and the digital elevation model. The author also thanks J.-Y. Huang in Department of Civil Engineering, Feng Chia University, Taiwan, for refining the figures in the article.

Conflicts of Interest: The author declares no conflicts of interest.

References

1. Chen, W.W.; Shen, J.-P.; Wang, J.-A.; Tsai, F. Scripting STABL with PSO for analysis of slope stability. *Neurocomputing* **2015**, *148*, 167–174. [CrossRef]
2. Duncan, J.M.; Wright, S.G. The accuracy of equilibrium methods of slope stability analysis. *Eng. Geol.* **1980**, *16*, 5–17. [CrossRef]
3. Pawluszek, K. Landslide features identification and morphology investigation using high-resolution DEM derivatives. *Nat. Hazards* **2019**, *96*, 311–330. [CrossRef]
4. Prakash, N.; Manconi, A.; Loew, S. Mapping landslides on EO data: Performance of deep learning models vs. traditional machine learning models. *Remote Sens.* **2020**, *12*, 346. [CrossRef]
5. Shirvani, Z.; Abdi, O.; Buchroithner, M. A synergetic analysis of sentinel-1 and -2 for mapping historical landslides using object-oriented random forest in Hyrcanian forests. *Remote Sens.* **2019**, *11*, 2300. [CrossRef]
6. Samodra, G.; Chen, G.; Sartohadi, J.; Kasama, K. Generating landslide inventory by participatory mapping: An example in Purwosari Area, Yogyakarta, Java. *Geomorphology* **2018**, *306*, 306–313. [CrossRef]
7. Yu, B.; Chen, F.; Xu, C. Landslide detection based on contour-based deep learning framework in case of national scale of Nepal in 2015. *Comput. Geosci.* **2020**, *135*, 104338. [CrossRef]
8. Tsai, F.; Hwang, J.-H.; Chen, L.-C.; Lin, T.-H. Post-disaster assessment of land-slides in southern Taiwan after 2009 Typhoon Morakot using remote sensing and spatial analysis. *Nat. Hazards Earth Syst. Sci.* **2010**, *10*, 2179–2190. [CrossRef]
9. Deng, Y.C.; Tsai, F.; Hwang, J.H. Landslide characteristics in the area of Xiaolin Village during Morakot typhoon. *Arab. J. Geosci.* **2016**, *9*, 332. [CrossRef]
10. Lu, P.; Bai, S.; Casagli, N. Spatial relationships between landslide occurrences and land cover across the Arno river basin (Italy). *Environ. Earth Sci.* **2015**, *74*, 5541–5555. [CrossRef]
11. Mondini, A.C.; Viero, A.; Cavalli, M.; Marchi, L.; Herrera, G.; Guzzetti, F. Comparison of event landslide inventories: The Pogliaschina catchment test case, Italy. *Nat. Hazards Earth Syst. Sci.* **2014**, *14*, 1749–1759. [CrossRef]
12. Tsou, C.-Y.; Feng, Z.-Y.; Chigira, M. Catastrophic landslide induced by Typhoon Morakot, Shiaolin, Taiwan. *Geomorphology* **2011**, *127*, 166–178. [CrossRef]
13. Wu, C.-H.; Chen, S.-C.; Chou, H.-T. Geomorphologic characteristics of catastrophic landslides during typhoon Morakot in the Kaoping Watershed, Taiwan. *Eng. Geol.* **2011**, *123*, 13–21. [CrossRef]
14. Dai, F.C.; Lee, C.F.; Ngai, Y.Y. Landslide risk assessment and management: An overview. *Eng. Geol.* **2002**, *64*, 65–87. [CrossRef]
15. Zhao, X.; Chen, W. GIS-based evaluation of landslide susceptibility models using certainty factors and functional trees-based ensemble techniques. *Appl. Sci.* **2020**, *10*, 16. [CrossRef]

16. Pawluszek-Filipiak, K.; Oreńczak, N.; Pasternak, M. Investigating the Effect of Cross-Modeling in Landslide Susceptibility Mapping. *Appl. Sci.* **2020**, *10*, 6335. [CrossRef]
17. Wu, Y.; Ke, Y.; Chen, Z.; Liang, S.; Zhao, H.; Hong, H. Application of alternating decision tree with AdaBoost and bagging ensembles for landslide susceptibility mapping. *Catena* **2020**, *187*, 104396. [CrossRef]
18. van Westen, C.J.; Castellanos, E.; Kuriakose, S.L. Spatial data for landslide susceptibility, hazard, and vulnerability assessment: An overview. *Eng. Geol.* **2008**, *102*, 112–131. [CrossRef]
19. Fu, S.; Chen, L.; Woldai, T.; Yin, K.; Gui, L.; Li, D.; Du, J.; Zhou, C.; Xu, Y.; Lian, Z. Landslide hazard probability and risk assessment at the community level: A case of western Hubei, China. *Nat. Hazards Earth Sys. Sci.* **2020**, *20*, 581–601. [CrossRef]
20. Chang, K.-T.; Chiang, S.-H.; Chen, Y.-C.; Mondini, A.C. Modeling the spatial occurrence of shallow landslides triggered by typhoons. *Geomorphology* **2014**, *208*, 137–148. [CrossRef]
21. Chen, Y.-C.; Chang, K.-T.; Lee, H.-Y.; Chiang, S.-H. Average landslide erosion rate at the watershed scale in southern Taiwan estimated from magnitude and frequency of rainfall. *Geomorphology* **2015**, *228*, 756–764. [CrossRef]
22. Claessens, L.; Heuvelink, G.B.M.; Schoorl, J.M.; Veldkamp, A. DEM resolution effects on shallow landslide hazard and soil redistribution modeling. *Earth Surf. Process. Landf.* **2005**, *30*, 461–477. [CrossRef]
23. Gariano, S.L.; Brunetti, M.T.; Iovine, G.; Melillo, M.; Peruccacci, S.; Terranova, O.; Vennari, C.; Guzzetti, F. Calibration and validation of rainfall thresholds for shallow landslide forecasting in Sicily, Southern Italy. *Geomorphology* **2015**, *228*, 653–665. [CrossRef]
24. Lainas, S.; Sabatakakis, N.; Koukis, G. Rainfall thresholds for possible landslide initiation in wildfire-affected area of western Greece. *Bull. Eng. Geol. Environ.* **2016**, *75*, 883–896. [CrossRef]
25. Guzzetti, F.; Mondini, A.C.; Cardinali, M.; Fiorucci, F.; Santangelo, M.; Chang, K.T. Landslide inventory maps: New tools for an old problem. *Earth Sci. Rev.* **2012**, *112*, 42–66. [CrossRef]
26. Harp, E.L.; Keefer, D.K.; Sato, H.P.; Yagi, H. Landslide inventories: The essential part of seismic landslide hazard analyses. *Eng. Geol.* **2011**, *122*, 9–21. [CrossRef]
27. Malamud, B.D.; Turcotte, D.L.; Guzzetti, F.; Reichenbach, P. Landslide inventories and their statistical properties. *Earth Surf. Process. Landf.* **2004**, *29*, 687–711. [CrossRef]
28. Shao, X.; Ma, S.; Xu, C.; Shen, L.; Lu, Y. Inventory, distribution and geometric characteristics of landslides in Baoshan City, Yunnan province, China. *Sustainability* **2020**, *12*, 2433. [CrossRef]
29. Galli, M.; Ardizzone, F.; Cardinali, M.; Guzzetti, F.; Reichenbach, P. Comparing landslide inventory maps. *Geomorphology* **2008**, *94*, 268–289. [CrossRef]
30. Wang, H.; Liu, G.; Xu, W.; Wang, G. GIS-based landslide hazard assessment: An overview. *Prog. Phys. Geog.* **2005**, *29*, 548–567.
31. Lee, C.-T.; Huang, C.-C.; Lee, J.-F.; Pan, K.-L.; Lin, M.-L.; Dong, J.J. Statistical approach to storm event-induced landslide susceptibility. *Nat Hazards Earth Syst. Sci.* **2008**, *8*, 941–960. [CrossRef]
32. Dai, F.C.; Lee, C.F. Landslide characteristics and slope instability modeling using GIS, Lantau Island, Hong Kong. *Geomorphology* **2002**, *42*, 213–228. [CrossRef]
33. Mondini, A.C.; Chang, K.-T.; Yin, H.-Y. Combing multiple change detection indices for mapping landslides triggered by typhoons. *Geomorphology* **2011**, *134*, 440–451. [CrossRef]
34. Lai, J.-S.; Chiang, S.-H.; Tsai, F. Exploring influence of sampling strategies on event-based landslide susceptibility modeling. *ISPRS Int. J. Geo Inf.* **2019**, *8*, 397. [CrossRef]
35. Tsai, F.; Lai, J.-S.; Chen, W.W.; Lin, T.-H. Analysis of topographic and vegetative factors with data mining for landslide verification. *Ecol. Eng.* **2013**, *61*, 669–677. [CrossRef]
36. Pham, V.D.; Nguyen, Q.H.; Nguyen, H.-D.; Pham, V.-M.; Vu, V.M.; Bui, Q.T. Convolutional neural network-optimized moth flame algorithm for shallow landslide susceptible analysis. *IEEE Access* **2020**, *8*, 32727–32736. [CrossRef]
37. Mehrabi, M.; Pradhan, B.; Moayedi, H.; Alamri, A. Optimizing an adaptive neuro-fuzzy inference system for spatial prediction of landslide susceptibility using four state-of-art metaheuristic techniques. *Sensors* **2020**, *20*, 1723. [CrossRef]
38. Hong, H.; Tsangaratos, P.; Ilia, I.; Loupasakis, C.; Wang, Y. Introducing a novel multi-layer perceptron network based on stochastic gradient decent optimized by a meta-heuristic algorithm for landslide susceptibility mapping. *Sci. Total Environ.* **2020**, *742*, 140549. [CrossRef] [PubMed]

39. Arabameri, A.; Saha, S.; Roy, J.; Chen, W.; Blaschke, T.; Bui, D.T. Landslide susceptibility evaluation and management using different machine learning methods in the Gallicash river watershed, Iran. *Remote Sens.* **2020**, *12*, 475. [CrossRef]
40. Luo, X.; Lin, F.; Zhu, S.; Yu, M.; Zhang, Z.; Meng, L.; Peng, J. Mine landslide susceptibility assessment using IVM, ANN and SVM models considering the contribution of affecting factors. *PLoS ONE* **2019**, *14*, e0215134. [CrossRef] [PubMed]
41. Huang, F.; Cao, Z.; Guo, J.; Jiang, S.-H.; Li, S.; Guo, Z. Comparisons of heuristic, general statistical and machine learning models for landslide susceptibility prediction and mapping. *Catena* **2020**, *191*, 104580. [CrossRef]
42. Chen, W.; Li, Y. GIS-based evaluation of landslide susceptibility using hybrid computational intelligence models. *Catena* **2020**, *195*, 104777. [CrossRef]
43. Meena, S.R.; Mishra, B.K.; Piralilou, S.T. A hybrid spatial multi-criteria evaluation method for mapping landslide susceptibility areas in Kullu Valley, Himalayas. *Geosciences* **2019**, *9*, 156. [CrossRef]
44. Breiman, L. Random forests. *Mach. Learn.* **2001**, *45*, 5–32. [CrossRef]
45. Belgiu, M.; Dragut, L. Random forest in remote sensing: A review of applications and future directions. *ISPRS J. Photogramm. Remote Sens.* **2016**, *114*, 24–31. [CrossRef]
46. Stumpf, A.; Kerle, N. Object-oriented mapping of landslides using Random Forests. *Remote Sens. Environ.* **2011**, *115*, 2564–2577. [CrossRef]
47. Díaz-Uriarte, R.; De Andres, S.A. Gene selection and classification of microarray data using random forest. *BMC Bioinform.* **2006**, *7*, 3. [CrossRef]
48. Lai, J.-S.; Tsai, F. Improving GIS-based landslide susceptibility assessments with multi-temporal remote sensing and machine learning. *Sensors* **2019**, *19*, 3717. [CrossRef]
49. Desai, A.; Jadav, P.M. An empirical evaluation of adaboost extensions for cost-sensitive classification. *Int. J. Comput. Appl.* **2012**, *44*, 34–41.
50. Dou, J.; Paudel, U.; Oguchi, T.; Uchiyama, S.; Hayakawa, Y.S. Shallow and deep-seated landslide differentiation using support vector machines: A case study of the Chuetsu Area, Japan. *Terr. Atmos. Ocean. Sci.* **2015**, *26*, 227–239. [CrossRef]
51. Chiang, S.-H.; Chang, K.-T.; Mondini, A.C.; Tsai, B.-W.; Chen, C.-Y. Simulation of event-based landslides and debris flows at watershed level. *Geomorphology* **2012**, *138*, 306–318. [CrossRef]
52. National Disaster Prevention and Protection Commission (NDPPC). *Disaster Response Disposition Report of Typhoon Morakot*; National Disaster Prevention and Protection Commission: New Taipei City, Taiwan, 2009; p. 74.
53. Guo, L.; Chehata, N.; Mallet, C.; Boukir, S. Relevance of airborne lidar and multispectral image data for urban scene classification using random forests. *ISPRS J. Photogramm. Remote Sens.* **2011**, *66*, 56–66. [CrossRef]
54. Tsai, F.; Lai, J.-S.; Lu, Y.-H. Land-cover classification of full-waveform LiDAR point cloud with volumetric texture measures. *Terr. Atmos. Ocean. Sci.* **2016**, *27*, 549–563. [CrossRef]
55. Witten, I.H.; Frank, E.; Hall, M.A. *Data Mining: Practical Machine Learning Tools and Techniques*, 3rd ed.; Morgan Kaufmann: San Francisco, CA, USA, 2011.
56. Elkan, C. The foundations of cost-sensitive learning. In Proceedings of the 7th International Joint Conference on Artificial Intelligence, Seattle, VA, USA, 4–10 August 2001.
57. Mountrakis, G.; Im, J.; Ogole, C. Support vector machines in remote sensing: A review. *ISPRS J. Photogramm. Remote Sens.* **2011**, *66*, 247–259. [CrossRef]
58. Rossi, M.; Guzzetti, F.; Reichenbach, P.; Mondini, A.C.; Peruccacci, S. Optimal landslide susceptibility zonation based on multiple forecasts. *Geomorphology* **2010**, *114*, 129–142. [CrossRef]
59. Pham, B.T.; Nguyen-Thoi, T.; Qi, C.; Phong, T.V.; Dou, J.; Ho, L.S.; Le, H.V.; Prakash, I. Coupling RBF neural network with ensemble learning techniques for landslide susceptibility mapping. *Catena* **2020**, *195*, 104805. [CrossRef]
60. Du, P.; Samat, A.; Waske, B.; Liu, S.; Li, Z. Random forest and rotation forest for fully polarized SAR image classification using polarimetric and spatial features. *ISPRS J. Photogramm. Remote Sens.* **2015**, *105*, 38–53. [CrossRef]
61. Saleem, N.; Huq, M.E.; Twumasi, N.Y.D.; Javed, A.; Sajjad, A. Parameters derived from and/or used with digital elevation models (DEMs) for landslide susceptibility mapping and landslide risk assessment: A review. *ISPRS Int. J. Geo Inf.* **2019**, *8*, 545. [CrossRef]

62. Glenn, N.F.; Streutker, D.R.; Chadwick, D.J.; Thackray, G.D.; Dorsch, S.J. Analysis of LiDAR-derived topographic information for characterizing and differentiating landslide morphology and activity. *Geomorphology* **2006**, *73*, 131–148. [CrossRef]
63. Goetz, J.N.; Bell, R.; Brenning, A. Could surface roughness be a poor proxy for landslide age? Results from the Swabian Alb, Germany. *Earth Surf. Process. Landf.* **2014**, *39*, 1697–1704. [CrossRef]
64. Van Den Eeckhaut, M.; Kerle, N.; Poesen, J.; Hervás, J. Object-oriented identification of forested landslides with derivatives of single pulse LiDAR data. *Geomorphology* **2012**, *173*, 30–42. [CrossRef]
65. Pawluszek, K.; Borkowski, A.; Tarolli, P. Sensitivity analysis of automatic landslide mapping: Numerical experiments towards the best solution. *Landslides* **2018**, *15*, 1851–1865. [CrossRef]

applied sciences

MDPI

Article

Study on the Early Warning Methods of Dynamic Landslides of Large Abandoned Rockfill Slopes

Nan Qiao [1,*], Yun-Ling Duan [1], Xiao-Meng Shi [2], Xue-Fei Wei [3] and Jin-Ming Feng [1]

1 State Key Laboratory of Hydroscience and Engineering, Tsinghua University, Beijing 100084, China; yduan@tsinghua.edu.cn (Y.-L.D.); jmfeng126@126.com (J.-M.F.)
2 School of Civil Engineering, Beijing Jiaotong University, Beijing 100044, China; shixm@bjtu.edu.cn
3 CIECC Engineering Company Limited, Beijing 100048, China; wxf0341@163.com
* Correspondence: qn13@mails.tsinghua.edu.cn

Received: 22 July 2020; Accepted: 28 August 2020; Published: 2 September 2020

Abstract: The excavation of large-scale underground projects produces a large amount of rubble waste material that is temporarily deposited near the project site, which forms a large-scale waste rockfill artificial slope. The slope has a granular structure, thus, during excavation and trans-shipment, surface shallow landslides may frequently occur. Existing contact monitoring methods such as buried sensors and GPS (Global Position System) are difficult to apply to the monitoring of rockfill landslides. Therefore, there are no appropriate early warning methods for waste rockfill slope landslides during dynamic transfer. Here, we used ground-based interferometric synthetic aperture radar to monitor the deformation of a rockfill slope during the excavation and transfer processes as a proposed method for the early warning against landslides on rockfill slopes during dynamic construction based on the radar interference measurement results. Through data cleaning and data interpolation, the line of equal displacement was generated, and the cross-sectional area of the equal displacement bodies of landslides was calculated. In addition, we established a four-level early warning grading standard, with the rate of change of the cross-sectional area of the equal displacement body as the early warning index, and realized real-time dynamic early warning of waste rockfill landslides during excavation and transportation. Finally, five landslide examples were used to verify the proposed warning method. The results show that the warning method can make an early warning 8–14 min before the occurrence of landslide, which can effectively avoid the appearance of catastrophic events.

Keywords: rockfill; ground-based interferometric synthetic aperture radar; construction; cross-sectional area of equal displacement body; landslide warning method

1. Introduction

In recent years, the construction volume of large underground projects in China has grown [1]. During underground engineering excavation, a large amount of rubble waste material is temporarily piled up near the project area to form a large-scale waste rockfill body. With the development and utilization of rubble waste material, the temporary waste material dump needs to be excavated and transported. Excavation disturbance inevitably causes slope failure. Therefore, the safety of waste rockfill slopes during excavation and transportation has become an important factor [2].

Slope deformation monitoring and landslide warning rank high among the most effective safety prevention and control measures [3]. Displacement monitoring forms the basis of landslide deformation monitoring. Slope displacement and its related parameters are important reference indicators for landslide early warning research [4]. In 1968, Saito proposed the "three-segment" change rule of slope displacement–time curves [5]. Since then, scholars have conducted a vast amount of research on slope deformation monitoring and landslide warning from different perspectives and with different means. For instance, Hoek (1970) analyzed the displacement–time curve measured at the Chuquicanata mine and proposed an extension method to extrapolate the displacement–time curves and to predict the time of landslides according to the curve trend [6–8]. Zhong proposed the Verhulst inverse function model according to the characteristics of displacement–time curves [9]. Qiang et al. analyzed and studied many typical landslide displacement monitoring results and proposed a landslide early warning method using acceleration as the evaluation index [10]. In addition, many scholars have tried to combine various mathematical models with slope displacement–time curves. Such examples include the golden section method for early warning of the time of landslides [11–13], the markov early warning model [14,15], the fuzzy mathematics early warning model [16,17], the poisson cycle early warning model [18–21], and the gradient-sinusoidal early warning model [20,22,23].

For natural landslides, the displacement vectors at various points on the landslide body can generally be kept consistent. Even under special circumstances, the deformation of each point in the initial stage of slope deformation is disordered, but, once entering deformation acceleration, the displacement of each point on the landslide body and the time curve also tend to be the same. Therefore, the displacement data of one or more monitoring points can be used as the basis for slope deformation analysis and landslide warning research [24]. However, the waste rockfill are massive and unconsolidated and feature dynamic excavation and transportation processes. In slope deformation monitoring, it is difficult to apply contact monitoring methods such as embedded sensors and GPS. Because of the granular structure characteristics of a rockfill body, the rubble particles on the slope will rotate during slope excavation, which leads to inconsistent deformation among the rubble. It is irrational to describe the displacement and warn of the deformation of the slope by a single point. Therefore, traditional slope monitoring and landslide warning methods are not applicable for large abandoned rockfill slopes.

To effectively solve the problem of catastrophic events that may occur on the slopes of the rockfills, it is necessary to select appropriate monitoring methods, to describe the deformation process of slopes with scientific quantitative indicators, and to provide early warning for slope landslides. GB-InSAR (ground-based interferometric synthetic aperture radar) can provide full-field deformation information within the monitored area [25], which makes it possible to use the change of deformation region on the slope as the characteristic value to study the slope deformation. Therefore, the GB-InSAR system (IBIS-L system from Innovative Interferometric Radar for Environmental and Civil Engineering Applications, Italy) was used to carry out on-site monitoring tests on the slope of China's first temporary large-scale groundwater-sealed oil storage dumping site in Huangdao, Shandong Province. Through data processing of the full-field displacement results obtained from monitoring (data cleaning, data interpolation, generating equal displacement lines, etc.), we propose to describe the deformation process of the slope by using the curve of the cross-sectional area of the constant displacement body with time. By summarizing the results of the data, a landslide early warning method with an equal displacement cross-sectional area as an early warning index is proposed to achieve the real-time monitoring and early warning of rockfill landslides during construction.

2. Project Overview

The waste rockfill is located about 500 m southwest of the underground project entrance, and the floor area is approximately 9.6×10^4 m^2. The ground where the waste rockfill body is located is high in the north and low in the south, and the difference of elevation between the south and north is about 20 m. The east and west directions are relatively gentle, and there is a slightly protruding depression in the middle part. The difference of elevation is about 4 m. The waste rockfill is dumped by a dump truck [26], and it is leveled by a bulldozer. The rockfill body slopes can be divided into two stages based on the formation process. The boundary between the two stages are shown in Figure 1 (the yellow dotted line).

Figure 1. Overview of the rockfill. The horizontal boundary width of the red curve is about 100 m, and the maximum vertical level difference is about 50 m.

The waste is mainly rubble produced by the blasting and crushing of granite gneiss. Table 1 shows the petrophysical and mechanical properties of granite gneiss in intact rock mass. The rockfill is highly porous and unconsolidated with high permeability of the rubble. The bottom of the rockfill body is composed of large rubbles with an average particle size of 1 m, and the rubble particles are independent of each other. The slope within the monitoring area (the red polygon area in Figure 1) is the residual slope after the occurrence of the previous landslides, and many large boulders are already broken, having an average size of 10–50 cm (Figure 2). Besides, due to the occurrence of landslide, the rock rolling and impact will be broken, resulting in much coarse-grained soil left on the slope.

Table 1. Physical and mechanical properties of intact rocks (from geological survey report).

Classification	Specific Gravity	Uniaxial Saturation Compressive Strength (MPa)	Elasticity Modulus (GPa)	Shearing Strength		Tensile Strength (MPa)
				Cohesion (MPa)	Internal Friction Angle (°)	
Granite gneiss below +20 m (the elevation of the underground cavern roof is −30 m)	2.6	90.7	52.7	10.2	71.1	11.7

Figure 2. Rock fragments left on the slope after the previous landslide.

Due to the stacking height limits and the economic value of rubble, rubble waste is transferred to a dumping site about 1 km south of the site of rubble transfer, as shown in Figure 3.

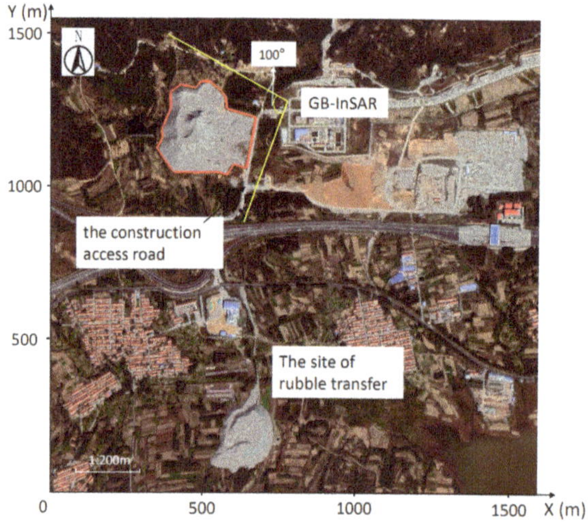

Figure 3. Relative position of project site (adapted from google maps). Where GB-InSAR is ground-based interferometric synthetic aperture radar.

To prevent uncontrollable large-scale slippage caused by an excavation-related disturbance, rubble excavation is carried out in two phases. In the first phase, a "C"-shaped platform is constructed above the platform of the original rockfill body. Excavation is carried out using construction equipment located on the platform (Figure 4) to control the location and slope height of the area affected by the excavation, thereby controlling the scale of the landslide and the direction of the sliding body. After the removal of most of the rubble via the above the platform, the second phase of the operation is to transport the rubble under the platform.

Figure 4. Loading and transportation.

3. Slope Monitoring of Large-Scale Abandoned Rockfill Based on Ground-Based Interferometric Synthetic Aperture Radar

3.1. Monitoring Equipment

This research used the IBIS-L system [2,27] for monitoring. The IBIS-L system is mainly composed of four parts: radar sensor, linear scanner, computer, and power supply module (Figure 5). When using

IBIS for field monitoring tests, in addition to ensuring the normal monitoring of the target, the power supply of the equipment should also be considered. IBIS has a power of about 100 W. Even though a battery can be used as a power source, it does not effectively guarantee the continuous working time of IBIS.

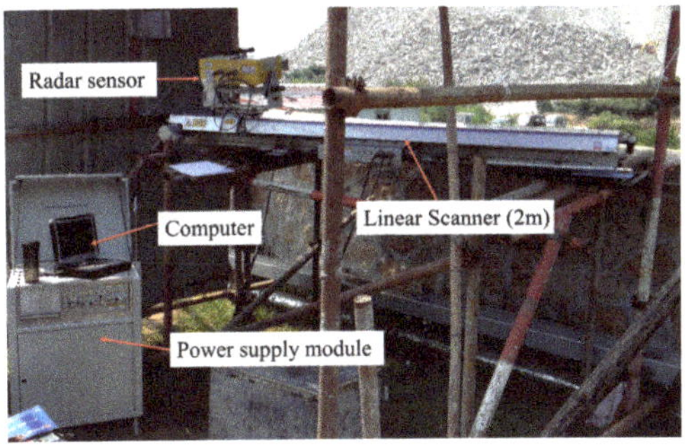

Figure 5. IBIS-L system (from Innovative Interferometric Radar for Environmental and Civil Engineering Applications, Italy) composition.

The theoretical monitoring range of the IBIS-L system is 0–4000 m. Due to the limitation of the radar antenna beam angle, the azimuth monitoring range is −50° to 50°. The specific system parameter settings are shown in Table 2. As for the working process of the system, the computer controls the radar sensor to move from the left end to the right end of the linear guide in an "off–on–off" manner. The length of the linear guide is 2 m, "off" every 5 mm, during which electromagnetic waves are emitted and echoes are received a total of 401 times. Then, the received signal is processed by synthetic aperture technology to realize azimuth focusing. Finally, according to the phase information of each pixel obtained by the synthetic aperture, the deformation information d of the target can be acquired by the interferometry of the phase difference between the echo signals of the targets at different times:

$$d = \frac{\lambda(\varphi_2 - \varphi_1)}{4\pi} \quad (1)$$

where φ_1 and φ_2 represent the phase information of each pixel obtained by synthetic aperture at different times, and λ is the radar wavelength. Thus, the displacement of each pixel in the whole field can be obtained.

Table 2. The parameters of IBIS-L.

System Parameters	
Range resolution (the range of the minimum resolution element)	0.5 m
Azimuth resolution	4.4 mrad
Accuracy	0.1 mm
Operating range	10–4000 m
Frequency band	17.1–17.3 GHz
Scan time	5 min
Linear scansion length	2 m
Power consumption	0.1 kW·h

3.2. Monitoring Plan

To make a preliminary judgment on the position of the site excavation and potential landslides, the optimal monitoring angle of GB-InSAR was first ensured, and, at the same time, the safety of the equipment and the power supply requirements were taken into account. The GB-InSAR was located at the project department entrance about 120 m to the northeast of the rockfill, as shown in

Figure 2. The range indicated by the red line in the figure is the area occupied by the piled rockfill, and the range indicated by the yellow line is the azimuth scanning range of the radar.

3.3. Monitoring Results

The monitoring lasted 69 days, during which time 11,376 observations were obtained and 72 shallow landslides (the sliding surface depth did not exceed 1 m) were recorded, including 41 small-area landslides that were destroyed within a short time by excavator disturbance. However, when the tops of some slopes were removed, it made it very easy to lose control, resulting in large-scale landslides. In total, 31 larger landslides were recorded, as shown in Figure 6, where the top of the slopes was removed. The records of the landslide area of the 72 landslides are shown in Figure 7, and the numbers represent the landslide areas, where the landslide area is the area of each landslide which was recorded during the field monitoring. The source of the final landslide area record was divided into two parts. Firstly, all the landslide area records that occurred during the day were recorded from the perspective of radar through visual estimation. Secondly, almost all the records of the landslide area from night to morning were obtained by the visual estimation of the construction workers standing near the bottom of the slope. Then, the landslide area visible to the radar could be calculated according to the relative position of the radar position, the construction personnel, and the landslide position. However, the landslide area records shown in this paper are estimated values of the landslide area from the perspective of radar. The landslide areas shown in Figure 7 are the final data that could be measured by radar.

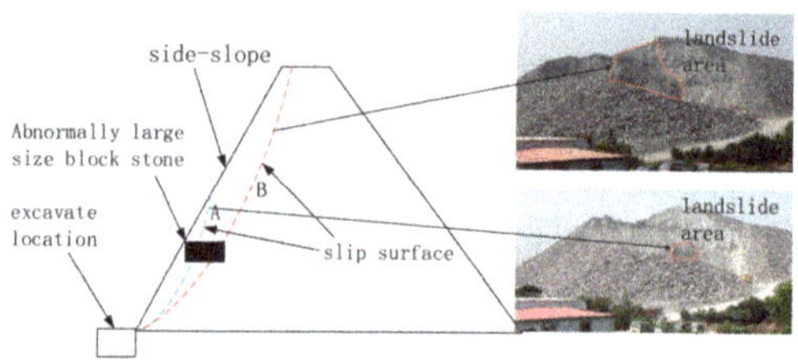

Figure 6. Position of the sliding surface. The 72 landslides records can be divided into two types according to the position of the sliding surface as A and B. Examples of the two types of landslide surfaces are shown on the right.

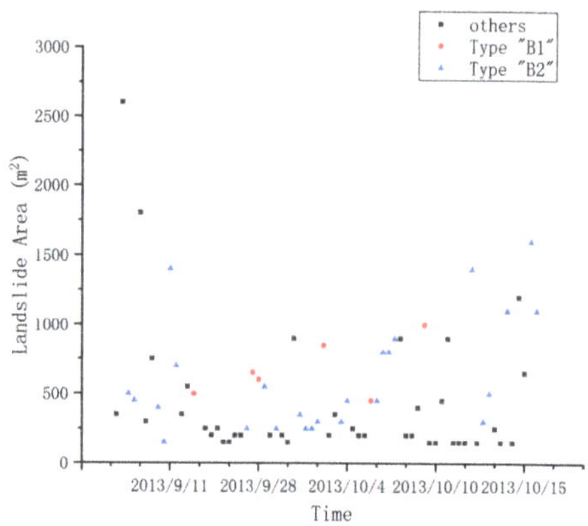

Figure 7. Landslide areas (72 landslides). The landslide area is the area of each landslide recorded during the field monitoring. These data are not accurate measurements but intuitive estimates. Since this paper is based on IBIS-L measurement results for analysis, the area data in the figure represent the area of the landslide area that can be seen by the corresponding radar equipment.

According to the field observation, the surface morphology of the 31 landslides can be divided into two types. One is the slope surface above the excavation location of the mechanical equipment, which has no apparent large particle size and dense agglomeration zone. The other is the slope with a significant dense zone (Figure 8). Due to rainfall, the rolling of construction vehicles, and the uneven particle sizes, some layers or areas of the rockfill have a higher density than the surrounding dense zone. Therefore, the 31 landslides were classified based on the surface shape of the slopes (Table 3).

Figure 8. Slope with a significant dense zone.

Table 3. Classification of landslides.

Landslide Classification	Classification Basis	Number of Landslides
Type "B1"	There is a dense belt, the slope surface is uneven.	6
Type "B2"	No dense belt, the slope surface is even.	25

This article focuses on the abovementioned 31 landslides. First, according to the monitoring results of IBIS-L, the echo signals of each pixel were extracted. Then, we dealt with the pixels that could not meet the precision requirement. The displacement values of the remaining pixels in each monitoring period were calculated by Equation (1). According to the displacement values of existing pixel points, the interpolation method could obtain the displacement values of all pixel points in the area around each cleaned point. Finally, the contour of equal displacement could be generated based on the values. Considering the limitation of displacement monitoring accuracy of IBIS-L, the displacement cross-sectional area of −0.5 mm was selected as the index for the landslide early warning research. The −0.5 mm equal displacement cross-sectional areas of 31 landslides at different times were extracted, as shown in Figure 9.

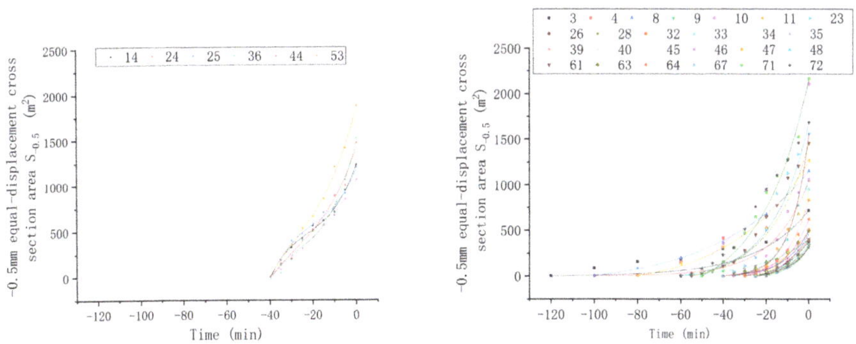

(a) Type "B1" (b) Type "B2"

Figure 9. The area–time curve of the equal displacement cross section. The numbers in the legend represent landslide numbers recorded in chronological order. The "0" time represents the last scan before the severe landslides.

The two types of different slope forms exhibit two distinct curves under the same excavation construction conditions. The −0.5 mm equal displacement cross-sectional area of a Type B1 landslide lasts about 30–40 min before a landslide, and it is a cubic curve. The reason for this is the dense zone on the slope has a significant supporting effect on the upper rubble during the deformation of the slope. The −0.5 mm equal displacement cross-sectional area of a Type B2 landslide varies with time for about 30–120 min, and most last less than 60 min. The curve conforms to the exponential form with the accelerated deformation stage in the three sections of the slope displacement–time curve.

4. Early Warning Method for Landslides Based on the Rate of Change of the Cross-Sectional Area of Equal Displacement Bodies

The uniqueness of waste rockfill bodies makes it unsuitable to use single-point displacement as an indicator for slope deformation analysis or empirical judgment. Therefore, in this study, the deformation characteristics of waste rockfill slopes were characterized by the change of the cross-sectional area of the equal displacement body. On this basis, a warning method for waste rockfill slope landslides during excavation is proposed.

4.1. Ideas on Early Warning

(1A) Based on the results in [2] and the 26 landslides monitored in the previous period (the remaining five landslides were used to verify the early warning method), the surface characteristics of waste rockfill slopes can be identified, and a preliminary judgment can be made about the possible landslide patterns in the excavation process. During the accumulation of waste rockfill, the density of some accumulation layers and areas caused by rainfall, the rolling of construction vehicles, etc. is quite different, or there might be a certain amount of abnormally large-sized rubble on the slope surface and in the shallow layer. Both conditions play a great role in supporting the rubble particles above the dense zone. Therefore, combined with the surface characteristics of the rockfill slope, a possible landslide pattern can be determined through the analysis of the characteristics of the cross-sectional area of different displacement bodies over time.

(1B) For possible landslides being monitored, based on the real-time constant displacement cross-sectional area data obtained from slope deformation monitoring, a time series prediction model such as ARIMA-SVR [28] (Autoregressive Integrated Moving Average model-Support Vector Regression: a hybrid ARIMA and SVR model of time series prediction mode) can be used. Based on the existing data and the prediction of future data, the future trend of constant displacement cross-sectional areas can be predicted.

(2) Real-time monitoring data can be extracted, the real-time changing process of the cross-sectional area of equal displacement can be studied and analyzed, and the mode of slope deformation can be determined according to the change trend of the cross-sectional area of the equal displacement body.

(3) According to the measured (1A) and predicted values of the cross-sectional area of the equal displacement body (1B), the relationship between the cross-sectional area change of the equal displacement body in different modes and landslides of the rockfill slope can be analyzed, and an early warning judgment of such landslides can be determined. Then, the change process of the cross-sectional area of different landslides and equal displacement bodies can be independently analyzed.

(4) On the basis of the determined early warning criteria of the rockfill slope, an early warning grade division standard of the rockfill slope can be established.

The detailed process is shown in Figure 10.

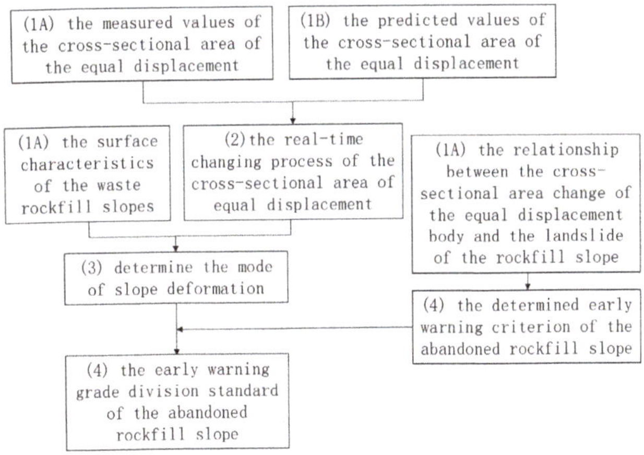

Figure 10. Early warning method of landslides.

4.2. Early Warning Indicators and Grading Standard

Considering the needs of on-site early warning in terms of the number of data calculations and the data extraction speed, under the condition of ensuring the accuracy of radar displacement monitoring, the deformation of −0.5 mm on the slope is easier to identify in time. Therefore, the cross-sectional area of the −0.5 mm equal displacement body was used as an index to validate the early warning method of the rockfill slope during excavation and transportation [2]. The on-site conditions of waste rockfill slope landslides are complicated, and both different and similar types of landslides often occur in the same area. Therefore, the cross-sectional area of the −0.5 mm equal displacement body was normalized, and then the rate of change was used as a warning indicator of landslides. The cross-sectional area of the −0.5 mm constant displacement body at each moment after normalization is

$$S^{*}_{-0.5}(t) = \frac{S_{-0.5}(t)}{S_{max}} \tag{2}$$

where $S_{-0.5}(t)$ is the measured or predicted value of the cross-sectional area of the −0.5 mm constant displacement body at time t and S_{max} is the cumulative value of the cross-sectional area of the −0.5 mm equal displacement body obtained from the last measurement before the slope starts sliding.

The normalized −0.5 mm equal displacement body cross-sectional area change rate is

$$V^{*}_{-0.5}(t) = \frac{S_{-0.5}(t)'}{S_{max}} \tag{3}$$

Based on the field measurement results and the analogy between landslide cases, the characteristics of the cross-sectional area of the −0.5 mm equal displacement body of the two types of 29 landslides that were previously monitored in the rockfill were statistically analyzed. Then, early warning standards were developed according to two different types of landslides. Since the data selected were actually measured landslide data, the lower limit of the 95% confidence band of the mean value of the cross-sectional area of the two types of landslides of the −0.5 mm equal displacement body was selected as the standard to formulate the landslide warning level. According to the characteristic values of 50%, 30%, and 10% of the lower limit of the 95% confidence band of the average displacement rate of the cross-sectional area of the −0.5 mm equal displacement body before sliding, the landslide warning was divided into five levels, as shown in Figure 6. Note that the rate of change of the cross-sectional area of the "B1"-type landslide of the −0.5 mm equal displacement body shows a decreasing trend within a certain period of the initial stage of the development of the displacement field. Therefore, the maximum value of $V^*_{-0.5}(t)$ before the start of deformation acceleration is defined as the upper limit of the yellow warning for class B landslides. To make the early warning method more convenient and to ensure the compatibility of the two types of landslides, the early warning standard was uniformly divided into four levels based on the statistical results of the early warning thresholds of the two types of landslides, as shown in Figure 11. The grading results and construction regulations were then established, as shown in Table 4.

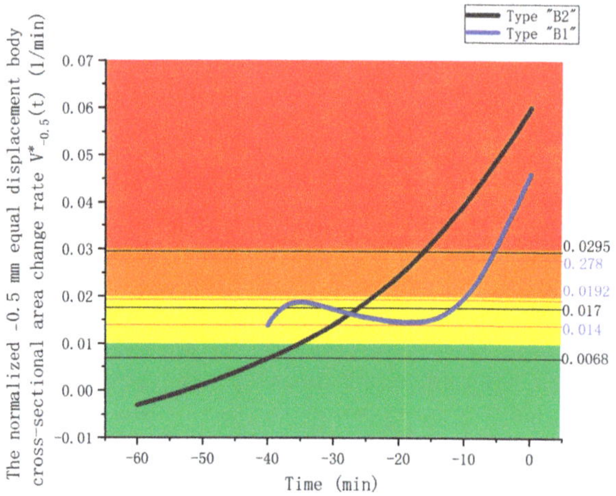

Figure 11. The lower limit of the 95% confidence band for $V^*_{-0.5}(t)$

Table 4. Early warning classification and dynamic control method of the waste rockfill slope during excavation and transportation.

Warning Level		Level 1	Level 2	Level 3	Level 4
		Green	**Yellow**	**Orange**	**Red**
Slope Status		Stable	Relatively Stable	Relatively Unstable	Unstable
Early-warning Index	Area change rate index $V^*_{-0.5}(t)$ (1/min)	<0.01	0.01–0.02	0.02–0.03	≥0.03
Evaluation		Stable	Relatively stable	Relatively dangerous	Dangerous
Dynamic regulation		Construct as normal	Construct as normal while paying attention to the slopes	Pay close attention to the slope anomalies and adjust the construction position appropriately	Stop construction immediately and evacuate, evacuate the original excavation location, and avoid landslides

5. Method Validation

To check the feasibility and reliability of the landslide warning method of rockfill slopes, the last five landslides of two different types of landslides recorded during field monitoring were used to verify the landslide warning method.

The waste rockfill slope was taken as the remaining broken surface after the previous landslide, and the slope angle was approximately 35–38°. The excavation and loading speed during construction was approximately a 10-t load muck truck loaded every 6–9 min. During the five landslide monitoring periods, there was no rainfall, the daytime temperature was about 20–25 °C, and the nighttime temperature was about 15–20 °C.

The landslides occurred at 4:55 on 5 October 2013, 2:45 on 10 October 2013, 3:00 on 15 October 2013, 21:00 on 15 October 2013, and 9:38 on 16 October 2013. The final landslide areas were 450, 1000, 1100, 1600, and 1100 m^2, respectively. The specific records are shown in Table 5.

Table 5. Characteristics of the five test landslides.

Landslide Number	Types of Slopes and Landslides	Duration of Displacement to Failure	Moment of Landslide	Final Landslide Area
Landslide 44	Type "B1"	40 min	5 October 2013 4:55	450 m^2
Landslide 53	Type "B1"	40 min	10 October 2013 2:45	1000 m^2
Landslide 67	Type "B2"	40 min	15 October 2013 3:00	1100 m^2
Landslide 71	Type "B2"	50 min	15 October 2013 21:00	1600 m^2
Landslide 72	Type "B2"	60 min	16 October 2013 9:38	1100 m^2

First, the slope was initially classified according to whether there was a dense belt formed by the rolling of the slope or whether there was abnormally large-sized rubble in the middle of the slope. Landslides 44 and 53 conformed to the B1-type failure mode, while Landslides 67, 71, and 72 conformed to the B2-type failure mode. Then, ARIMA-SVR model was used to predict the variation trend of −0.5 mm equal displacement cross-section area of these five landslides, and the results are shown in Figure 12. The average deviations between the predicted value and the measured value of five test landslides are shown in Table 6.

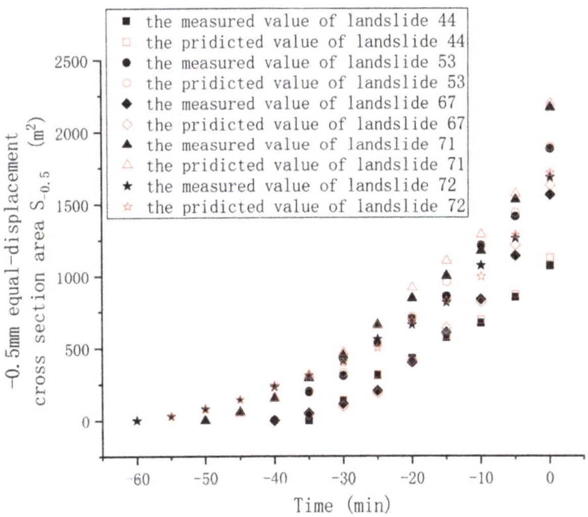

Figure 12. The measured and predicted values of the cross-sectional area of the −0.5 mm equal displacement bodies.

Table 6. The average deviations between the predicted value and the measured value of the five test landslides.

Landslide Number	Landslide 44	Landslide 53	Landslide 67	Landslide 71	Landslide 72
Average deviations (m^2)	9.8	11.9	15.3	21.2	10.5

According to Equation (2), the prediction results of five −0.5 mm equal displacement bodies of five landslides were normalized. Then, the deformation rates of the cross-sectional area of the −0.5 mm equal displacement bodies were calculated after normalization using Equation (3). The results are shown in Figure 13.

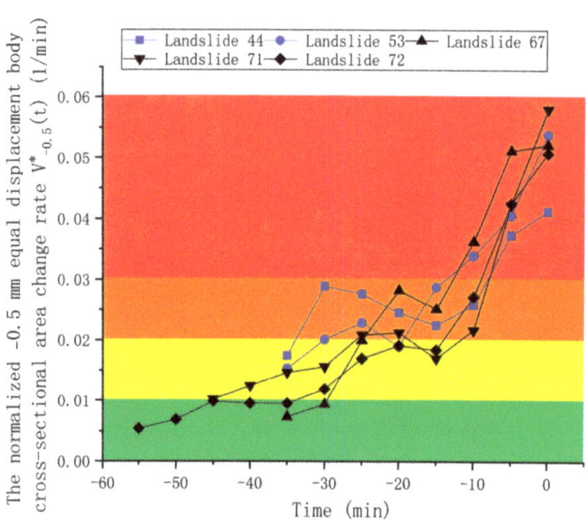

Figure 13. The normalized −0.5 mm equal displacement body cross-sectional area change rate.

From the analysis results of the above examples, the landslide early warning method and the early warning classification established in this paper could provide a warning to avoid risks before the occurrence of all five tested landslides.

Compared with the traditional landslide warning method based on slope deformation, it mainly depends on the displacement of some points on the slope. For a loose rockfill slope, it is feasible to express the deformation of the slope by the displacement of a certain point, but it is difficult to find a suitable point. However, as described in Section 3.3, the variation curve of a −0.5 mm equal displacement cross-sectional area over time conforms to the form of a "three-section" curve. Combined with previous research experience, this will make it easier for the equal displacement cross-sectional area to be used as an early warning indicator.

It should be noted that this verification was based on the last instance of landslide of each type on the site, and the actual working conditions were similar to the previous landslides. The establishment of the landslide warning standard also depends on the statistical results of these previous similar landslide cases. Therefore, these four successful landslide warnings are limited by certain working conditions, such as rock lithology, rubble particle gradation, rolling of vehicles during the accumulation of rock mass, rainfall, and so on. These conditions will directly affect the repose angle of the rockfill, and the repose angle is an important factor affecting rockfill landslides. Generally, when the angle of the slope is less than the repose angle, the slope is unstable. Rainfall has a direct impact on the water content of rockfill, which is directly proportional to the angle of internal friction. For loose rockfill, due to the structural characteristics of the large pores, leakage and evaporation are rapid, so water content is usually small. Particle gradation will also directly affect the repose angle of the rockfill body. Small-particle-size rubble plays a particular "lubrication" role in the rockfill body. When the content of small-particle-size rubble is relatively low, large-particle-size rubble relies on friction and the "aggregate interlock capacity of rubble" to keep the slope in a stable state. When the content of small-particle-size rubble is high, due to the "lubricating" effect of small-particle-size rubble particles, the relative movement between large-particle-size rubble particles is more likely to occur, resulting in slope instability. Therefore, for the slopes of other rockfill bodies under similar working conditions, the actual situation should be considered to make appropriate adjustments to the early warning standard.

In general, most landslide warning models are based on the displacement–time curve of a single point on the slope. According to the curve change that conforms to the three stages of "initial deformation–uniform deformation–accelerated deformation", the landslide warning levels are divided into "caution–vigilance–danger". For example, Wang [29], He [30], and Qiang [31], among others, all studied the landslide early warning index and landslide early warning classification standard based on the "three stages" of the single point displacement–time curve. However, which point of displacement data should be selected as the basis in the process of landslide warning is a fundamental question. To solve this problem, Qiang [24] proposed that the deformation of each point on the slope is disordered before the basic penetration of the sliding surface. However, after the basic penetration of the sliding surface, the slope begins to slide as a whole. As mentioned above, the rockfill body is loosely piled, and the deformation of adjacent positions on the slope is disordered before the occurrence of landslide. Moreover, under the condition of excavation disturbance, a shallow landslide is likely to occur in a short time, and the time between the connection of the sliding surface and the occurrence of landslide is generally quick. The landslide early warning method established in this paper describes the deformation process of the slope through the area of equal displacement cross-section, avoiding the problem of selecting appropriate points on the slope, which has significant importance for landslide warning of loose accumulation.

6. Conclusions

Due to the granular structure of waste rockfills, it is difficult to find a suitable single-point displacement as an early warning index. The advantage of GB-InSAR full-field scanning is that it helps to obtain slope displacement and other displacement information (the electromagnetic wave reflection intensity of the measured target, etc.) through data processing. In this paper, a landslide early warning system based on the rate of change of the cross-sectional area of equal displacement bodies was proposed for the landslides of unconsolidated coarse sedimentary rocks.

By summarizing the previous landslide monitoring and data processing results, using the normalized rate of change of the cross-sectional area of a −0.5 mm equal displacement body as the early warning criteria, two types of landslides and four levels of early warning classifications were established as follows: (a) Level 1 (green) warning area, $V^*_{-0.5}(t) < 0.01$; (b) Level 2 (yellow) warning area, $0.01 \le V^*_{-0.5}(t) < 0.02$; (c) Level 3 (orange) warning area, $0.02 \le V^*_{-0.5}(t) < 0.03$; and (d) Level 4 (red) warning area, $V^*_{-0.5}(t) \ge 0.03$.

The validity of the early warning method was verified based on five landslides to compare the predicted and real values in this paper. Further research is needed to take into consideration comprehensive factors such as rock mechanics and rainfall for early warning of rockfill landslides.

Author Contributions: N.Q. contributed to conceptualization, numerical and experimental investigations, and preparation of the paper; Y.-L.D. contributed to conceptualization, numerical and experimental studies, and preparation of the paper; X.-M.S. contributed to conceptualization, numerical and experimental investigations, and preparation of the paper; X.-F.W. contributed to conceptualization, numerical and experimental investigations, and preparation of the paper; and J.-M.F. contributed to experimental investigations. All authors have read and agreed to the published version of the manuscript.

Funding: This research received no external funding.

Conflicts of Interest: The authors declare no conflict of interest.

References

1. Kairong, H. State of art and Prospect of Tunnels and Underground Works in China. *Tunn. Constr.* **2015**, *35*, 95.
2. Qiao, N.; Duan, Y.; Wei, X.; Shi, X. Study on Landslide Monitoring Experiment of Underground Chambers Spoil Area. *E&ES* **2020**, *455*, 012212.
3. Chae, B.G.; Park, H.J.; Catani, F.; Simoni, A.; Berti, M. Landslide prediction, monitoring and early warning: A concise review of state-of-the-art. *Geosci. J.* **2017**, *21*, 1033–1070. [CrossRef]
4. Xu, F.; Wang, Y.; Du, J.; Ye, J. Study of displacement prediction model of landslide based on time series analysis. *Chin. J. Rock Mech. Eng.* **2011**, *30*, 746–751.

5. Saito, M. Forecasting the time of occurrence of a slope failure. In Proceedings of the 6th International Conference on Soil Mechanics and Foundation Engineering, Montreal, QC, Canada, 8–15 September 1965; pp. 537–541.
6. Hoek, E. Estimating the stability of excavated slopes in opencast mines. *Inst. Min. Metall. A* **1970**, *105*, A132.
7. Hencher, S.R.; Liao, Q.H.; Monaghan, B.G. Modelling slope behaviour for open-pits. *Trans. Inst. Min. Metall. Sect. A Min. Ind.* **1996**, *105*, A37.
8. Faquan, W.; Niansheng, W. An Embryo of Dynamical Analysis Method on Landslide Displacement. *Chin. J. Geol. Hazard. Control.* **1996**, *7*, 38–41.
9. Zhong, Z.; Baochen LI, U. Verhulst Inverse-Function Residual Correction Model for Landslide Prediction and Forecast. *China Railw. Sci.* **2009**, *4*, 28–36.
10. Xu, Q.; Huang, R.Q.; Li, X.Z. Research progress in time forecast and prediction of landslides. *Adv. Earth Sci.* **2004**, *3*, 479–483.
11. Shitian, W.; Zhuoyuan, Z.; Zheng, Z. On the characteristics and dynamics of the catastrophic mount sale landslide, Gansu (kansu), China. *J. Chengdu Univ. Technol.* **1988**, *2*, 58–63.
12. Iading, W.; Zhuoyuan, Z. A study on the mechanism of high-speed loess landslide induced by earthquake. *Chin. J. Geotech. Eng. Chin. Ed.* **1999**, *21*, 670–674.
13. Siqing, Q. The nonlinear dynamics method to forecast landslide disaters. *Hydrogeol. Eng. Geol.* **1993**, *5*, 1–14, 58.
14. Gui-lan, L.; Wei-ya, X. Application of fuzzy Markov chains model in landslide stability prediction. *Chin. J. Geol. Hazard. Control.* **2006**, *4*, 14.
15. Xu, X.; Shang, Y.; Wang, Y. Research on comprehensive evaluation decision system for landslide disaster. *Rock Soil Mech.* **2010**, *10*, 22.
16. Huo, Z.; Liang, S. Application of Fuzzy Mathematics Method on Landslide Stability Assessment. *Northwest. Seismol. J.* **2007**, *1*, 35–39.
17. Ercanoglu, M.; Gokceoglu, C. Use of fuzzy relations to produce landslide susceptibility map of a landslide prone area (West Black Sea Region, Turkey). *Eng. Geol.* **2004**, *75*, 229–250. [CrossRef]
18. Tongzhen, Y. Prediction on Stages of Single Process Circle for Landslide and Ground Subsidence. *Chin. J. Geol. Hazard. Control.* **1990**, *2*, 11–21.
19. Li, C.; Tang, H.; Hu, X.; Li, D.; Hu, B. Landslide prediction based on wavelet analysis and cusp catastrophe. *J. Earth Sci.* **2009**, *20*, 971. [CrossRef]
20. Ren, D.; Leslie, L.M.; Fu, R.; Dickinson, R.E.; Xin, X. A storm-triggered landslide monitoring and prediction system: Formulation and case study. *Earth Interact.* **2010**, *14*, 1–24. [CrossRef]
21. Clare, M.A.; Talling, P.J.; Challenor, P.; Malgesini, G.; Hunt, J. Distal turbidites reveal a common distribution for large (>0.1 km^3) submarine landslide recurrence. *Geology* **2014**, *42*, 263–266. [CrossRef]
22. Dietrich, W.E.; Bellugi, D.; De Asua, R.R. Validation of the shallow landslide model, SHALSTAB, for forest management. *Water Sci. Appl.* **2001**, *2*, 195–227.
23. Pope, E.L.; Talling, P.J.; Urlaub, M.; Hunt, J.E.; Clare, M.A.; Challenor, P. Are large submarine landslides temporally random or do uncertainties in available age constraints make it impossible to tell. *Mar. Geol.* **2015**, *369*, 19–33. [CrossRef]
24. Qiang, X.U. Understanding the Landslide Monitoring and Early Warning: Consideration to Practical Issues. *J. Eng. Geol.* **2020**, *28*, 360–374.
25. Liu, L.; Jiang, L.; Sun, Y.; Wang, H.; Sun, Y.; Xu, H. Diurnal fluctuations of glacier surface velocity observed with terrestrial radar interferometry at Laohugou No. 12 Glacier, western Qilian mountains, China. *J. Glaciol.* **2019**, *65*, 239–248. [CrossRef]
26. Shouren, Z. Major technical problems about river closure and second stage cofferdam design at Three Gorges Project. *Yangtze River* **1997**, *4*, 129–138.
27. Rödelsperger, S.; Läufer, G.; Gerstenecker, C.; Becker, M. Monitoring of displacements with ground-based microwave interferometry: IBIS-S and IBIS-L. *J. Appl. Geod.* **2010**, *4*, 41–54. [CrossRef]
28. Zhang, Y.; Luo, L.; Yang, J.; Liu, D.; Kong, R.; Feng, Y. A hybrid ARIMA-SVR approach for forecasting emergency patient flow. *J. Ambient Intell. Human Comput.* **2019**, *10*, 3315–3323. [CrossRef]
29. Shang-qing, W.; Jin-jun, X.U.; Mian, L. Study of warning of dangerous state of Baishuihe Landslide in Three Gorges Reservoir Area. *J. Geomat. Inf. Sci. Wuhan Univ.* **2009**, *34*, 1218–1221.

30. He, K.; Chen, W.; Zhang, P. Real-time monitoring of dynamic stability coefficient and displacement criterion of the creep slope. *J. Chin. J. Rock Mech. Eng.* **2016**, *35*, 1377–1385.
31. Qiang, X.U.; Dalei, P.; Chaoyhang, H.; Xing, Q.; Kuanyao, Z.; Dehao, X. Theory and method of monitoring and early warning for sudden loess landslide-a case study at HeiFangTai Terrace. *J. Eng. Geol.* **2020**, *28*, 111–121.

applied sciences

MDPI

Article

Assessment of Landslide-Induced Geomorphological Changes in Hítardalur Valley, Iceland, Using Sentinel-1 and Sentinel-2 Data

Zahra Dabiri [1,*], Daniel Hölbling [1], Lorena Abad [1], Jón Kristinn Helgason [2], Þorsteinn Sæmundsson [3] and Dirk Tiede [1]

1 Department of Geoinformatics-Z_GIS, University of Salzburg, Schillerstrasse 30, 5020 Salzburg, Austria; daniel.hoelbling@sbg.ac.at (D.H.); lorenacristina.abadcrespo@sbg.ac.at (L.A.); dirk.tiede@sbg.ac.at (D.T.)
2 Icelandic Meteorological Office, Bústaðavegur 7–9, 105 Reykjavík, Iceland; jonkr@vedur.is
3 Department of Geography and Tourism, University of Iceland, Sturlugata 7, 101 Reykjavik, Iceland; steinis@hi.is
* Correspondence: zahra.dabiri@sbg.ac.at

Received: 17 July 2020; Accepted: 20 August 2020; Published: 24 August 2020

Abstract: Landslide mapping and analysis are essential aspects of hazard and risk analysis. Landslides can block rivers and create landslide-dammed lakes, which pose a significant risk for downstream areas. In this research, we used an object-based image analysis approach to map geomorphological features and related changes and assess the applicability of Sentinel-1 data for the fast creation of post-event digital elevation models (DEMs) for landslide volume estimation. We investigated the Hítardalur landslide, which occurred on the 7 July 2018 in western Iceland, along with the geomorphological changes induced by this landslide, using optical and synthetic aperture radar data from Sentinel-2 and Sentinel-1. The results show that there were no considerable changes in the landslide area between 2018 and 2019. However, the landslide-dammed lake area shrunk between 2018 and 2019. Moreover, the Hítará river diverted its course as a result of the landslide. The DEMs, generated by ascending and descending flight directions and three orbits, and the subsequent volume estimation revealed that—without further post-processing—the results need to be interpreted with care since several factors influence the DEM generation from Sentinel-1 imagery.

Keywords: object-based image analysis; Sentinel-1; Sentinel-2; digital elevation model; InSAR; landslide; landslide-dammed lake; river; Iceland

1. Introduction

Landslide mapping and analysis are essential aspects of hazard and risk analysis, and the accurate detection of land surface changes is crucial for understanding processes and interactions between human and natural phenomena [1,2]. Landslides can be triggered by earthquakes, snowmelt, severe rainfall, human activities, or a combination of these factors. Landslides can partially or entirely block rivers and subsequently create landslide-dammed lakes, whereby potential dam failures pose a substantial risk for downstream areas [3]. Large, rapid mass movements are a common geomorphological process in Iceland and represent a significant threat to people and infrastructure [4–6]. Examples of such landslides are a rock avalanche onto the Morsárjökull outlet glacier in 2007 [7], a debris slide at the Móafellshyrna Mountain in northern Iceland in 2012 [8], and the landslide at the Askja caldera in July 2014, which caused a displacement wave [9–11].

Earth Observation (EO) data from optical and synthetic aperture radar (SAR) sensors have proven valuable for mapping and monitoring geomorphological features and, in particular, different types of landslides [1,12–14]. Optical imagery from different sources such as satellites, or unmanned aerial vehicles (UAVs) [15,16] has been used for landslide mapping and analysis [15,17]. However, mapping landslides using only optical imagery is challenging because of the spectral, spatial, and temporal characteristics of landslides [18]. SAR imagery has the potential to be used for hazard assessments by providing large-scale two-dimensional high spatial and temporal resolution images of the Earth's surface [19]. Radar pulses can penetrate through clouds (nearly weather independent), and they can provide data during the day and night (sun independent) [20]. According to Lee and Pottier [19], the surface reflectivity measured by radar imagery (also known as the radar backscatter coefficient σ^0) is a function of the radar system parameters (such as the frequency, polarization, incident angle) and the surface parameters (such as the topography, roughness, dielectric properties of the medium, moisture). These parameters can be used to extract features from SAR imagery. Moreover, different polarizations provide different information about features of interest on the ground, and depending on the structure and position of the feature of interest, they might appear differently with varying polarizations [21]. The combined interpretation of optical and SAR imagery provides promising opportunities for landslide investigation while existing approaches still need to be improved [22–25].

Several studies have shown that object-based image analysis (OBIA) can achieve better classification accuracies than pixel-based techniques [26,27]. OBIA has been used for more than two decades as a framework for feature extraction from different imagery [28,29]. OBIA allows the use of spectral, textural and spatial information and thus provides a suitable framework for landslide mapping [30]. The integration of optical and SAR imagery within an OBIA approach for mapping geomorphological features is promising since the properties of both data types can be exploited in one interlinked workflow [24].

The estimation of landslide volumes is another key parameter for hazard analysis. There are several methods for estimating the volume of a landslide, of which the use of pre- and post-event digital elevation models (DEMs) can provide a good estimation [31]. However, accurate post-event DEMs are rarely freely available. Thus, the straightforward generation of post-event DEMs based on freely available data is important. There are several techniques to create DEMs, such as using optical imagery and stereoscopy or using SAR imagery and interferometric SAR (InSAR) techniques. The InSAR technique is based on the processing of at least two complex SAR images covering the same area and acquired from slightly different points of view [32]. InSAR allows the extraction of three-dimensional information using the phase difference and the along- and cross-track location of targets on the image and can be used for measuring the topography of a surface and its changes over time [33–35]. The generation of DEMs from SAR imagery using InSAR techniques is not new; for example, the well-known Shuttle Topography Mission (SRTM) provides near-global DEM data produced by InSAR [36]. However, there is a need for higher temporal and spatial resolution DEM data, particularly for assessing landslide events.

In this study, Sentinel-2 optical data and Sentinel-1 SAR data were used for geomorphological feature mapping. Sentinel-2 is a multispectral optical sensor from the European Union's Earth Observation programme, Copernicus, with 13 spectral bands combining different spatial resolutions (up to 10 m for Red, Green, Blue (RGB) and Near-infrared (NIR) bands) and a repeat cycle of 5 days [37]. The Sentinel-1 A SAR instrument was the first satellite of the Copernicus Sentinel missions [38], which provides radar data following the retirement of the ERS-2 and Envisat missions. The Sentinel-1 mission comprises a constellation of two polar-orbiting satellites (A and B). The Sentinel-1 instrument provides high-resolution C-band (wavelength of 5.6 cm) data with a short revisit time of 12 days for each satellite, and six days with Sentinel-1 A and B [39]. The satellite has a new type of ScanSAR mode, known as Terrain Observation with Progressive Scan (TOPS) SAR [40]. It has a similar pixel resolution (about 3.5 m in the range direction and 14 m in the azimuth direction) to ScanSAR but with an enhanced signal-to-noise ratio distribution [41]. The high temporal and spatial resolution of the Sentinel-1 A and B SAR has been used in various InSAR applications [42]. However, using Sentinel-1 data for generating DEMs and its application in landslide analysis is still in the early stages [43,44]. In this research, we focused on a large landslide that occurred on the 7 July 2018 on the eastern side of the Fagraskógarfjall mountain in Hítardalur valley, western Iceland, leading to a landside-dammed lake and changes in the watercourse [45].

The main objectives of this research are to semi-automatically assess the landslide-induced geomorphological changes in Hítardalur valley by jointly using Sentinel-1 and Sentinel-2 data within an OBIA workflow and to assess the potential of post-event DEMs generated from Sentinel-1 image pairs for the estimation of the Hítardalur landslide volume.

2. Materials and Methods

2.1. Study Area

Our study area is the Hítardalur valley and, in particular, the area surrounding the Hítardalur landslide, which is located on the eastern side of the Fagraskógarfjall mountain in the Vesturland region, western Iceland (Figure 1). The elevation in the Hítardalur valley ranges from 89 to 743 m above sea level. The valley floor is partly covered by a lava field (Hagahraun) [46]. Thick sedimentary layers can be found within the stratigraphic sequence. The rock is rather dense as the cavities have been filled with minerals. Quaternary volcanic formations, created during the last glaciation and Holocene, can be found within the vicinity of the landslide. Faults and fractures are prominent within the region and can be divided into three groups, i.e., NE–SW, N–S and NW–SE oriented [47]. The faults and fractures that belong to the NE–SW and N–S groups formed mainly during the formation of the lava pile and are older than 8 million years. NW–SE oriented fractures belong to the so-called Snæfells fracture zone which extends from Kerlingarskarð in the north to Borgarfjördur valley in the south. The NW–SE fracture zone was very active 8 to 4.5 million years ago and is considered to be still active [48].

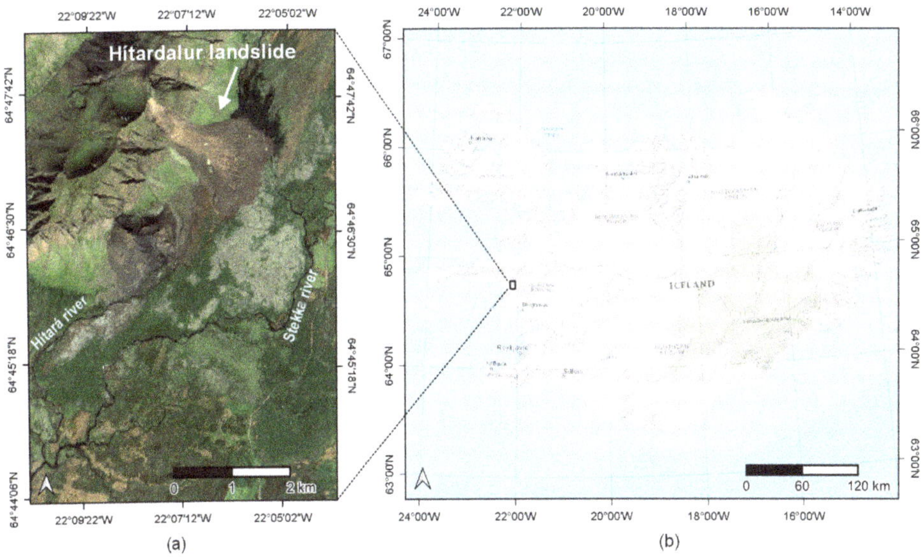

Figure 1. (**a**) The study area around the Hítardalur landslide shown on a Sentinel-2 image acquired on 17 July 2018 and (**b**) its location in western Iceland; background data © ESRI.

On the morning of the 7 July 2018, a large landslide blocked the Hítará river and led to the creation of a landslide-dammed lake. Water from the landslide-dammed lake found a new way through the highly porous lava field. The landslide originated in an area of the Fagraskógarfjall mountain that shows evidence of earlier displacements [45]. The year 2018 had more rainfall than usual in Iceland [49], likely leading to a higher water pressure in cracks and fractures that further weakened the rocks, and thus contributed to landslide initiation [50]. Helgason et al. [45] estimated that approximately 7 million m^3 of material was released and that a total of 10–20 million m^3 of material was displaced. The landslide deposition covers an area of approximately 1.5 km^2 on the valley floor and is up to 30 m thick. The fall height was about 450 m over a runout length of approximately 2.3 km at a runout angle of approximately 12°. The length of the riverbed covered with debris was estimated to be around 1.6 km. The Hítardalur landslide is considered to be one of the largest landslides in Iceland in recorded history [51]. Figure 2 shows field photographs of the Hítardalur landslide, the landslide-dammed lake and the river finding a new path after emerging from the dammed lake.

(a) (b) (c) (d)

Figure 2. (**a**) Photograph showing the Hítardalur landslide and the landslide-dammed lake soon after the event occurred, i.e., taken on the afternoon of 7 July 2018 (photograph: © S. Asgeirsson). (**b**) View from the landslide deposition area towards the landslide source area (photograph: © J. K. Helgason; 7 July 2018). (**c**) Water from the dammed lake finds a new way (photograph: © J. K. Helgason; 7 July 2018). (**d**) View from the Fagraskógarfjall mountain towards the Hítardalur valley showing the landslide deposition area and the new watercourse flowing out from the lake into the Stekka river (photograph: © M. Olafsson; 23 November 2018).

2.2. Data and Data Preparation

2.2.1. Optical Data

We used pre- and post-landslide event Sentinel-2 data (Table 1). The data selection was made based on the cloud cover estimates (<30%) provided by the European Space Agency (ESA) for pre- and post-event datasets, using the Google Earth Engine (GEE) platform.

Table 1. Sentinel-2 datasets used in this research. MSI stands for Multispectral Instrument.

Date	Sensor, Product	Pre-/Post-Event
20 June 2018	Sentinel-2 B MSI, Level-1C	Pre
17 July 2018	Sentinel-2 B MSI, Level-1C	Post
1 August 2019	Sentinel-2 B MSI, Level-1C	Post

2.2.2. Radar Data

We used the Sentinel-1 Interferometric Wide Swath (IWS) Level-1 Single Look Complex (SLC) product. The SLC is the main product of the IWS mode. It is mainly used for interferometric applications [38,52]. The Sentinel-1 IWS SLC product includes three sub-swaths and each of them includes one intensity image per polarization channel, and several bursts over each of the sub-swaths [40,52]. However, for mapping purposes, the SLC data should be post-processed and converted to the Ground Range Detected Geo-referenced product (GRD). For mapping geomorphological features and related changes, backscatter coefficients (also known as sigma naught, or sigma 0) were created by converting the Sentinel-1 SLC images with slant-range geometry to Sentinel-1 GRD products. The GRD data consist of focused SAR data that have been corrected using SLC data by applying a calibration and speckle noise filtering using the Lee refined filter [53,54], and a ground geometry correction using an Earth ellipsoid model. The thermal noise of the GRD products was removed using a noise look-up table provided for each image to derive the calibrated noise profiles matching the calibrated GRD data and to improve the quality of the TOPSAR images [55,56]. A visual inspection of several Sentinel-1 scenes from different orbit tracks and the two available flight directions (ascending with the satellite moving north, and descending with the satellite moving south) showed that ascending images were more suitable than the descending images for mapping the geomorphological features and related changes in Hítardalur valley due to their geometry. The whole landslide, including both the landslide source and the deposition areas, is usually well visible on the ascending image, whereas only the deposition area is visible on the descending image. Therefore, one pre-event (5 July 2018) and two post-event (17 July 2018 and 5 August 2019) ascending Sentinel-1 TOPSAR scenes (SLC type) in the interferometric wide (IW) mode, with vertical-vertical (VV) and vertical-horizontal (VH) dual polarizations from track 16 were selected for the geomorphological mapping.

Figure 3 shows the combination of Sentinel-1 sigma naught VV polarization and the Sentinel-2 datasets for each year, which were used for geomorphological feature mapping.

Moreover, we used six additional Sentinel-1 IWS SLC datasets for generating post-event DEMs using InSAR. The quality of DEMs generated from InSAR analysis essentially depends on the perpendicular component of the spatial baseline (known as the perpendicular baseline or B_{perp}) between the radar antenna locations, divided by the distance to the ground. The distance to the ground is similar for the orbiting satellites, so the spatial baseline is the important variable. If the perpendicular baseline is too small, the topographic effects on the differential phase are not pronounced enough. On the other hand, if the baselines are too big, the coherent phase is increasingly different, leading to decorrelation. For example, for the ERS satellite, a suitable B_{perp} for DEM generation is recommended to be between 150 and 300 m at the time of the image acquisition [57]. This grants an angle between both acquisitions, which allows the retrieval of topographic variations from parallel-like effects. This B_{perp} was originally recommended for the ERS data, while for Sentinel-1 a lower range is usually used according to the literature. For example, Geudtner et al. [58] described that due to the orbit maintenance strategy, the baselines for the Sentinel-1 A and B are mostly in the range of ±150 m. Kyriou et al. [59] used a B_{perp} between 96 and 170 m for landslide mapping with Sentinel-1. For this study, we downloaded the available Sentinel-1 images with a B_{perp} range according to these recommendations for 2018 and 2019 (only from June to September to avoid potential snow cover) and processed them to generate post-event DEMs (Section 2.4).

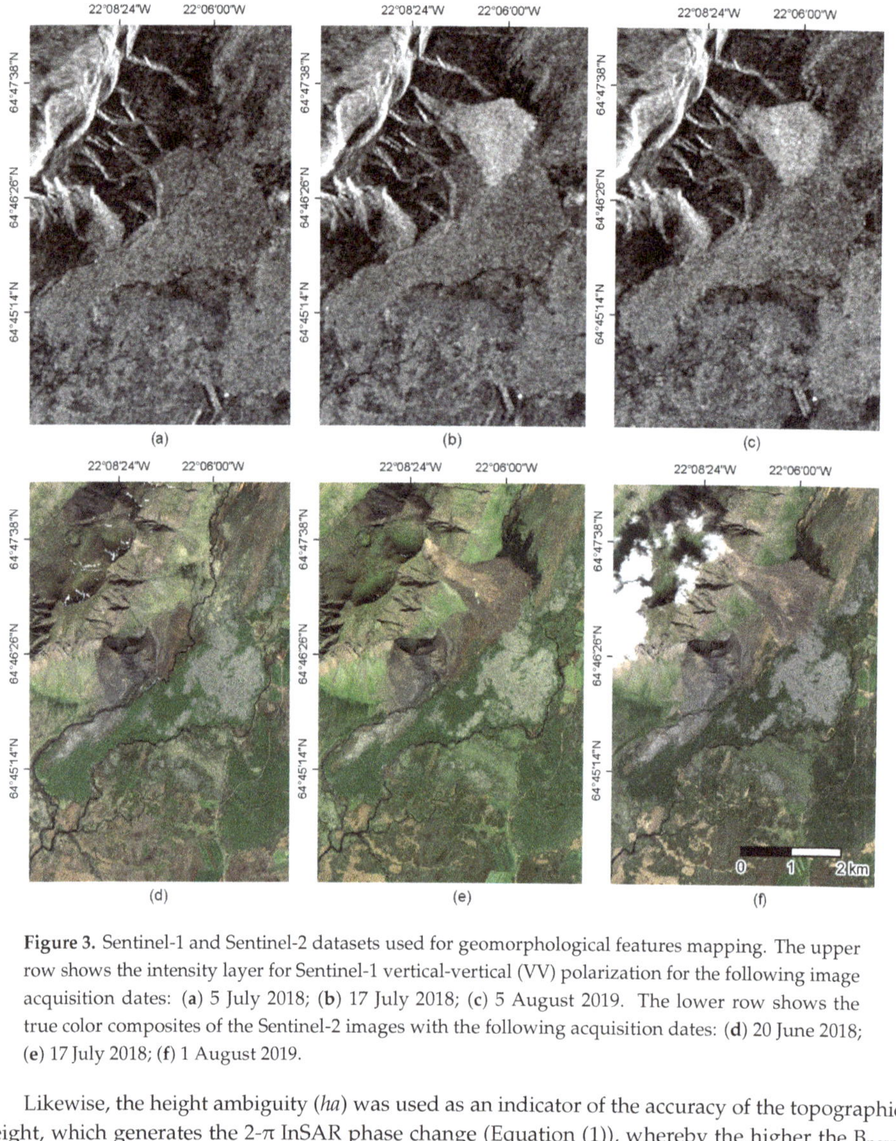

Figure 3. Sentinel-1 and Sentinel-2 datasets used for geomorphological features mapping. The upper row shows the intensity layer for Sentinel-1 vertical-vertical (VV) polarization for the following image acquisition dates: (**a**) 5 July 2018; (**b**) 17 July 2018; (**c**) 5 August 2019. The lower row shows the true color composites of the Sentinel-2 images with the following acquisition dates: (**d**) 20 June 2018; (**e**) 17 July 2018; (**f**) 1 August 2019.

Likewise, the height ambiguity (*ha*) was used as an indicator of the accuracy of the topographic height, which generates the 2-π InSAR phase change (Equation (1)), whereby the higher the B_{perp}, the more accurate the altitude measurement [57,60]:

$$ha = \frac{\lambda\, R \sin\theta}{2B_{perp}} \quad (1)$$

where *ha* is the height ambiguity, λ is the radar wavelength, *R* is the range from satellite to ground, and θ is the look angle.

From the above equation, it is clear that *ha* is proportionally related to the B_{perp}, and a higher B_{perp} theoretically should result in a more accurate altitude measurement. However, for a longer baseline, it is more difficult to solve the 2π phase ambiguity by phase unwrapping than for a shorter baseline [61].

The temporal interval between two SAR images was another factor to consider for the image selection. It influences the coherence, which determines how accurately the phase can be measured. A low coherence indicates a higher phase noise, which can cause larger elevation errors and phase unwrapping errors. A maximum relative coherency is recommended to be considered for the selection of SAR data pairs for DEM generation. Moreover, the interferometric phase depends on the orbit geometry, scene topography, line-of-sight surface displacements, and atmospheric path delays [62].

We used the Alaska Satellite Facility Baseline Tool to identify and select image pairs. From the available orbit tracks (i.e., number 16, number 118, and number 155), and flight directions (descending and ascending), we selected three post-event Sentinel-1 (A and B) image pairs (Table 2) for DEM generation and volume estimation. We used the InSAR stack overview function implemented in the SNAP (Sentinel Application Platform) toolbox to estimate the expected coherency (called "modeled coherency"), and the height ambiguity between two TOPSAR pair images.

Table 2. Sentinel-1 data and their characteristics used to generate the digital elevation models (DEMs). The B_{perp} stands for the spatial perpendicular baseline, which was used as the main criterion for the selection of Sentinel-1 A and B image pairs. The modeled coherency and height ambiguity measurements were derived using the interferometric synthetic aperture radar (SAR) (InSAR) stack overview function implemented in the Sentinel Application Platform (SNAP) toolbox.

Sentinel-1 Image Pairs	Orbit Track	Flight Direction	Time Baseline (Days)	B_{perp} (m)	Modeled Coherency	Height Ambiguity (m)
11 July 2018 and 17 July 2018	16	Ascending	6	134	0.88	115
5 August 2018 and 11 August 2018	118	Ascending	6	142	0.87	110
4 July 2019 and 10 July 2019	155	Descending	6	159	0.86	98

All SAR processing was completed using the open-source SNAP toolbox provided by ESA. Moreover, we used existing global earth topography and sea surface elevation data at 30 arc-second resolution (GETASSE30), which was freely available and accessible in the SNAP toolbox in the DEM generation process.

2.2.3. Topographic Data

We used the freely available ArcticDEM (spatial resolution of 2 m), provided by the Polar Geospatial Center, for the validation of the generated DEMs. The ArcticDEM dataset is generated based on overlapping very high-resolution (VHR) optical satellite images with sub-meter resolution and stereoscopic imagery [63,64].

2.3. *Geomorphological Features Mapping*

In the mapping of geomorphological features, we focused on the Hítardalur landslide, the landslide-dammed lake, and the riverbed with flowing water. Additionally, we differentiated between the landslide source and deposition area and identified changes in the watercourse based on the pre-and post-event images. We used an OBIA approach and created a knowledge-based classification ruleset in eCognition (Trimble) software, integrating Sentinel-1 and Sentinel-2 datasets. We used spectral indices derived from the Sentinel-2 data, including a brightness layer (average of the three visible spectral bands blue (B2), green (B3), red (B4)) and the normalized difference vegetation index (NDVI). For the detection of riverbeds, we calculated an edge extraction layer based on the Sentinel-2 NIR (B8) spectral band using the Lee sigma [65] algorithm. We also used the intensity information from the pre- and post-event Sentinel-1 VV, and VH polarization intensity images to create subtraction layers for 2018 and 2019. The Sentinel-1 intensity subtraction layers and the edge extraction layers derived from band B8 were used for the object-based classification for 2018 and 2019 and are shown in Figure 4.

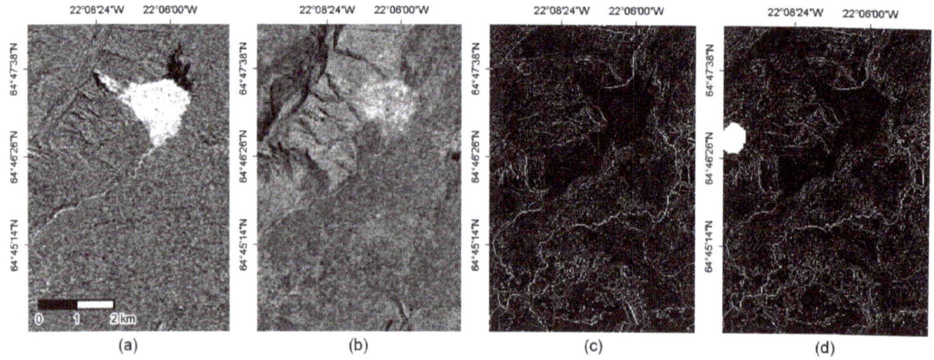

Figure 4. Examples of the layers extracted from the Sentinel-1 and Sentinel-2 datasets. (**a**) and (**b**) show the subtraction layers using pre- and post-event Sentinel-1 VV polarization intensity images for 2018 and 2019, respectively. The Hítardalur landslide and the landslide-dammed lake are distinguishable. (**c**) and (**d**) show the edge extraction layers derived from the post-event Sentinel-2 images (Near-infrared (NIR) band (B8)) for 2018 and 2019, respectively.

The Hítardalur landslide and landslide-dammed lake are distinguishable in the Sentinel-1 subtraction layer of the year 2018 (Figure 4a), but they are less obvious on the Sentinel-1 subtraction layer from 2019 (Figure 4b). Due to the SAR amplitude changes, the Hítardalur landslide appears very bright, and the landslide-dammed lake very dark on the Sentinel-1 subtraction layer of 2018 (Figure 4a). The white path visible on the Sentinel-1 VV polarization subtraction layer of 2018 (Figure 4a) is the Hítará riverbed (i.e., towards the south-west of the landslide). In the Sentinel-1 subtraction layer of the year 2019, the Hítará river change is hardly visible (Figure 4b). The Sentinel-2 edge extraction layer of the year 2018 clearly shows the new watercourse between the dammed lake and Stekka river (i.e., the river towards the south-east of the landslide) (Figure 4c). This connection, however, is not readily distinguishable on the edge extraction layer of the year 2019. In both images (Figure 4c,d), the outlines of the lake are partly visible, but, it is difficult to differentiate them from the edges of other features. The white circular area in Figure 4d is a cloud mask.

The OBIA classification of the landslide was based on the pre- and post-event VV polarization subtraction layer, the brightness layer was used for the classification of the landslide-dammed lake, and the riverbed with water was classified using the edge extraction layer and refined based on spectral and spatial parameters. The first step in the object-based mapping was the creation of image objects using the multiresolution segmentation algorithm. Then, we created a knowledge-based classification ruleset which includes thresholds based on the backscatter information from Sentinel-1 and the spectral information from the Sentinel-2 images for each year. The size of the segmentation-derived image segments is controlled by the scale parameter (SP), which is directly related to the local variations of the bands used in the multiresolution segmentation algorithm [66,67]. In general, the higher the value of the SP, the larger the resulting segments, and vice versa. First, multiresolution segmentation was applied at the pixel level to create suitable objects for the subsequent classification of the landslide, using the Sentinel-1 VV intensity change layer, and the landslide-dammed lake, using the Sentinel-1 VV intensity change and NDVI layers. The same features were used for classification in both years, whereby some adaptations in the thresholds for the 2019 mapping were required. The difference in the values of the Sentinel-1 VV polarization change layers between the year 2018 and 2019 could be related to the different time of the Sentinel-1 image acquisitions, and potential changes in the surface structure of the landslide, as well as different dielectric properties (e.g., moisture content) of the landslide. Another multiresolution segmentation was performed based on the unclassified segments from the first segmentation and using only the Sentinel-2 post-event images to create suitable objects for the classification of the riverbeds with water. We excluded the Sentinel-1 datasets from this classification since riverbeds were barely visible in the datasets.

Moreover, we classified the original riverbeds with water using only the pre-event Sentinel-2 image (20 June 2018). Therefore, we used the edge extraction layer sigma >5 followed by a manual refinement, merging, and using a threshold for the object length of >470 pixels.

Figure 5 gives an overview of the OBIA mapping workflow and the features and thresholds which were used for the post-event geomorphological feature mapping for 2018 and 2019.

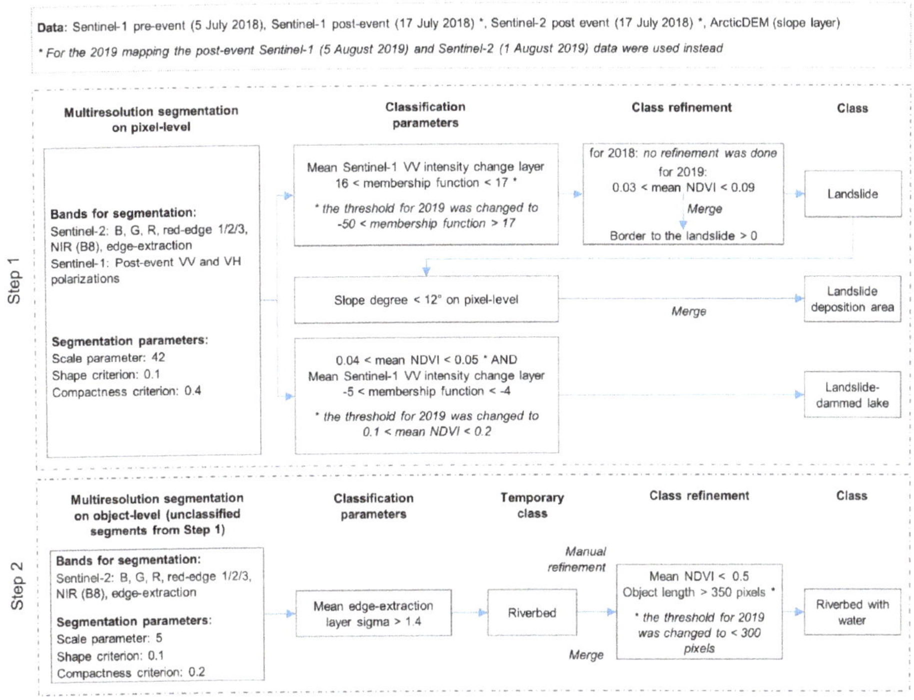

Figure 5. The classification framework used for the post-event geomorphological feature mapping for 2018 and 2019.

Moreover, we semi-automatically delineated the part of the landslide where the material was deposited in the year 2018 as an input for the subsequent landslide volume estimation. Therefore, we re-segmented the landslide class. Then the slope information derived from the Arctic DEM was used to differentiate the landslide deposition area from the landslide source area with a slope threshold of <12°. The value of 12° was selected based on the work of Helgason et al. [45], impressions from the field, and considering the visual comparison of the field photographs with the remote sensing data.

2.4. DEM Generation

DEMs are an important component of landslide modelling, volume calculation, and landform classification. In this research, we performed an exploratory assessment of the usability of Sentinel-1 SLC TOPSAR datasets for the generation of post-event DEMs within an automated workflow. The three SLC TOPSAR image pairs were used to create post-event DEMs using InSAR. Each TOPSAR image was calibrated using precise orbit ephemerides (POD) products provided by the ESA. The POD products contain auxiliary information about the position of the satellite during the data acquisition.

For computational efficiency, the split TOPSAR function was applied to each TOPSAR image to select the swath sub-bursts covering the area of interest. The TOPSAR images require a precise coregistration to ensure each ground target contributes to the same (range and azimuth) pixel in both images [57]. The coregistration of SLC TOPSAR images can be completed on a pixel by pixel basis, with an accuracy in the order of one-tenth of the resolution. Therefore, the process of DEM-assisted coregistration was applied using the back-geocoding function. Due to the steep azimuth spectrum ramp in each burst [42] the correction of the shift in the azimuthal direction was performed by enhanced spectral diversity [68]. Then, an interferogram generation was carried out by cross multiplying the complex conjugate of two coregistered TOPSAR images. During this process, the amplitude of both images is multiplied while the phase represents the phase difference between the two TOPSAR images. The flat-Earth phase component was then subtracted to remove the phase present in the interferometric signal due to the curvature of the reference surface. The flat-Earth phase component was estimated using the orbital- and metadata information [69]. The resulting interferometric fringes represent a full 2 π cycle, with arbitrary colors representing half the sensor's wavelength. In the case of repeat-pass acquisitions, such as in the case of Sentinel-1, the interferograms are rather noisy, mainly due to temporal decorrelation. To improve the quality of the interferograms and reduce the phase noise of the adjacent pixels, the Goldstein [70] filtering with fast Fourier transformation (FFT) was used. We tried different settings and window sizes of the FFT filter and finally selected 64 for the FFT filter definition and 3 × 3 for the window size. The multi-look speckle filtering was applied to increase the radiometric accuracy by reducing the local variability of the signal [71]. The multi-look speckle filtering improves the signal-to-noise ratio at the expense of the spatial resolution [72]. Moreover, applying a multi-look speckle filtering on the interferogram increases the computational efficiency for phase unwrapping. The phase unwrapping is necessary to resolve the unknown multiple-of-wavelength ambiguity in the interferometric phase. We used the statistical-cost network-flow algorithm for phase unwrapping (SNAPHU), which is a two-dimensional phase unwrapping algorithm [73–75]. The SNAPHU algorithm is written in C, can be used within the SNAP toolbox, or as a standalone program on a Linux platform. The unwrapped phase values were converted into elevation values with respect to a reference ellipsoid [76]. This process specifies each pixel in the unwrapped phase image with respect to a Cartesian reference system using radar coordinates (range, azimuth, phase variation). The last step in the DEM generation was data geocoding to allow for the comparison of the results with a reference DEM.

The OBIA-delineated landslide deposition area for 2018 was used for the calculation of the landslide volume. The volume estimation was conducted by subtracting each of the three post-event DEMs from the pre-event ArcticDEM.

2.5. Validation

The accuracy of the geomorphological feature mapping was assessed by comparing it to mapping results available from the National Land Survey of Iceland (NLSI) [77], and information provided by Helgason et al. [45].

As for the accuracy assessment of the DEMs generated using Sentinel-1 image pairs, we used the high-resolution ArcticDEM as a reference. This is in line with the literature, which recommends that the resolution of the reference data should be at least three times higher than the resolution of the DEM elevation being evaluated [78,79]. The ArcticDEM has known biases of several meters due to errors in the sensor models due to the satellite position and different sensors in the ArcticDEM constellation. However, its vertical accuracy is corrected by registering it to the Ice, Clouds, and Land Elevation Satellite mission (ICESat) altimetry information and the results can be used without further corrections [80]. The DEM quality assessment was conducted by comparing the statistical measures, such as the minimum, maximum, mean, and standard deviation, of the generated DEMs with those of the reference ArcticDEM. The vertical quality assessment was completed using the root mean square error (RMSE), whereas the horizontal quality assessment was completed using autocorrelation and the Moran's I index [81,82].

The RMSE includes both random and systematic errors introduced during data production [83], and is expressed by the following equation:

$$RMSE = \sqrt{\frac{\sum(y_i - y_{ti})^2}{N}} \quad (2)$$

where y_i refers to the ith interpolated elevation, yt_i refers to the ith known or measured elevation of a sample point in a reference dataset, and N is the number of sample points.

While the RMSE is a valuable quality-control statistic, it does not provide an accurate assessment of how well each cell in the DEM represents the true elevation. The RMSE only provides an assessment of how well the DEM corresponds to the data to which it is compared [84]. The Moran's I index expresses local homogeneity by comparing the difference between neighboring pixels to the standard deviation. The Moran's I index ranges between +1 and −1, where +1 indicates a strong spatial autocorrelation, 0 a spatially uncorrelated data, and −1 a strong negative spatial autocorrelation. The Moran's I was calculated by adopting the "Queen's" rule, which takes into account the eight neighboring pixels. The generated DEMs from the InSAR analysis were first subtracted from the ArcticDEM, and then the Moran's I index was calculated on the difference map. We consider only the "stable area" for applying the Moran's I and RMSE measures, i.e., the area without the landslide and the landslide-dam lake.

3. Results

3.1. Geomorphological Features Mapping Using OBIA

The results of the geomorphological features mapping are shown in Figure 6. The pre-event result shows the original Hítará riverbed with water (Figure 6a). The post-event results (Figure 6b for 2018 and Figure 6c for 2019) show the landslide, the landslide-induced changes in the watercourse, and the landslide-dammed lake. The reference map from NLSI [77], which was used for comparison, is shown in Figure 6d. The water flowing out of the dammed lake found a new route through a lava field towards the south and then merged with the Stekka river, resulting in more water flowing in this riverbed than prior to the landslide. However, the upper part of the connection between the lake and the Stekka river is not easily distinguishable in the Sentinel-2 image taken in August 2019, probably due to less water flowing there at the time of the image acquisition one month later in the year 2019 (July 2018 versus August 2019). It seems that the Hítará river partly passes through the landslide and appears as a spring in the river bed downstream, i.e., approximately 2.5 km downstream from the lake and approximately 1 km downstream from the closest part of the landslide deposition. This is confirmed by the manually created reference data from NLSI [77]. The landslide area was estimated to be approximately 2000 ha in 2018 and 2019. Slight differences between the landslide areas likely result from variations in the segmentation. We estimated a lake area of 58 ha in 2018 and 47.1 ha in 2019. The reference values for the lake area are reported to be approximately 47 ha by Helgason et al. [45], but, the lake area from the shapefile provided by NLSI is 39.8 ha [77]. A comparison of our results with these numbers reveals an overestimation of the OBIA result for the year 2018. This could be associated with a high moisture content near the lakeshore and shallow water areas with high sediment load, which introduces uncertainty in the segmentation and classification [17]. The lower value for 2019 is probably due to the warm summer in 2019, which caused the discharge of some rivers to be lower for a few weeks [85].

The maximum width of the landslide deposition is approximately 1.5 km, and the overall runout length of the landslide is approximately 2.3 km. These results are in line with the numbers reported by Helgason et al. [45].

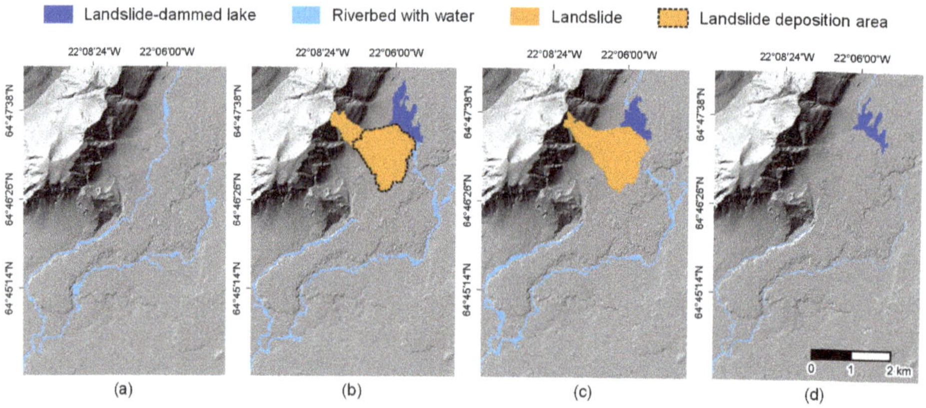

Figure 6. Object-based geomorphological feature mapping using Sentinel-1 and Sentinel-2 datasets. (**a**) shows the riverbed with water before the Hítardalur landslide using the pre-event Sentinel-2 image only; (**b**) shows the landslide, the landslide deposition area, the landslide-dammed lake, and the (now partly interrupted) riverbeds with water after the landslide event in 2018; (**c**) shows the landslide, the landslide-dammed lake, and the riverbeds with water in 2019; (**d**) shows the reference map from National Land Survey of Iceland IS 50V 24 September 2019 Vatnafar Flakar (© NLSI) [77]. The hillshade derived from the ArcticDEM was used as a background layer.

3.2. DEMs from Sentinel-1 and Landslide Volume Estimation

Figure 7 shows the results of the post-event DEMs generated using the Sentinel-1 image pairs from different orbit tracks, and the pre-event ArcticDEM for comparison.

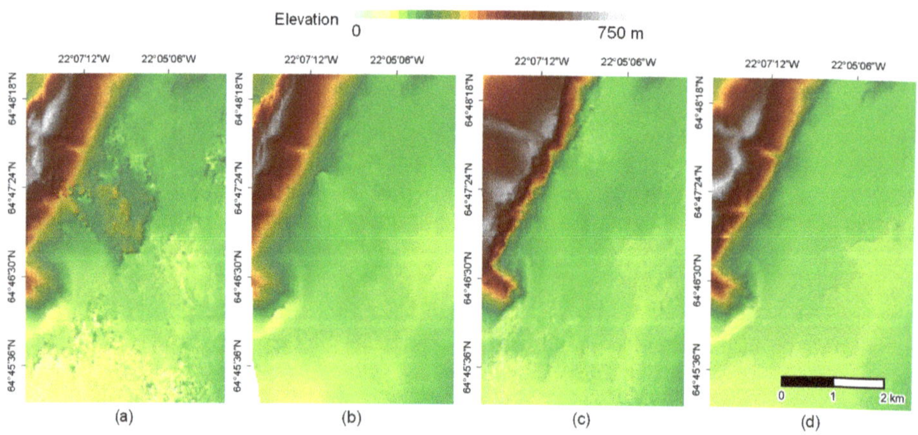

Figure 7. Illustration of the three post-event DEMs derived from the Sentinel-1 image pairs. (**a**) for orbit track number 16 (11 July 2018 and 17 July 2018), ascending; (**b**) orbit track number 118 (5 August 2018 and 11 August 2018), ascending; (**c**) orbit track number 155 (4 July 2019 and 10 July 219), descending; (**d**) ArcticDEM. For visualization purposes, each DEM is overlaid with the hillshade derived from it.

For the volume estimation of the landslide, each of the three generated DEMs was subtracted from the ArcticDEM. Then a spatial subset for the landslide was created using the OBIA delineation of the landslide deposition area for 2018 (Figure 8). The DEM from track 16 shows maximum elevation differences to the ArcticDEM of more than 80 m, whereas the other two DEMs (track number 118 and 155) show lower, more homogenous and realistic elevation differences in the landslide deposition area.

However, both DEMs from tracks 118 and 155 partly show negative values upslope, which indicate the removal of material and potential issues in the delineation of the deposition area.

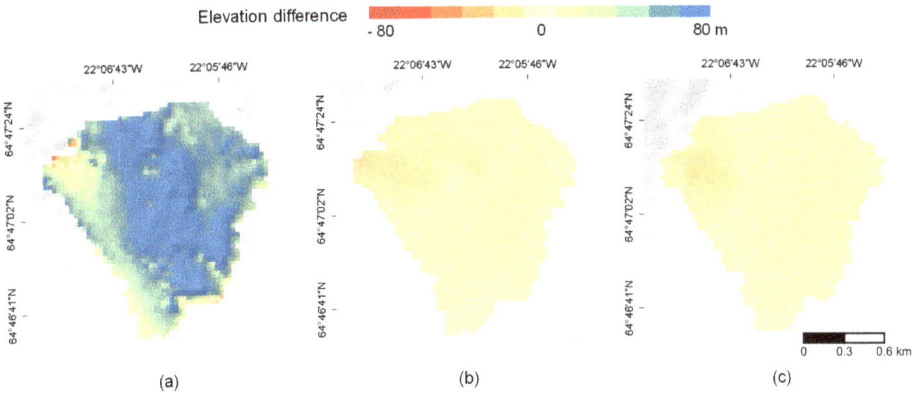

Figure 8. Elevation differences of the three Sentinel-1 DEMs compared to the ArcticDEM for the landslide deposition area used for calculation of the landslide volume. (**a**) elevation differences of DEM from track 16 (11 July 2018 and 17 July 2018); (**b**) elevation differences of DEM from track 118 (5 August 2018 and 11 August 2018); (**c**) elevation differences of DEM from track 155 (4 July 2019 and 10 July 2019). The respective hillshade derived from the Sentinel-1 DEMs was overlaid for each result.

Table 3 shows the landslide deposition volumes. The DEM from track 16 gives a much higher volume than the other two DEMs. The landslide volume estimations using the DEMs from track 118 and 155 are in a similar range.

Table 3. Volume estimation of the landslide based on the Sentinel-1 derived post-event DEMs and the pre-event ArcticDEM.

Sentinel-1 DEM	Orbit Track	Volume (Million m^3)
11 July 2018 and 17 July 2018	16	109
5 August 2018 and 11 August 2018	118	7
4 July 2019 and 10 July 2019	155	12

3.3. DEM Validation

Table 4 shows the measures used for the accuracy assessment of the generated DEMs. The comparison of the statistical measures shows an underestimation of the minimum DEM value for the track 16 DEM, while the other two DEMs show similar minimum statistics as the ArcticDEM. In contrast, the maximum value for the track 16 DEM is the closest compared to the ArcticDEM while the other two DEMs (tracks 118 and 155) show lower values. The mean and standard deviation values for the track 16 DEM are closer to the values of the ArcticDEM.

The RMSE for the track 16 DEM is the lowest of the three DEMs, which indicates less vertical variation between track 16 and the ArcticDEM, but at the same time it has the lowest Moran's I autocorrelation (0.92), which could be a result of the existence of more outliers compared to the track 118 and 155 DEMs, both of which have a Moran's I of 0.97. The latter DEMs have better autocorrelation due to more homogenous values.

It is worth mentioning that both quality assessment indicators used (i.e., RMSE and Moran's I) are global indicators, and they don't take into account local variations.

Table 4. Validation of the Sentinel-1 DEMs with respect to the ArcticDEM. Comparison is based on the area not influenced by the landslide. The Moran's I statistic is calculated on the difference map.

Acquisition Date	Orbit Tracks	Min Elevation (m)	Max Elevation (m)	Mean Elevation (m)	Std. Dev. (m)	RMSE (m)	Moran's I
			Arctic DEM				
-		89	743	180	117	-	-
			Post-event DEMs				
11 July 2018 and 17 July 2018	16	18	750	176	115	30	0.92
5 August 2018 11 August 2018	118	78	654	173	100	37	0.97
4 July 2019 and 10 July 2019	155	87	681	192	130	46	0.97

From the statistics in Table 4, it is challenging to choose the most suitable DEM for volume estimation. Despite a clear indication of which DEM should give the best results, we see major differences in the measured volumes for the three DEMs (track 16, 118, and 155). According to the available literature [45], approximately 7 million m^3 of material was released from the landslide source area, and the volume of the debris in the depositions area is about 10–20 million m^3. According to these statistics, the track 16 DEM shows an unrealistic value of 109 million m^3, whereas the landslide volumes calculated from the other two DEMs are closer to the estimated reference value.

4. Discussion

During the last two decades, OBIA has been used for many applications [28,29]. This research shows the advantage of integrating backscatter information from Sentinel-1 SAR data and spectral information from Sentinel-2 optical data within an expert knowledge-based classification approach in an OBIA environment to assess landslide-induced geomorphological changes. We used Hítardalur as a case study, where a large landslide blocked the river, and, consequently, a lake formed and the river changed its direction. The OBIA classification results show no significant change in the landslide area in the year after the event, but changes in the area of the landslide-dammed lake were detected. However, this might be partly explained by uncertainties in image segmentation and classification and seasonal variations due to different acquisition dates (July 2018 versus August 2019). Moreover, the spatial resolution of Sentinel-1 and Sentinel-2 images are limiting factors for river classification. Therefore, we first identified the existing riverbeds based on an edge detection layer and then assessed whether water was present in the riverbeds.

The Hítardalur landslide has several interesting characteristics that make it suitable to be considered as a case study for creating DEMs using Sentinel-1 datasets. It is a large landslide with a deposition area spreading over the rather flat valley. Therefore, both ascending and descending Sentinel-1 datasets can be used. The resolution and accuracy of DEMs have a direct influence on subsequent analyses such as the landform classification and hydrological modeling [86]. We applied a straightforward workflow for generating DEM using Sentinel-1 image pairs from three different tracks. Our results are heterogeneous and indicate that the generation of DEMs has several limitations in this situation. InSAR techniques have been extensively used for DEM generation using SAR datasets, but the accuracy of the resulting DEMs depends on several factors related to the data and processing workflows and techniques. For selecting suitable SAR datasets, the orbit indetermination and B_{perp} range, atmospheric conditions, and temporal decorrelation (due to scene changes between the two passes) need to be considered. To achieve a good vertical accuracy, the SAR image pair requires a large B_{perp} [76]. The Sentinel-1 mission was mainly designed for the retrieval of surface deformations using differential InSAR (DInSAR) which requires a low B_{perp} between consecutive SAR images, and DEM generation was not the primary goal of the mission [58]. The recommended B_{perp} for the

DEM generation using the ERS satellite is between 150 and 300 m [57,87], but the ERS satellite is known to have high B_{perp}, whereas the common B_{perp} for Sentinel-1 datasets is below 30 m, which is very suitable for interferometry analysis but challenging for DEM generation [88,89]. In this study, we selected datasets in the B_{perp} range of 130 to 160 m, with a short temporal baseline of 6 days between each image pair. The temporal baseline is important to reduce phase noise. Although radar imagery is known for its all-weather measurement capability, changes in the atmospheric conditions (mainly water vapor) cause a variable path delay which results in atmospheric distortions and is the main error source in repeat-pass SAR interferometry [62]. In Iceland, the atmospheric and weather conditions make it difficult to find Sentinel-1 image pairs with a suitable B_{perp} and short temporal interval [90]. Therefore, we limited our search to the summer months (June to September), also to avoid snow cover. We also considered the coherency and the height ambiguity in our data selection. We selected three SAR image pairs from different orbit tracks, i.e., track 16 (11 July 2018 and 17 July 2018) and track 118 (5 August 2018 and 11 August 2018) for the ascending flight direction and track 155 (4 July 2019 and 10 July 2019) for the descending flight direction. One of the selected image pairs was in the range of the recommended B_{perp}, i.e., track number 155 with a B_{perp} of 159. This image pair also showed the lowest height ambiguity (98) compared to the others. The other two image pairs had a slightly lower B_{perp} (134 for track 16 and 142 for track 118) than recommended, but when considering the other selection criteria, they seemed to be the best datasets available.

We faced challenges in validating our classification and DEM-generated results. One issue was the availability of reference data. We used the values provided by Helgason et al. [45] as a reference for our classification and landslide volume estimation. Our volume estimations need to be considered with care since there are several uncertainties associated with them, and certain limitations make the comparison difficult. For example, the area used for the calculation of the landslide volume is not entirely clear and most likely differs from the delineation we used. We used the landslide deposition area, which was semi-automatically extracted using OBIA. Therefore, a one to one comparison of our results with other results reported in literature is not possible. Another challenge is directly related to the reliability of using the generated DEMs for volume calculations. We directly used the generated DEM from the InSAR processing workflow for volume estimation without any further post-processing, while the literature suggests that the InSAR-produced DEMs should be further post-processed to be ready for the end-user [42]. However, since we aimed to test the suitability of DEMs for landslide volume estimation created through a straightforward workflow that could be directly transferred to other areas, we did not apply any post-pressing to improve the volume estimation results. In this study, we demonstrated the potential and limitations inherent to the approach and the Sentinel-1 datasets for DEM generation. Thus, we did not apply any vertical correction on the DEMs generated from Sentinel-1 datasets. We did not use any pre-event DEM generated with this workflow. This was mainly to avoid introducing further errors to the volume estimation. However, there are several possibilities to deal with error distribution in the InSAR-generated DEMs, such as the fusion of the final products from the ascending and descending flight directions [32,91] and the horizontal and vertical adjustment of pixels in the post- and pre-event DEMs, to match them better and make the direct comparison more reliable [92]. Further work is needed to fully evaluate such approaches and the potential of using Sentinel-1 data to generate post-landslide event DEMs for volume estimation.

5. Conclusions

The combined use of Sentinel-1 and Sentinel-2 data offers opportunities for assessing landslides and landslide-induced geomorphological changes. Although many applications use EO datasets for landslide mapping [18,93], studies on integrating intensity information derived from Sentinel-1 SAR data with spectral information from Sentinel-2 optical data are still rare. In this study, we presented a semi-automated workflow to map the landslide-induced geomorphological changes in Hítardalur valley, Iceland, by jointly using Sentinel-1 and Sentinel-2 data within an OBIA framework and assessed the potential of post-event DEMs generated from Sentinel-1 image pairs from different orbit tracks for the estimation of the Hítardalur landslide volume.

It is plausible that we will face more frequent and large landslides in the future due to climate change. The high spatial and temporal resolution, free availability, and the integrated and effective use of Sentinel-1 and Sentinel-2 datasets can contribute enormously to analyzing such events and their consequences. However, the quality of post-event DEMs derived from Sentinel-1 data depends on several factors. These include the SAR image pair selection considering the perpendicular baseline, high coherency, and atmospheric conditions, as well as the process for DEM generation, for example, precise coregistration, phase unwrapping, and conversion of the unwrapped phase into elevation values. These are still active research areas [42]. Further research needs to be conducted to systematically assess the accuracy of the DEMs generated using Sentinel-1.

Author Contributions: Conceptualization: Z.D., D.H. and L.A.; Methodology: Z.D., D.H., L.A. and D.T.; Validation: Z.D., D.H., L.A. and J.K.H.; Formal Analysis: Z.D.; Investigation: Z.D., D.H., L.A., J.K.H. and Þ.S.; Data Creation, Z.D., D.H. and L.A.; Writing—Original Draft Preparation: Z.D. and D.H.; Writing—Review and Editing: Z.D., D.H., L.A., D.T., J.K.H. and Þ.S.; Visualization: Z.D., D.H., L.A.; Supervision: D.H.; Project Administration: D.H.; Funding Acquisition: D.H. and D.T. All authors have read and agreed to the published version of the manuscript.

Funding: This research has been supported by the Austrian Science Fund (FWF) through the project MORPH (Mapping, monitoring and modelling the spatio-temporal dynamics of land surface morphology; FWF-P29461-N29) and the Doctoral Collage GIScience (DK W 1237-N23), as well as by the Austrian Academy of Sciences (ÖAW) through the project RiCoLa (Detection and analysis of landslide-induced river course changes and lake formation).

Acknowledgments: The authors would like to thank S. Asgeirsson and M. Olafsson for providing field photographs.

Conflicts of Interest: The authors declare no conflict of interest.

Data Availability: Figure 6d shows the reference map from the National Land Survey of Iceland (NLSI) vector layer, available online at https://www.lmi.is (accessed on 15 May 2020); © NLSI (license: https://www.lmi.is/en/licence-for-national-land-survey-of-iceland-free-data (accessed on 15 May 2020)).

References

1. Guzzetti, F.; Mondini, A.C.; Cardinali, M.; Fiorucci, F.; Santangelo, M.; Chang, K.-T. Landslide inventory maps: New tools for an old problem. *Earth Sci. Rev.* **2012**, *112*, 42–66. [CrossRef]
2. Tacconi Stefanelli, C.; Casagli, N.; Catani, F. Landslide damming hazard susceptibility maps: A new GIS-based procedure for risk management. *Landslides* **2020**, *17*, 1635–1648. [CrossRef]
3. Ermini, L.; Casagli, N. Prediction of the behaviour of landslide dams using a geomorphological dimensionless index. *Earth Surf. Process Landf.* **2003**, *28*, 31–47. [CrossRef]
4. Sæmundsson, Þ.; Petursson, H.G.; Kneisel, C.; Beylich, A. Monitoring of the Tjarnardalir Landslide, in Central North Iceland. In Proceedings of the First North America Landslide Conference, Vail, CO, USA, 3–8 June 2007; Schaefer, V., Schuster, R., Turner, A., Eds.; AEG Publication: Madison, WI, USA, 2007; Volume 23, pp. 1029–1040.
5. Decaulne, A. Slope processes and related risk appearance within the Icelandic Westfjords during the twentieth century. *Nat. Hazards Earth Syst. Sci.* **2005**, *5*, 309–318. [CrossRef]
6. Sæmundsson, Þ.; Petursson, H. Causes and triggering factors for large scale displacements in the Almenningar landslide area, in central North Iceland. In Proceedings of the Geophysical Research Abstracts, Vienna, Austria, 8–13 April 2018; Volume 20.
7. Sæmundsson, Þ.; Sigurðsson, I.; Pétursson, H.; Jónsson, H.; Decaulne, A.; Roberts, M.J.; Jensen, E.H. Bergflóðið sem féll á Morsárjökull 20. mars 2007—Hverjar hafa afleiðingar þess orðið? (The Morsárjökull rock avalanche in the southern part of the Vatnajökull glacier, south Iceland, in Icelandic). *Náttúrufræðingurinn* **2011**, *81*, 131–141.
8. Sæmundsson, Þ.; Morino, C.; Helgason, J.K.; Conway, S.J.; Pétursson, H.G. The triggering factors of the Móafellshyrna debris slide in northern Iceland: Intense precipitation, earthquake activity and thawing of mountain permafrost. *Sci. Total Environ.* **2018**, *621*, 1163–1175. [CrossRef]

9. Helgason, J.K.; Gylfadóttir, S.S.; Brynjólfsson, S.; Grímsdóttir, H.; Höskuldsson, Á.; Sæmundsson, Þ.; Hjartardóttir, Á.R.; Sigmundsson, F.; Jóhannesson, T. Berghlaupið í Öskju. 21. júlí 2014 (Rockslide in Askja on 21 July 2014, in Icelandic). *Náttúrufræðingurinn* **2019**, *89*, 5–21.
10. Gylfadóttir, S.S.; Kim, J.; Helgason, J.K.; Brynjólfsson, S.; Höskuldsson, Á.; Jóhannesson, T.; Harbitz, C.B.; Løvholt, F. The 2014 Lake Askja rockslide-induced tsunami: Optimization of numerical tsunami model using observed data. *J. Geophys. Res. Oceans* **2017**, *122*, 4110–4122. [CrossRef]
11. Jóhannesson, T.; Helgason, J.K.; Gylfadóttir, S.S. Comment on "Dynamics of the Askja caldera July 2014 landslide, Iceland, from seismic signal analysis: Precursor, motion and aftermath" by Schöpa et al. (2018). *Earth Surf. Dyn.* **2020**, *8*, 173–175. [CrossRef]
12. Scaioni, M.; Longoni, L.; Melillo, V.; Papini, M. Remote Sensing for Landslide Investigations: An Overview of Recent Achievements and Perspectives. *Remote Sens.* **2014**, *6*, 9600–9652. [CrossRef]
13. Joyce, K.E.; Belliss, S.E.; Samsonov, S.V.; McNeill, S.J.; Glassey, P.J. A review of the status of satellite remote sensing and image processing techniques for mapping natural hazards and disasters. *Prog. Phys. Geogr. Earth Environ.* **2009**, *33*, 183–207. [CrossRef]
14. Casagli, N.; Cigna, F.; Bianchini, S.; Hölbling, D.; Füreder, P.; Righini, G.; Del Conte, S.; Friedl, B.; Schneiderbauer, S.; Iasio, C.; et al. Landslide mapping and monitoring by using radar and optical remote sensing: Examples from the EC-FP7 project SAFER. *Remote Sens. Appl. Soc. Environ.* **2016**, *4*, 92–108. [CrossRef]
15. Hervás, J.; Barredo, J.I.; Rosin, P.L.; Pasuto, A.; Mantovani, F.; Silvano, S. Monitoring landslides from optical remotely sensed imagery: The case history of Tessina landslide, Italy. *Geomorphology* **2003**, *54*, 63–75. [CrossRef]
16. Fiorucci, F.; Cardinali, M.; Carlà, R.; Rossi, M.; Mondini, A.C.; Santurri, L.; Ardizzone, F.; Guzzetti, F. Seasonal landslide mapping and estimation of landslide mobilization rates using aerial and satellite images. *Geomorphology* **2011**, *129*, 59–70. [CrossRef]
17. Hölbling, D.; Abad, L.; Dabiri, Z.; Prasicek, G.; Tsai, T.; Argentin, A.-L. Mapping and Analyzing the Evolution of the Butangbunasi Landslide Using Landsat Time Series with Respect to Heavy Rainfall Events during Typhoons. *Appl. Sci.* **2020**, *10*, 630. [CrossRef]
18. Zhong, C.; Liu, Y.; Gao, P.; Chen, W.; Li, H.; Hou, Y.; Nuremanguli, T.; Ma, H. Landslide mapping with remote sensing: Challenges and opportunities. *Int. J. Remote Sens.* **2020**, *41*, 1555–1581. [CrossRef]
19. Lee, J.-S.; Pottier, E. *Polarimetric Radar Imaging*; CRC Press: Boca Raton, FL, USA, 2017; ISBN 9781315219332.
20. Jensen, J.R. *Introductory Digital Image Processing: A Remote Sensing Perspective*, 4th ed.; Pearson Education, Incorporated: New York City, NY, USA, 2016; ISBN 978-0134058160.
21. Jensen, J.R. *Remote Sensing of the Environment: An Earth Resource Perspective*, 2nd ed.; Prentice Hall series in Geographic Information Science; Pearson Prentice Hall: Upper Saddle River, NJ, USA, 2009; ISBN 9788131716809.
22. Shirvani, Z. A Holistic Analysis for Landslide Susceptibility Mapping Applying Geographic Object-Based Random Forest: A Comparison between Protected and Non-Protected Forests. *Remote Sens.* **2020**, *12*, 434. [CrossRef]
23. Plank, S.; Hölbling, D.; Eisank, C.; Friedl, B.; Martinis, S.; Twele, A. Comparing object-based landslide detection methods based on polarimetric SAR and optical satellite imagery—A case study in Taiwan. In Proceedings of the 7th International Workshop on Science and Applications of SAR Polarimetry and Polarimetric Interferometry, POLinSAR 2015, Frascati, Italy, 26–30 January 2015; pp. 1–5.
24. Hölbling, D.; Friedl, B.; Dittrich, J.; Cigna, F.; Pedersen, G. Combined interpretation of optical and SAR data for landslide mapping. In *Advances in Landslide Research, Proceedings of the 3rd Regional Symposium on Landslides the Adriatic-Balkan Region, Ljubljana, Slovenia, 11–13 October 2017*; Jemec Auflic, M., Mikos, M., Verbovsek, T., Eds.; Geological Survey of Slovenia: Ljubljana, Slovenia, 2018; pp. 11–13.
25. Plank, S.; Twele, A.; Martinis, S. Landslide Mapping in Vegetated Areas Using Change Detection Based on Optical and Polarimetric SAR Data. *Remote Sens.* **2016**, *8*, 307. [CrossRef]
26. Moosavi, V.; Talebi, A.; Shirmohammadi, B. Producing a landslide inventory map using pixel-based and object-oriented approaches optimized by Taguchi method. *Geomorphology* **2014**, *204*, 646–656. [CrossRef]
27. Keyport, R.N.; Oommen, T.; Martha, T.R.; Sajinkumar, K.S.; Gierke, J.S. A comparative analysis of pixel- and object-based detection of landslides from very high-resolution images. *Int. J. Appl. Earth Obs. Geoinf.* **2018**, *64*, 1–11. [CrossRef]

28. Chen, G.; Weng, Q.; Hay, G.J.; He, Y. Geographic object-based image analysis (GEOBIA): Emerging trends and future opportunities. *GISci. Remote Sens.* **2018**, *55*, 159–182. [CrossRef]
29. Blaschke, T.; Hay, G.J.; Kelly, M.; Lang, S.; Hofmann, P.; Addink, E.; Queiroz Feitosa, R.; van der Meer, F.; van der Werff, H.; van Coillie, F.; et al. Geographic Object-Based Image Analysis—Towards a new paradigm. *ISPRS J. Photogramm. Remote Sens.* **2014**, *87*, 180–191. [CrossRef] [PubMed]
30. Hölbling, D.; Eisank, C.; Albrecht, F.; Vecchiotti, F.; Friedl, B.; Weinke, E.; Kociu, A. Comparing Manual and Semi-Automated Landslide Mapping Based on Optical Satellite Images from Different Sensors. *Geosciences* **2017**, *7*, 37. [CrossRef]
31. Tsutsui, K.; Rokugawa, S.; Nakagawa, H.; Miyazaki, S.; Cheng, C.-T.; Shiraishi, T.; Yang, S.-D. Detection and Volume Estimation of Large-Scale Landslides Based on Elevation-Change Analysis Using DEMs Extracted From High-Resolution Satellite Stereo Imagery. *IEEE Trans. Geosci. Remote Sens.* **2007**, *45*, 1681–1696. [CrossRef]
32. Crosetto, M. Calibration and validation of SAR interferometry for DEM generation. *ISPRS J. Photogramm. Remote Sens.* **2002**, *57*, 213–227. [CrossRef]
33. Rosen, P.A.; Hensley, S.; Joughin, I.R.; Li, F.K.; Madsen, S.N.; Rodriguez, E.; Goldstein, R.M. Synthetic aperture radar interferometry. *Proc. IEEE* **2000**, *88*, 333–382. [CrossRef]
34. Gens, R.; van Genderen, J.L. Review Article SAR interferometry—Issues, techniques, applications. *Int. J. Remote Sens.* **1996**, *17*, 1803–1835. [CrossRef]
35. Haghighi, M.H.; Motagh, M. Sentinel-1 InSAR over Germany: Large-scale interferometry, atmospheric effects, and ground deformation mapping. *ZFV—Z. Geodasie Geoinf. Landmanag.* **2017**, *142*, 245–256.
36. Yang, L.; Meng, X.; Zhang, X. SRTM DEM and its application advances. *Int. J. Remote Sens.* **2011**, *32*, 3875–3896. [CrossRef]
37. Drusch, M.; Del Bello, U.; Carlier, S.; Colin, O.; Fernandez, V.; Gascon, F.; Hoersch, B.; Isola, C.; Laberinti, P.; Martimort, P.; et al. Sentinel-2: ESA's Optical High-Resolution Mission for GMES Operational Services. *Remote Sens. Environ.* **2012**, *120*, 25–36. [CrossRef]
38. Torres, R.; Snoeij, P.; Geudtner, D.; Bibby, D.; Davidson, M.; Attema, E.; Potin, P.; Rommen, B.; Floury, N.; Brown, M.; et al. GMES Sentinel-1 mission. *Remote Sens. Environ.* **2012**, *120*, 9–24. [CrossRef]
39. Ottavianello, G. *Copernicus Space Component Data Access Portfolio: Data Warehouse 2014–2020*; European Space Agency: Frascati, Italy, 2019.
40. De Zan, F.; Monti Guarnieri, A. TOPSAR: Terrain Observation by Progressive Scans. *IEEE Trans. Geosci. Remote Sens.* **2006**, *44*, 2352–2360. [CrossRef]
41. Farkas, P.; Hevér, R.; Grenerczy, G. Geodetic integration of Sentinel-1A IW data using PSInSAR in Hungary. In Proceedings of the Geophysical Research Abstracts, Vienna, Austria, 12–17 April 2015; Volume 17.
42. Dai, K.; Li, Z.; Tomás, R.; Liu, G.; Yu, B.; Wang, X.; Cheng, H.; Chen, J.; Stockamp, J. Monitoring activity at the Daguangbao mega-landslide (China) using Sentinel-1 TOPS time series interferometry. *Remote Sens. Environ.* **2016**, *186*, 501–513. [CrossRef]
43. Kyriou, A.; Nikolakopoulos, K. Landslide mapping using optical and radar data: A case study from Aminteo, Western Macedonia Greece. *Eur. J. Remote Sens.* **2019**, 1–11. [CrossRef]
44. Barra, A.; Monserrat, O.; Mazzanti, P.; Esposito, C.; Crosetto, M.; Scarascia Mugnozza, G. First insights on the potential of Sentinel-1 for landslides detection. *Geomat. Nat. Hazards Risk* **2016**, *7*, 1874–1883. [CrossRef]
45. Helgason, J.K.; Sæmundsson, Þ.; Drouin, V.; Jóhannesson, T. The Hítardalur landslide in West Iceland in July 2018. In Proceedings of the Geophysical Research Abstracts, Vienna, Austria, 7–12 April 2019; Volume 21.
46. Jordan, B.T.; Carley, T.L.; Banik, T.J. Iceland: The Formation and Evolution of a Young, Dynamic, Volcanic Island—A Field Trip Guide. In *Iceland: The Formation and Evolution of a Young, Dynamic, Volcanic Island—A Field Trip Guide*; Geological Society of America: Boulder, CO, USA, 2019; Volume 54, ISBN 9780813700540.
47. Ragnarsdóttir, K.V. *Jarðlagaskipan Fagraskógarfjalls og Vatnshlíðar í Hítardal (Geology of Mt. Fagraskógarfjall and Mt. Vatnshlíðarfjall in Hítardalur Valley, in Icelandic)*; University of Iceland: Reykjavík, Iceland, 1979.
48. Johannesson, H. Jarlagaskipan og róun rekbelta á Vesturlandi (Stratigraphy and development of spreading centers in Western Iceland, in Icelandic). *Náttórufrœingurinn* **1980**, *50*, 13–31.
49. Larkin, H.; Magnall, N.; Thomas, A.; Holley, R.; Mccormack, H. Utilising satellite-based techniques to identify and monitor slope instabilities: The Fagraskógarfjall and Limnes landslides. In Proceedings of the 2020 International Symposium on Slope Stability in Open Pit Mining and Civil Engineering, Australian Centre for Geomechanics, Perth, Australia, 12–14 May 2020; pp. 1455–1466.

50. Iceland Met Office (IMO). A Large Landslide Falls in Hítardalur Valley. Available online: https://en.vedur.is/about-imo/news/a-large-landslide-falls-in-hitardalur-valley (accessed on 11 August 2020).
51. Pétursson, H.G. Large Landslides Since the Middle of the Last Century. Available online: https://en.vedur.is/media/frettir-myndasafn-2018/Big_landslides_Table_ENG.pdf (accessed on 17 June 2020).
52. Wegnüller, U.; Werner, C.; Strozzi, T.; Wiesmann, A.; Frey, O.; Santoro, M. Sentinel-1 Support in the GAMMA Software. *Procedia Comput. Sci.* **2016**, *100*, 1305–1312. [CrossRef]
53. Lee, J.-S. Refined filtering of image noise using local statistics. *Comput. Graph. Image Process.* **1981**, *15*, 380–389. [CrossRef]
54. Yommy, A.S.; Liu, R.; Wu, A.S. SAR Image Despeckling Using Refined Lee Filter. In Proceedings of the 2015 7th International Conference on Intelligent Human-Machine Systems and Cybernetics, Hangzhou, China, 26–27 August 2015; Volume 2, pp. 260–265.
55. Park, J.-W.; Korosov, A.A.; Babiker, M.; Sandven, S.; Won, J.-S. Efficient Thermal Noise Removal for Sentinel-1 TOPSAR Cross-Polarization Channel. *IEEE Trans. Geosci. Remote Sens.* **2018**, *56*, 1555–1565. [CrossRef]
56. Filipponi, F. Sentinel-1 GRD Preprocessing Workflow. *Proceedings* **2019**, *18*, 11. [CrossRef]
57. Ferretti, A.; Monti Guarnieri, A.; Prati, C.; Rocca, F. *InSAR Principles: Guidelines for SAR Interferometry Processing and Interpretation*; Fletcher, K., Ed.; ESA Publications: Noordwijk, The Netherlands, 2007; ISBN 92-9092-233-8.
58. Geudtner, D.; Prats, P.; Yague-Martinez, N.; Navas-Traver, I.; Barat, I.; Torres, R.R. Sentinel-1 SAR interferometry performance verification. In Proceedings of the 11th European Conference on Synthetic Aperture Radar, EUSAR, Berlin, Germany, 6–9 June 2016; VDE VERLAG GMBH: Berlin, Germany, 2016; pp. 65–68.
59. Kyriou, A.; Nikolakopoulos, K. Assessing the suitability of Sentinel-1 data for landslide mapping. *Eur. J. Remote Sens.* **2018**, *51*, 402–411. [CrossRef]
60. Bürgmann, R.; Rosen, P.A.; Fielding, E.J. Synthetic Aperture Radar Interferometry to Measure Earth's Surface Topography and Its Deformation. *Annu. Rev. Earth Planet. Sci.* **2000**, *28*, 169–209. [CrossRef]
61. Rott, H.; Nagler, T. The contribution of radar interferometry to the assessment of landslide hazards. *Adv. Space Res.* **2006**, *37*, 710–719. [CrossRef]
62. Wegmüller, U.; Santoro, M.; Werner, C.; Strozzi, T.; Wiesmann, A.; Lengert, W. DEM generation using ERS–ENVISAT interferometry. *J. Appl. Geophys.* **2009**, *69*, 51–58. [CrossRef]
63. Porter, C.; Morin, P.; Howat, I.; Noh, M.-J.; Bates, B.; Peterman, K.; Keesey, S.; Schlenk, M.; Gardiner, J.; Tomko, K.; et al. *ArcticDEM. Harvard Dataverse*; Polar Geospatial Center, University of Minnesota: Minneapolis, MN, USA, 2018; Volume 1.
64. Schumann, G.J.-P.; Bates, P.D. The Need for a High-Accuracy, Open-Access Global DEM. *Front. Earth Sci.* **2018**, *6*, 225. [CrossRef]
65. Lee, J.-S. NOTE Digital Image Smoothing and the Sigma Filter. *Comput. Vision Graph. Image Process.* **1983**, *24*, 255–269. [CrossRef]
66. Baatz, M.; Schäpe, A. Multiresolution Segmentation: An optimization approach for high quality multi-scale image segmentation. In *Proceedings of the Angewandte Geographische Informationsverarbeitung XII*; Strobl, J., Blaschke, T., Griesebner, G., Eds.; Wichmann: Heidelberg, Germany, 2000; pp. 12–23.
67. Eisank, C.; Smith, M.; Hillier, J. Assessment of multiresolution segmentation for delimiting drumlins in digital elevation models. *Geomorphology* **2014**, *214*, 452–464. [CrossRef]
68. Yague-Martinez, N.; Prats-Iraola, P.; Rodriguez Gonzalez, F.; Brcic, R.; Shau, R.; Geudtner, D.; Eineder, M.; Bamler, R. Interferometric Processing of Sentinel-1 TOPS Data. *IEEE Trans. Geosci. Remote Sens.* **2016**, *54*, 2220–2234. [CrossRef]
69. Lazecký, M.; Hlaváčová, I.; Martinovič, J.; Ruiz-Armenteros, A.M. Accuracy of Sentinel-1 Interferometry Monitoring System based on Topography-free Phase Images. *Procedia Comput. Sci.* **2018**, *138*, 310–317. [CrossRef]
70. Goldstein, R.M.; Werner, C.L. Radar interferogram filtering for geophysical applications. *Geophys. Res. Lett.* **1998**, *25*, 4035–4038. [CrossRef]
71. Lee, J.-S.; Hoppel, K.W.; Mango, S.A.; Miller, A.R. Intensity and phase statistics of multilook polarimetric and interferometric SAR imagery. *IEEE Trans. Geosci. Remote Sens.* **1994**, *32*, 1017–1028.
72. Schmitt, A. Multiscale and Multidirectional Multilooking for SAR Image Enhancement. *IEEE Trans. Geosci. Remote Sens.* **2016**, *54*, 5117–5134. [CrossRef]

73. Chen, C.W.; Zebker, H.A. Network approaches to two-dimensional phase unwrapping: Intractability and two new algorithms. *J. Opt. Soc. Am. A* **2000**, *17*, 401. [CrossRef]
74. Chen, C.W.; Zebker, H.A. Phase unwrapping for large SAR interferograms: Statistical segmentation and generalized network models. *IEEE Trans. Geosci. Remote Sens.* **2002**, *40*, 1709–1719. [CrossRef]
75. Chen, C.W.; Zebker, H.A. Two-dimensional phase unwrapping with use of statistical models for cost functions in nonlinear optimization. *J. Opt. Soc. Am. A* **2001**, *18*, 338. [CrossRef]
76. Richards, M.A. A Beginner's Guide to Interferometric SAR Concepts and Signal Processing [AESS Tutorial IV]. *IEEE Aerosp. Electron. Syst. Mag.* **2007**, *22*, 5–29. [CrossRef]
77. National Land Survey of Iceland IS 50V 24/12 2019 Vatnafar Flakar. Available online: https://gatt.lmi.is/geonetwork/srv/eng/catalog.search#/metadata/83E61CBF-8498-4259-A40C-3B628EA34FB7 (accessed on 15 May 2020).
78. Maune, D.F. *Digital Elevation Model Technologies and Applications: The DEM Users Manual*, 2nd ed.; American Society for Photogrammetry and Remote Sensing: Bethesda, MD, USA, 2007; ISBN 9781570830822.
79. Höhle, J.; Höhle, M. Accuracy assessment of digital elevation models by means of robust statistical methods. *ISPRS J. Photogramm. Remote Sens.* **2009**, *64*, 398–406. [CrossRef]
80. Candela, S.G.; Howat, I.; Noh, M.-J.; Porter, C.C.; Morin, P.J. ArcticDEM Validation and Accuracy Assessment. In Proceedings of the American Geophysical Union, Fall Meeting, New Orleans, LA, USA, 11–15 December 2017.
81. Griffith, D. *Spatial Autocorrelation: A Primer*; Association of American Geographers: Washington, DC, USA, 1987; ISBN 9780892911974.
82. Moran, P.A.P. The Interpretation of Statistical Maps. *J. R. Stat. Soc. Ser. B* **1948**, *10*, 243–251. [CrossRef]
83. Wechsler, S.P. Perceptions of digital elevation model uncertainty by DEM users. *URISA J.* **2003**, *15*, 57–64.
84. Wechsler, S.P.; Kroll, C.N. Quantifying DEM Uncertainty and its Effect on Topographic Parameters. *Photogramm. Eng. Remote Sens.* **2006**, *72*, 1081–1090. [CrossRef]
85. Iceland Met Office (IMO). The Weather in Iceland in 2019. Available online: https://en.vedur.is/about-imo/news/the-weather-in-iceland-in-2019 (accessed on 15 May 2020).
86. Kramm, T.; Hoffmeister, D. Evaluation of Digital Elevation Models for Geomorphometric Analyses on Different Scales for Northern Chile. *ISPRS Int. Arch. Photogramm. Remote Sens. Spat. Inf. Sci.* **2019**, *XLII-2/W13*, 1229–1235. [CrossRef]
87. Braun, A. *DEM Generation with Sentinel-1 Workflow and Challenges*; European Space Agency: Paris, France, 2020.
88. Berardino, P.; Fornaro, G.; Lanari, R.; Sansosti, E. A new algorithm for surface deformation monitoring based on small baseline differential SAR interferograms. *IEEE Trans. Geosci. Remote Sens.* **2002**, *40*, 2375–2383. [CrossRef]
89. Geudtner, D.; Torres, R.; Snoeij, P.; Davidson, M.; Rommen, B. Sentinel-1 System capabilities and applications. In Proceedings of the 2014 IEEE Geoscience and Remote Sensing Symposium, Quebec City, QC, Canada, 13–18 July 2014; pp. 1457–1460.
90. Amitrano, D.; Guida, R.; Dell'Aglio, D.; Di Martino, G.; Di Martire, D.; Iodice, A.; Costantini, M.; Malvarosa, F.; Minati, F. Long-Term Satellite Monitoring of the Slumgullion Landslide Using Space-Borne Synthetic Aperture Radar Sub-Pixel Offset Tracking. *Remote Sens.* **2019**, *11*, 369. [CrossRef]
91. Sansosti, E.; Lanari, R.; Fornaro, G.; Franceschetti, G.; Tesauro, M.; Puglisi, G.; Coltelli, M. Digital elevation model generation using ascending and descending ERS-1/ERS-2 tandem data. *Int. J. Remote Sens.* **1999**, *20*, 1527–1547. [CrossRef]
92. Tang, C.; Tanyas, H.; van Westen, C.J.; Tang, C.; Fan, X.; Jetten, V.G. Analysing post-earthquake mass movement volume dynamics with multi-source DEMs. *Eng. Geol.* **2019**, *248*, 89–101. [CrossRef]
93. Zhao, C.; Lu, Z. Remote Sensing of Landslides—A Review. *Remote Sens.* **2018**, *10*, 279. [CrossRef]

applied sciences

MDPI

Article

Bayesian Updating of Soil–Water Character Curve Parameters Based on the Monitor Data of a Large-Scale Landslide Model Experiment

Chengxin Feng [1], Bin Tian [1,*], Xiaochun Lu [1], Michael Beer [2,3,4,*], Matteo Broggi [2], Sifeng Bi [2], Bobo Xiong [1] and Teng He [1]

1 College of Hydraulic and Environment Engineering, China Three Gorges University, Yichang 443002, China; 2014101517@ctgu.edu.cn (C.F.); luxiaochun1014@ctgu.edu.cn (X.L.); zbqxbb19870@hhu.edu.cn (B.X.) 201708150021013@ctgu.edu.cn (T.H.)
2 Institute for Risk and Reliability, Leibniz University Hannover, Callinstr. 34, 30167 Hannover, Germany; broggi@irz.uni-hannover.de (M.B.); sifeng.bi@irz.uni-hannover.de (S.B.)
3 Institute for Risk and Uncertainty, University of Liverpool, Peach Street, Liverpool L69 7ZF, UK
4 International Joint Research Center for Engineering Reliability and Stochastic Mechanics, Tongji University, Shanghai 200092, China
* Correspondence: eudiltb@ctgu.edu.cn (B.T.); beer@irz.uni-hannover.de (M.B.)

Received: 30 June 2020; Accepted: 6 August 2020; Published: 10 August 2020

Abstract: It is important to determine the soil–water characteristic curve (SWCC) for analyzing landslide seepage under varying hydrodynamic conditions. However, the SWCC exhibits high uncertainty due to the variability inherent in soil. To this end, a Bayesian updating framework based on the experimental data was developed to investigate the uncertainty of the SWCC parameters in this study. The objectives of this research were to quantify the uncertainty embedded within the SWCC and determine the critical factors affecting an unsaturated soil landslide under hydrodynamic conditions. For this purpose, a large-scale landslide experiment was conducted, and the monitored water content data were collected. Steady-state seepage analysis was carried out using the finite element method (FEM) to simulate the slope behavior during water level change. In the proposed framework, the parameters of the SWCC model were treated as random variables and parameter uncertainties were evaluated using the Bayesian approach based on the Markov chain Monte Carlo (MCMC) method. Observed data from large-scale landslide experiments were used to calculate the posterior information of SWCC parameters. Then, 95% confidence intervals for the model parameters of the SWCC were derived. The results show that the Bayesian updating method is feasible for the monitoring of data of large-scale landslide model experiments. The establishment of an artificial neural network (ANN) surrogate model in the Bayesian updating process can greatly improve the efficiency of Bayesian model updating.

Keywords: large-scale landslide model experiment; soil–water characteristic curve; Bayesian updating; Markov chain Monte Carlo; artificial neural networks

1. Introduction

The prediction of a landslide and its stability performance have always been key difficulties of engineering. Particularly for the sensitive and landslide-prone areas in reservoir areas or mountain areas, the water migration inside a landslide is very prominent for its stability [1–3]. The soil–water characteristic curve (SWCC) is the cornerstone of the seepage calculation of unsaturated soil. In the analysis of saturated-unsaturated seepage, it is very important to fully describe the function of the slope of the moisture content [4,5]. The preliminary prediction based on the survey data often deviates from the actual monitoring value. As the real conditions are very complicated, the mechanism cannot be fully grasped through analytical models. Therefore, the trial algorithm is often used to repeatedly correct the soil parameters to obtain results that match the monitoring value [6,7]. The Bayesian method provides an effective way to solve such problems. By using the detection data, a priori information, and model assumptions, the parameter update is realized and the parameter uncertainty is quantified [8,9]. This technique uses a set of output measurement data (such as a point in a multi-dimensional output space) [10,11]. The purpose of this paper is to solve the problem of how to check the parameters of a SWCC model of a landslide using the Bayesian updating method.

The SWCC is a crucial input for modeling the geotechnical problems with unsaturated soil. It can be describing the relationship between matric suction and volumetric water content in unsaturated soils [12]. The main feature of SWCC is that the slope of the residual water content or saturation function in the pore water pressure range is from negative to positive [13]. As a basic equation in unsaturated soil mechanics, it plays an important role in the application of unsaturated soil mechanics for studying the strengths, deformations, and permeabilities of soils [14]. Past studies have shown that the shear strength of unsaturated soils can be estimated by the SWCC [15–17]. It can be seen that these parameters play important roles in the safety of geotechnical engineering [18–21]. Therefore, the SWCC parameters are evaluated as a method to determine the unsaturated soil properties of landslides. In engineering practice, the SWCC is usually measured by field or laboratory tests. The typical methods used are: the tensiometer method, the pressure plate meter method, the cold mirror hygrometer method, and the filter paper method. The limited number of data points are then used to estimate the SWCC by best fitting them with some parametric SWCC models, such as the van Genuchten–Mualem model (VGM), the van Genuchten-Burdine model (VGB), the van Genuchten model (VG), and the Fredlund and Xing model (FX) [14,22].

This paper adopts the Bayesian updating method, which is essentially a Bayesian statistical reasoning problem [23], to study the uncertainty of the parameters involved in the model. The Bayesian method is a probabilistic method based on the application of Bayesian rules. The resulting solution is called the posterior distribution, which is a multi-dimensional probability distribution function in the update parameter space. The posterior distribution is usually a complex distribution, and in principle it is integral in the parameter space. It can be approximated by statistical methods and using modern algorithms, namely, the Markov chain Monte Carlo (MCMC) method [24]. The distribution model, also known as the maximum posterior probability (MAP), defines the updated model, and the distribution around the model provides a measure of confidence in the model. The peak distribution often means a high degree of confidence in the updated model, and shallowness may make uncertain the choice of parameters [25,26]. Bayesian updating and prediction methods are widely used in geotechnical engineering, such as pile bearing capacity analysis, soft soil consolidation inversion analysis, jack-up drilling platform bearing capacity analysis, foundation pit support under normal use protection, and excavation problems and slope stability problems [27]. The key difficulty of the Bayesian method lies in the calculation of the MAP distribution. Because of the complexity of the MAP distribution and the possibility of involving high-dimensional integration problems, the direct numerical integration method is less efficient. To overcome this problem, the MCMC method avoids the computational complexity of directly solving the MAP distribution by generating a large number of random samples that obey the MAP distribution, and then it uses these random samples to characterize the MAP distribution [28,29]. Metropolis–Hastings (M–H) algorithm, as one of the commonly used MCMC methods, has been widely used in geotechnical engineering [30]. In this paper, the M–H method is used to generate a large number of MCMC random samples that follow the MAP distribution parameters, such as the posterior mean, posterior standard deviation, and posterior distribution. These statistics are the basic inputs for reliability analysis of geotechnical engineering [31].

Therefore, the purpose of this article is to show how to use MCMC simulations to effectively describe SWCC parameter uncertainty involving random variables. The structure of this article is as follows. First, the theory of the SWCC and the Bayesian theory of parameter uncertainty are introduced. Second, the hybrid Markov chain with emphasis on the uncertainty characterization of the geotechnical model is introduced, and a framework for updating the posterior distribution of the parameters of SWCC is introduced. Then, a large-scale landslide model experiment is used to illustrate how to obtain the posterior distribution of the parameters of the SWCC through Bayesian updating. Finally, we analyze the influencing factors of posterior distribution and characterize the uncertainty of SWCC parameters.

2. Theories and Methods

2.1. Soil–Water Characteristic Curve

Among the existing empirical equations for curve-fitting of the SWCC, the most commonly used VG model was adopted in the present study, which is given by [12]:

$$\theta_w = \theta_r + \frac{\theta_s - \theta_r}{[1 + \frac{\lambda}{a}^n]^{1-\frac{1}{n}}} \tag{1}$$

where θ_w is the volumetric water content; θ_r is the residual volumetric water content; θ_s is the saturated volumetric water content; λ is matric suction; a and n are non-negative curve fitting parameters.

To obtain the prior distribution of SWCC parameters, a simple sensitivity analysis of the modeling parameters sheds light on the possible range of the numeric values of the parameters [23]. The morphology of the SWCC between a from 0.1 to 10 and n from 1.1 to 5 was studied. Figure 1a describes the influence of the value of a on the SWCC model when n is 5. Figure 1b describes the effect of the value of n on the SWCC model when a is 2. Each model parameter has a certain physical meaning, except a and n. From Figure 1, it becomes clear that the parameters a and n are critical for SWCC, determining the shape of the curve.

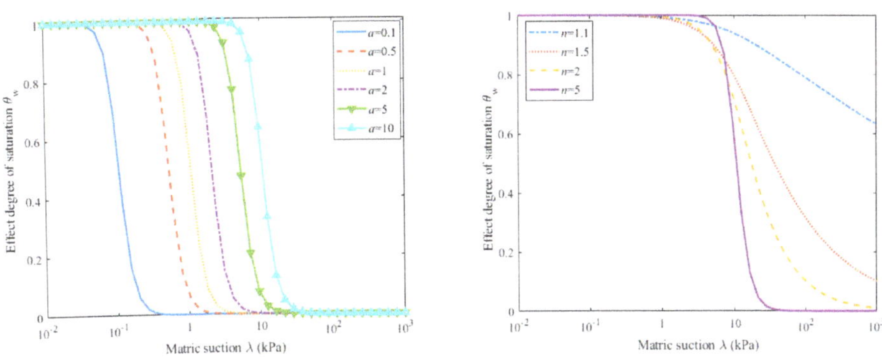

(a) Parameters a (b) Parameters n

Figure 1. (**a**) The influence of parameters a on the soil–water characteristic curve (SWCC), and the n is 5 in this figure; (**b**) The influences of parameters n on the SWCC equation, and the a is 2 in this figure.

2.2. Bayesian Model Updating

The principle of Bayesian model updating is the Bayes' theorem. The pioneering work in the late 1990s attracted the attention of the engineering research community and gave detailed and up-to-date instructions [32,33]. Then, the obstacle to computing in practical engineering applications was the high level of computer resources required. This problem still exists, but there has been great progress in solving it.

The Bayesian model's updating process is based on Bayes' theorem. In our case, likelihood is a function of the parameters in the statistical model, which represents the likelihood of model parameters. The prior distribution is based on existing research assumptions. The posterior distribution can be calculated by combining the prior distribution and the likelihood function. Its general formula is given by:

$$P(\theta|\mathbf{X}_{\text{obs}}) = \frac{P(\mathbf{X}_{\text{obs}}|\theta)P(\theta)}{P(\mathbf{X}_{\text{obs}})} \quad (2)$$

where $P(\theta)$ is the prior distribution representing the initial knowledge about the parameters θ; $P(\theta|\mathbf{X}_{\text{obs}})$ is the posterior distribution representing the updated knowledge based on the observation data; $P(\mathbf{X}_{\text{obs}})$ is the normalization factor; $P(\mathbf{X}_{\text{obs}}|\theta)$ is the likelihood function of $\mathbf{X}_{\text{obs}}$ for an instance of the parameters θ.

If it is possible to provide an update a prior distribution according the available knowledge, then the MAP distribution can be obtained. The Bayesian updating is a challenging problem, since it is very difficult to directly calculate the posterior probability density over the entire parameter space. However, the well-known M–H algorithm is an effective update tool [15]. The algorithm is essentially an iterative method, sampling from a series of intermediate partial differential equations that gradually converge to the MAP distribution. The jth intermediate probability density function (PDF) is expressed as a function of the likelihood probability function P_L as:

$$P_j = P_L(\mathbf{X}_{\text{obs}}|\theta)^{\beta_j} P(\theta) \quad (3)$$

where the index of likelihood β_j is the so-called reduction factor. Its value gradually increases from β_o= 0 in the first iteration until it reaches β_j = 1 in the last step [24]. β_j is adaptively calculated based on the sample from the previous step.

The likelihood is a key component of the Bayesian update framework because it quantifies how relevant the model is to a given parameter instance by representing the likelihood of observation. Under the assumption of independence between observations, the probability P_L is theoretically defined as:

$$P_L(\mathbf{X}_{\text{obs}}|\theta) = \prod_{k=1}^{N_{\text{obs}}} P(\mathbf{X}_k|\theta) \quad (4)$$

where $P(\mathbf{X}_k|\theta)$ is the probability density value at $\mathbf{X}_k$ for a given set of parameters θ; $\mathbf{X}_k$ is the kth $\mathbf{X}_{\text{obs}}$. θ is a set of parameters—in this analysis, the parameters a and n. In this process, the PDF should be estimated separately for each instance. However, in a complete Bayesian updating process, the required computation is costly, due to the large number of evaluations that are necessary to obtain to define a statistical sample.

3. The Bayesian Updating Procedure of Monitor Data in the Large-Scale Landslide Model Experiment

In this paper, the above-mentioned Bayesian updating method is used to calculate the SWCC model parameters. Bayesian updating is a method of statistical inference, in which as more evidence or information (in this paper is the observation samples) is obtained, Bayes' theorem (Equation (2)) is used to update the probability of a hypothesis. First, the prior distribution of the SWCC model parameters is determined through reference research data or data from a SWCC experiment, as done in this work. Then, one uses the M–H method to generate a defined (large) number of random MCMC samples following a iterative calculation until β_j = 1, and the posterior distribution is obtained. The M–H method is a powerful Markov chain method with which to simulate multivariate distributions. The main steps include initialization and iteration [34]. Figure 2 shows a flowchart for implementation of the proposed Bayesian approach schematically. The work done covers both the experimental results to feed the model (left branch on Figure 2), and the analysis methodology (right branch on Figure 2). The experimental data were obtained after basic soil mechanical experiments, to characterize the material of the landslide body, and a large-scale landslide model experiment, including water content sensors to measure in real-time the hydrodynamic process. The setup and results of the experiments are described in Sections 4 and 5. Each step and its detailed description are summarized as follows.

1. Regarding the left branch, the main goal was to obtain Bayesian updating evidence (water content observation samples X_{obs}) through the water content detection of large-scale landslide models:
   (a) Infiltration experiments and a triaxial experiment were carried out, and the material ratio of landslide body was determined to be 7:3 for soil:stone through these experiments.
   (b) Large-scale landslide models were built according to the layer-by-layer compaction method. The place where the sensor was buried was carefully compacted to avoid damage to the sensor.
   (c) After the landslide model and sensor were ready, the moisture content sensor was tested to determine the reliability of its data.
2. In the right branch, the main work was to obtain the Bayesian updating simulation samples (X_k) through the finite element model:
   (d) The SWCC model was selected (VG model), and we analyzed the two model parameters within its range to observe their influences on the curve shape.
   (e) We assumed the prior distribution as normal or lognormal according to existing research. Then we determined the prior distribution of SWCC for the model parameters a and n, based on experimental data and research results.
   (f) A finite element (FE) numerical model was developed according to the large-scale landslide test; we evaluated the seepage flow under the conditions of the reservoir water level.
   (g) In the Bayesian updating process, it is necessary to call the FE model, taking a long time to obtain the large number of output values. Therefore, this paper used an artificial neural network (ANN) as a surrogate model to speed up the analysis.
3. In the lower part of the flowchart, the main work was to perform Bayesian updating by combining the water content observation samples (X_{obs}) of the large landslide model of the left branch and the simulation samples (X_k) of the right branch:
   (h) The MCMC algorithm was used to generate N random samples that followed the posterior distribution of SWCC parameters and iteratively continued until $\beta_j = 1$. We got the posterior distribution of SWCC model parameters (posterior most likely value, posterior mean, posterior standard deviation, etc.).
   (i) We analyzed the factors influencing the posterior distribution results, such as the number of observation datums, the prior distribution, and the observation data that changed over time.
   (j) We characterized the uncertainty of the SWCC model. For the parameters of the SWCC, the N samples generated from the MCMC algorithm were ranked, and a 95% confidence interval was obtained.

These steps were programmed, and MATLAB 2019b was used in this process. The programming of the program is divided into three steps: firstly, import the simulated and observed values; then use the MCMC algorithm to generate a large number of samples (100,000 sets of data in this program) to iterate for Bayesian updating; finally, the posterior distribution of the parameters of the SWCC model is visualized. In a different analysis case, the same analysis tools can be used so that geotechnical practitioners only need to provide laboratory or in-situ test results, prior knowledge (e.g., reasonable ranges of model parameters), and candidate SWCC models as inputs. The main advantage of the above method is that it can quickly check distributions of the posterior parameters of the SWCC.

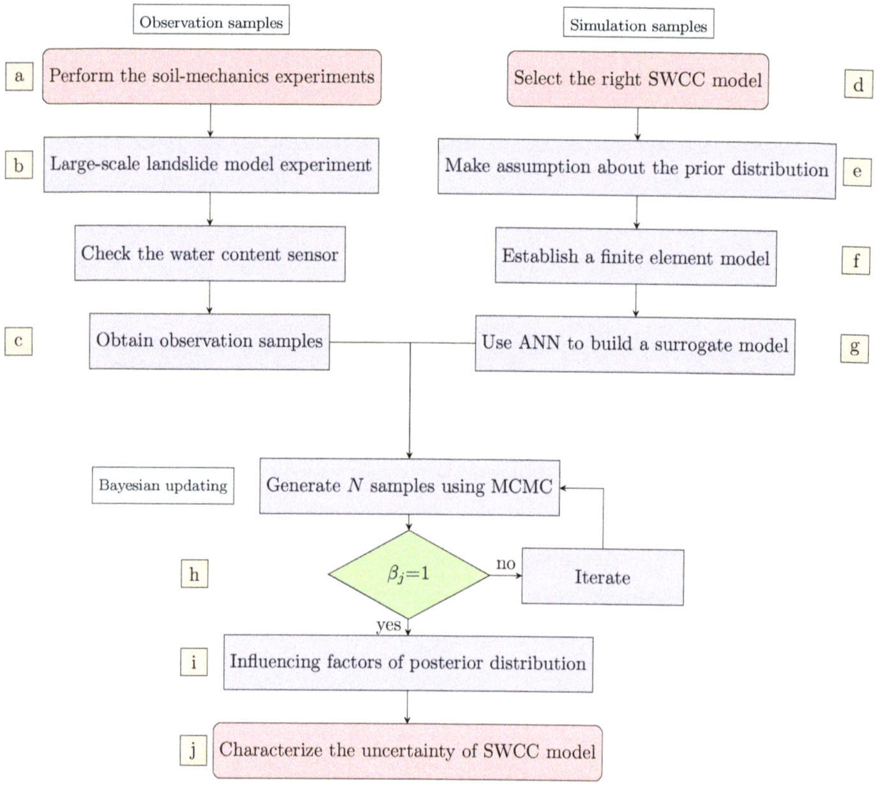

Figure 2. Flowchart of the Bayesian updating procedure.

4. The Large-Scale Landslide Model Experiment and Bayesian Updating

A large-scale landslide experiment was conducted using a soil–rock mixture and by monitoring the changes in water content to clarify the water migration during a landslide. The principle of numerical modeling, using an earlier developed model, is briefly explained. Simulations are presented and their results are compared to the experimental ones.

4.1. Experiment Setup

The experiments have been carried out in a large-scale landslide model experiment at The Key Laboratory of Geological Hazards and the Ministry of Education of the Three Gorges Reservoir Area of the Three Gorges University (China); the setup is shown in Figure 3. The model test platform was 6 m long, 0.8 m wide, and 1.5 m high. The door at the right end of the test platform could control the rise and fall of the reservoir water level. The landslide model was connected to external water supply pipes and drainage pipes. The speed at which the water level of the reservoir changed was adjusted by setting a flow valve.

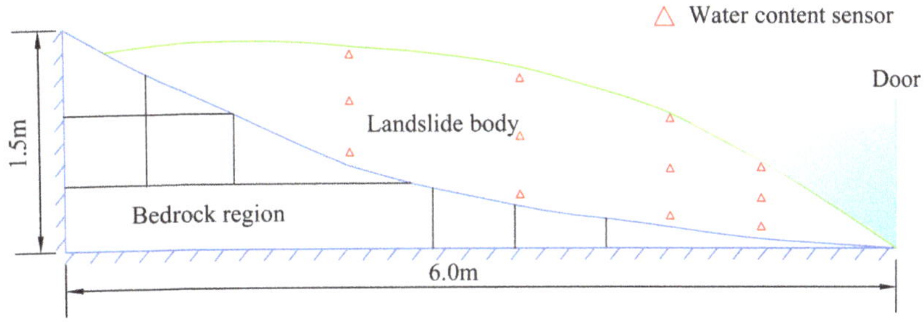

Figure 3. The side view of the landslide model and configuration of instruments in the experiment. A drawing of the configuration of the experimental model, showing the bedrock region (within blue lines), the landslide body region (below the green line), and the embedded sensors to monitor the water content.

According to the nature of the landslide body in the Three Gorges reservoir area, the material of the landslide body was designed. The material used in the landslide body was composed of soil and small stones, and the ratio was 7:3. The landslide model test was completed according to the principle of landslide model forming. Considering the size of the model and the need to embed sensors, it was decided to choose the tamping method to build the landslide model. The sensors were embedded in the model in advance. The moisture content sensors were arranged in four sections, with 3 sensors in each section. Then we increased the water level of the reservoir to monitor the change of its water content. The overall landslide model is shown in Figure 4. In this figure, instruments for displacement monitoring and high-density electrical methods are included (the results are not used in this article.) In order to obtain the material parameters of the landslide model, a triaxial test and a constant head permeability coefficient experiment were carried out. A pressure plate instrument was used to obtain the characteristic curve of unsaturated soil water, as shown in Figure 5a,b. In this figure, the pressure pump and pressure plate instruments are displayed.

Figure 4. A picture of the landslide model built indoors, from a frontal view: the model was 1.5 m (height) * 0.8 m (width). The water content sensor is inside the landslide body (not visible; the actual water content sensor is shown in Figure 6a).

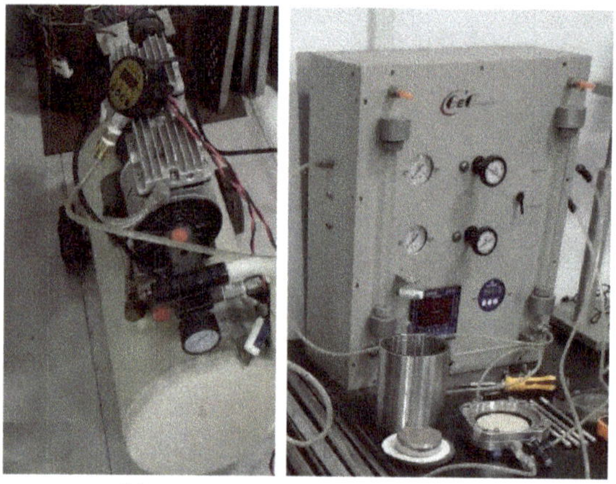

(a) (b)

Figure 5. Pressure plate instrument: (**a**) pump; (**b**) pressure gauge.

Sensors were used to obtain water content data in real time. Figure 6a shows the water content sensor. The power supply voltage of the sensor was 5 V, and the output signal type was RS485. The monitoring data of the water content sensor were saved when the water level was from 0.3 to 1 m. Figure 6b shows the sensor data interface. Its function was to convert RS485 signals into USB signals. Based on the Modbus protocol, the data acquisition software was developed using C# to save the water content data to the database. All sensors conducted data sampling and recording every 10 s. Figure 7 is a histogram representation of moisture sensor monitoring data. In this figure, the abscissa represents different moisture content, and the ordinate represents its frequency. These water content data were used as observation samples in Bayesian updating (specifically, as shown in Figure 2).

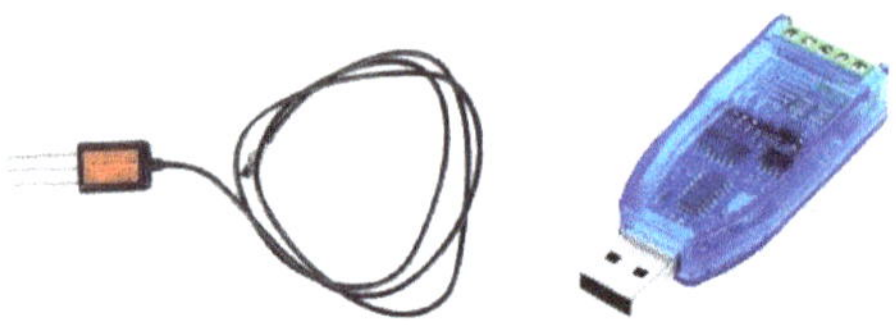

(a) (b)

Figure 6. (**a**) Water content sensor (the rated voltage was 5 V); (**b**) signal conversion device (converting the RS485 signal to USB signal).

4.2. FEM Simulation

In the Bayesian updating process, a numerical model of the experimental setup is needed to provide simulation results. The SEEP/W software, a part of Geo Studio is a numerical simulation software for seepage analysis. In this study, Geo Studio software (2007) was used. It can mathematically simulate the real physical process of water flowing through the soil medium based on the finite element model. The equation used in the process is a steady-state seepage equation. The work of this section aims to model the seepage analysis of the landslide model in Figure 3 through the FEM. For this reason, a geometric model of a large-scale landslide model was built in FEM software. The purpose of this research was to study the seepage characteristics of landslides through FEM to simulate the changes in water content of landslides when the water level rises. We used SEEP/W software to develop and calibrate computer models of landslide model, and compare observed data with simulated data.

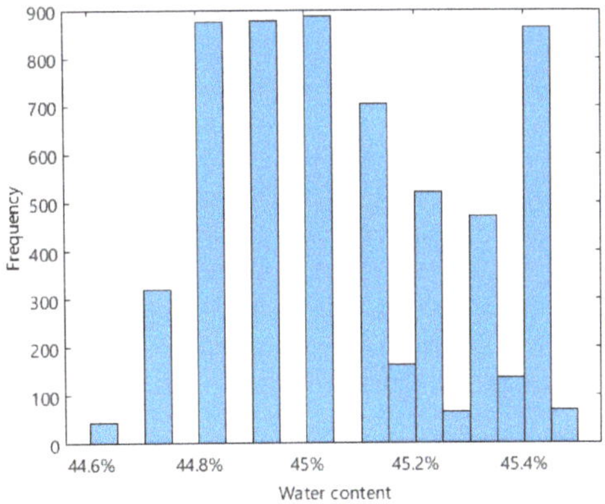

Figure 7. Monitoring data of water content.

In the FE model process, the main steps are to establish a landslide model, set material parameters, divide calculation units, and set boundary conditions [35]. Firstly, SEEP/W was used for geometric modeling of the landslide model according to the dimensions in Figure 3. The material models and properties of the soil were put into saturated-unsaturated seepage model, defined with a seepage constitutive relationship requiring two functions: the permeability coefficient function and water content function (SWCC) [35]. In this process, the saturated permeability coefficient was set to 1.6×10^6, and the SWCC parameters were set according to the prior distribution. Figure 8 shows the geometric model and the mesh used in this work. In this figure, the yellow area represents the landslide body. The unit accuracy of the finite element mesh was set to 0.1 m. The red part represents the boundary conditions. Two boundary conditions were set for the numerical model based on water level changes: water level boundary conditions and pressure boundary conditions. The boundary conditions of the model were a key component of the numerical analysis, and the ability to control the boundary conditions was also a direct reason for the powerful numerical analysis. In this process, the calculation method was the steady-state calculation, and the solver adopted the implicit solution method.

The FEM can evaluate water content values, depending on the input parameters (a,n) of the SWCC model. This results were the simulation samples used in Bayesian updating. The main purpose of this was to obtain the complete results of the FE model evaluation by SEEP/W software calculation, including water pressure, moisture content, etc. The results calculated by the SEEP/W software are analyzed and compared with the monitoring data of the landslide model in Section 4.1 to check for errors. The calculation of the SEEP/W software in this section also provides a model for the establishment of the surrogate model in the next section.

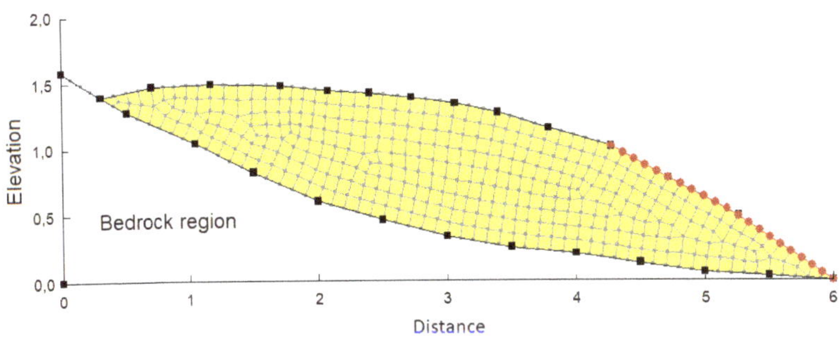

Figure 8. Mesh generation of finite element model (FEM).

4.3. Surrogate Model

Bayesian updating requires two sets of data: one of experimental data, and the other of simulation data. The experimental data were obtained through the above-mentioned large-scale landslide model experiment, and the data are shown in Figure 7. The simulation data were calculated according to the FEM in Section 4.2. It took 690 h (about 29 days) to perform a complete Bayesian update calculation with the FEM, and 2 h to use the surrogate model (Intel(R) Core(TM) i5-5200U CPU @ 2.20 GHz, RAM 8.00 GB). Therefore, a surrogate model was used in this study to obtain the simulation value. The surrogate model is a commonly used optimization method in engineering problems. When the actual problem is very computationally intensive and difficult to solve, a simplified model with less computational complexity and a rapid solution can be used to replace the original model to speed up the optimization process. In this study, an ANN was used to build the surrogate model.

In order to improve the efficiency of Bayesian updating, we adopted the method of the surrogate model. We used an ANN to train the surrogate model. The ANN is an algorithm whose design is driven by the function of the human brain and its components. ANNs use nonlinear techniques, so they can be used to model highly complex and nonlinear systems. In this study, a fully connected feedforward multi-layer structure using a backpropagation momentum learning algorithm was adopted. This kind of ANN architecture usually consists of an input layer, some hidden layers, and an output layer. For us there were 10 neuron nodes used in each layer. Each node was interconnected to the nodes of the upper layer through adjustable weights, and horizontal, self-direction, or backward connection was not allowed. The calculation and expression ability of a single neuron is limited. However, when they are connected to each other, the entire network can model complex functions. The training pattern consists of a set of matched input and output vectors. The logical sigmoid function was used as the activation function. In order to develop the artificial neural network model, Neural Network Fitting toolbox and MATLAB R2019b were used. The Neural Network toolbox provides various parameters for neural network development, which can be selected flexibly. We set the distribution of the parameters of the SWCC model, and then used the SEEP/W software to calculate the value of the water content. We saved the SWCC model parameters and the generated water content data for the establishment of the surrogate model. The input data selected for this part were a array (2×1000) of SWCC parameters a and n, and the target array was water content data. In the training process, 70% of the data were used for training, 15% of the data for verification, and 15% of the data for testing. The training data were presented to the network during the training process, and the network was adjusted according to its errors. The verification data were used to measure the generalization of the network and stop training when the generalization stopped improving. The test data had no effect on training, so they provide an independent measurement of network performance during and after training. In this section, the number of hidden neurons was 10. Finally, we checked the error of the surrogate model built using an ANN.

Figure 9 shows the regression diagram of the water content and SWCC parameters at specific points of the landslide model. In this figure, we can observe that all datasets were correctly fitted to the row. This shows that our neural network structure was correct, and it can also be used to predict the outputs of other input datasets. Figure 9a is the regression diagram of 70% of the training data. Figure 9b is the regression diagram of the 15% group of the validation data. Figure 9c is the regression diagram of the 15% group of test data. Figure 9d is the regression diagram of all data. The results of ANN prediction quality are detailed in Table 1. The goodness of fit (R) is close to 1 and the expected value of the mean square of the error (MSE) is small. The model results show that the ANN model can be used to predict the relationship between the water content at specific points of the landslide model and the parameters of the SWCC.

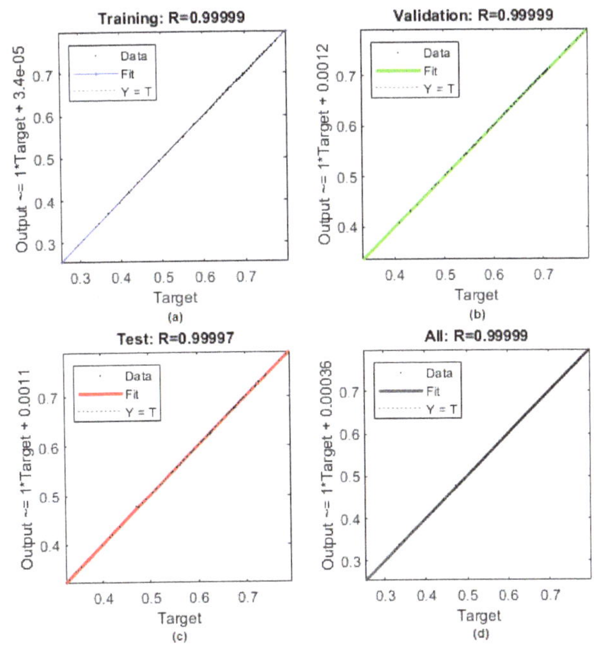

Figure 9. Neural network training regression: (**a**) The regression diagram of the 70% group of the training data; (**b**) The regression diagram of the 15% group of the validation data; (**c**) The regression diagram of the 15% group of test data; (**d**) The regression diagram of all data.

Table 1. The results of the prediction with the artificial neural networks (ANN).

	Samples	MSE	R
Training	700	1.43430×10^{-7}	9.99991×10^{-1}
Validations	150	1.67754×10^{-7}	9.99989×10^{-1}
Testing	150	5.56267×10^{-7}	9.99968×10^{-1}

5. Results

5.1. *Effect of Dataset Size on the Outcome of Bayesian Updating*

Using the surrogate model and the experimental data, the iterative Bayesian updating was done until the simulated value was close to the observed value. Bayesian updating is an iterative process. First, it sets the prior distribution of SWCC parameters and uses the MCMC algorithm to generate a large number of samples that obey the prior distribution. Then it performs iterative calculations, using surrogate models and experimental water cut monitoring data. Bayesian updating is performed until the simulated value is close to the observed value. For each iteration, a set of samples of SWCC parameters are generated. All the data of the iteration are presented into a plotmatrix, as shown in Figure 10a. In this figure, the top-left figure is the frequency diagram of parameter a, the bottom-right figure is the frequency diagram of parameter n, and the bottom-left and top-right figures represent the correlation. Figure 10b shows the distribution of SWCC parameters obtained by Bayesian updating, converged after 5 iterations (j = 5). This is the plotmatrix of the SWCC parameters after the fifth iteration.

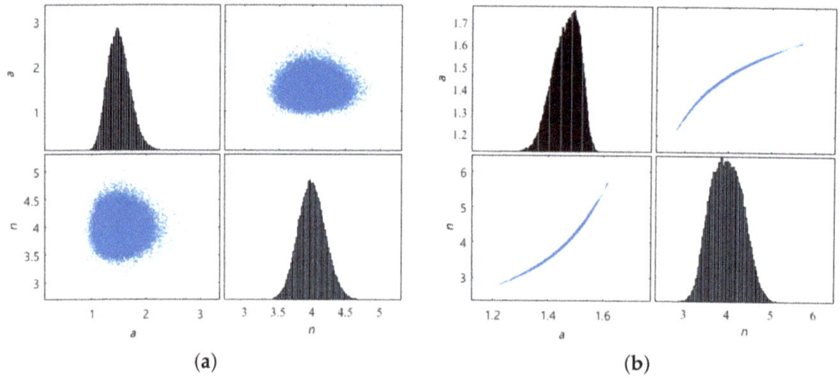

Figure 10. Results of using 1000 sets of water content monitoring data to update: (**a**) prior distribution; (**b**) posterior distribution.

In order to study the changes in the posterior distribution when monitoring the increase in the amount of data, we used 200, 500, 1000, 2000, and 5000 sets of monitoring data to calculate the posterior distribution and analyze the results. As the amount of data increased, the trends of Bayesian updating results are shown in Figure 11, and the values of Bayesian updating results are shown in Table 2. In that figure, the red line represents the median, the upper and lower lines of the blue rectangle represent the upper and lower quartiles, the short red lines represent outliers, and the short black lines represent the minimum and maximum values. The trend of the change of parameter a is shown in Figure 11a. It can be seen that the median value of parameter a fluctuates around 1.46 and has a tendency to increase slightly. Outliers are distributed below the minimum value, indicating that the posterior distribution is left skewed. The trend of the change of parameter n is shown in Figure 11b. It can be seen that the median value of the parameter n fluctuated around 3.98. Outliers are distributed over the maximum value, indicating that the posterior distribution skewed right. When the number of data was 1000, the data of the posterior distribution were relatively stable. Therefore, in the following analysis, 1000 sets of observation data were used as samples to study the influence of the distribution type on the posterior distribution.

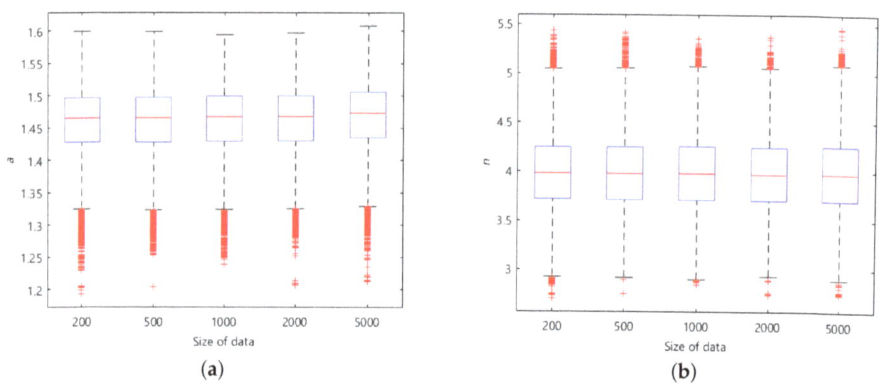

Figure 11. Impact of data volume on update results: (**a**) Parameter a; (**b**) parameter n.

Table 2. Impact of dataset size on update results.

	Most Probable VG Model Parameters		Posterior Mean		Posterior Standard Deviation	
Dataset Size	*a*	*n*	*a*	*n*	*a*	*n*
200	1.47	3.94	1.459	3.981	0.048	0.363
500	1.48	3.92	1.460	3.980	0.048	0.359
1000	1.49	3.85	1.462	3.985	0.048	0.364
2000	1.49	3.85	1.464	3.988	0.048	0.367
5000	1.50	3.84	1.468	3.978	0.048	0.364

5.2. The Effect of Prior Distribution on the Outcome of Bayesian Updating

In this Bayesian updating process, the prior distribution of SWCC parameters a and n needs to be set first. In the previous analysis (Section 5.1) the prior distribution of variables a and n was adjusted to a lognormal distribution. The converged posterior distribution was analyzed and the most likely value, mean, and standard deviation of the parameter were evaluated. In Table 3, the results for the analysis with dataset size value 1000 are shown.

To investigate the effect that a different prior distribution has on the results, the prior distribution was changed to a normal distribution. The results are plotted in Figure 12, and the distribution parameters are shown in Table 3. Comparing the results of Table 3, the values of the parameters a and n are compatible within the deviations obtained.

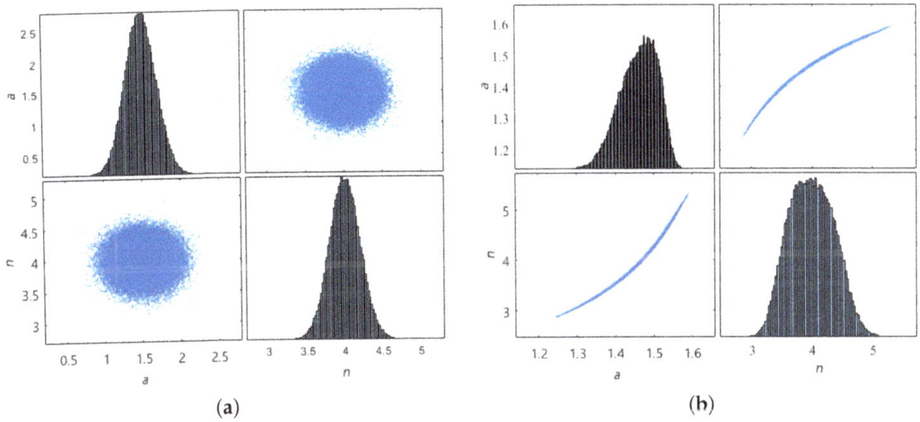

Figure 12. Update results when the prior distribution is a normal distribution: (**a**) prior distribution; (**b**) posterior distribution.

Table 3. Impact of prior distribution on update results.

	Most Probable VG Model Parameters		Posterior Mean		Posterior Standard Deviation	
Type	*a*	*n*	*a*	*n*	*a*	*n*
normal distribution	1.48	4.06	1.462	3.987	0.048	0.367
lognormal distribution	1.49	3.85	1.462	3.985	0.048	0.364

5.3. The Effects of Six Stages on the Outcome of Bayesian Updating

During the experiment (see Section 4.1 for details), we set the data collection time interval of the moisture content sensor to 10 s. When the water level changed from 0.3 m to 1 m, six stages were selected to study the trend of the posterior distribution of SWCC parameters when the water level changed. With the monitoring data obtained over time, the trend of Bayesian updating results is shown in Figure 13, and the value of Bayesian updating results is shown in Table 4. In this boxplot, the red line represents the median, the upper and lower lines of the blue rectangle represent the upper and lower quartiles, the red short lines represent the outliers, and the black short lines represent the minimum and maximum values. The trend of the change of parameter a is shown in Figure 13a. It can be seen that the median value of parameter a fluctuates around 1.47 and has a tendency to increase slightly. Outliers are distributed above the maximum value. The trend of the change of parameter n is shown in Figure 13b. It can be seen that the median value of parameter n fluctuates around 3.98 and has a slightly lower trend.

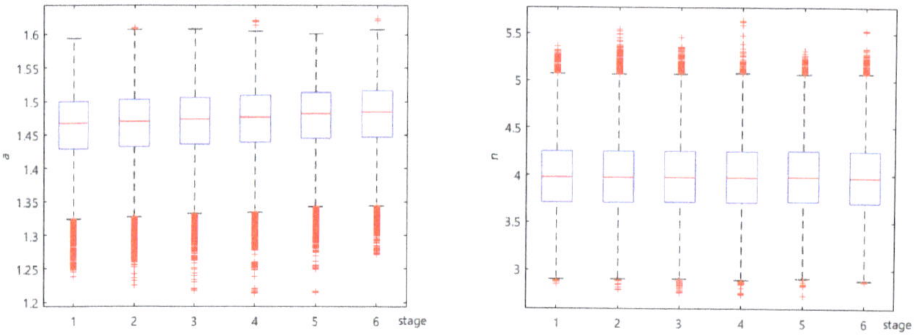

(a) (b)

Figure 13. The influence of the monitoring value at different time stages on the update result: (**a**) parameter a; (**b**) parameter n.

Table 4. Impacts of different stages on update results.

	Most Probable VG Model Parameters		Posterior Mean		Posterior Standard Deviation	
Stage	a	n	a	n	a	n
1	1.49	3.85	1.462	3.985	0.048	0.364
2	1.483	4.03	1.465	3.978	0.048	0.362
3	1.486	4.01	1.469	3.984	0.047	0.360
4	1.498	3.98	1.472	3.983	0.048	0.369
5	1.508	3.80	1.476	3.985	0.047	0.364
6	1.513	3.82	1.480	3.983	0.047	0.365

5.4. Characterization of Uncertainty

According to the previous results, it is possible to characterize the uncertainty of the model parameters. For a sufficiently large number of samples, and considering the minor effects caused by the selected prior distribution type and the time stage, the uncertainties obtained are reliable. Considering the VG model for the SWCC, the parameters a and n were evaluated with 100,000 MCMC samples and 95% confidence intervals (for 97.5% and 2.5% upper and lower bounds). The results are plotted on the curves in Figure 14. As can be seen in this figure, the most likely SWCC models were all within the 95% confidence interval. Therefore, the proposed method can reasonably determine the SWCC model based on limited experimental data and quantify the uncertainty of the SWCC model estimate.

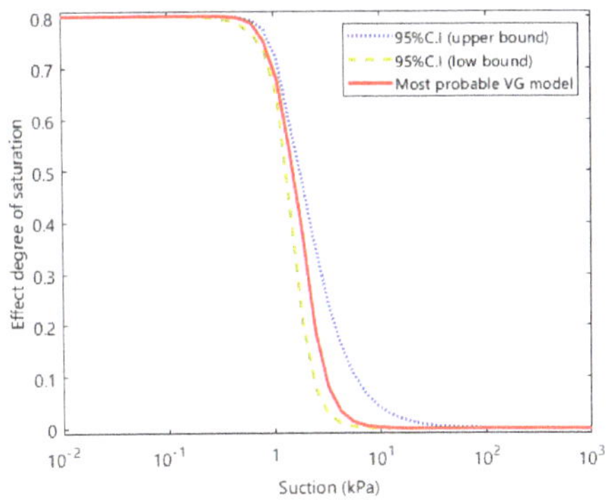

Figure 14. The 95% confidence interval of VG obtained from the Bayesian updating.

6. Discussion

Although slight changes of SWCC parameters have minor effects on the stability of a landslide, they improve the accuracy of landslide reliability calculation. It is very necessary to conduct in-depth research on SWCC parameters. In previous studies, the main focus was on the SWCC model. We used indoor test data to select SWCC model [36]. Compared with the deterministic SWCC model, this paper took into account the uncertain factors caused by soil variability and unstable monitoring data. This paper mainly checked the parameters of SWCC based on the monitoring data of large-scale model tests.

This paper mainly describes the impacts of monitoring data volume, prior distribution type, and different stages on posterior distribution. Figure 11 and Table 2 show the effect of data volume from 200 to 5000 on the posterior distribution of SWCC model parameters. The left figure represents the prior distribution, and it can be clearly seen that the prior distribution is the same. The figure on the right represents the posterior distribution. As the number of observation samples increases, it has little effect on the posterior distribution of SWCC parameters. It can be seen from Table 2 that the results of the posterior distribution are very stable. Therefore, it is feasible to perform Bayesian updating of parameters using monitoring data. This makes it possible to realize the online SWCC parameter monitoring and analysis system. In the research, this study selected 1000 sets of data for Bayesian updating. Figures 10 and 12 compare the effects of the prior distribution type on the results of the SWCC parameters' posterior distributions. It can be clearly seen that the coefficient of variation is smaller when the prior distribution is lognormal. Therefore, when using monitoring data to perform Bayesian updating of SWCC parameters, it is more reasonable to set the prior distribution to the lognormal distribution to obtain the posterior distribution. Figure 13 and Table 4 show the changes in the posterior distributions of SWCC parameters with different stages. In this figure, it can be seen that the mean value of parameter a in the SWCC posterior distribution increases from 1.462 to 1.480, while the mean value of parameter n fluctuates from 3.985 to 3.983 (the fluctuation range is within 0.1). Due to the short monitoring time, the change to the posterior distribution is very subtle. In the follow-up research, the monitoring time could be extended to obtain the SWCC parameter change rule. Figure 14 shows the 95% confidence interval of the posterior distribution of SWCC. The most likely SWCC falls within the 95% confidence interval of its posterior distribution.

This paper studied the method of using landslide monitoring data to study SWCC parameters. However, due to the influences of many uncertain factors, the simulation model's prediction and the real structural response often have certain errors. Therefore, it is necessary to verify the prediction accuracy of the simulation model in future research.

7. Conclusions

This paper presents a SWCC parameter calculation framework based on Bayesian landslide monitoring data theory. This method makes full use of prior information and limited monitoring data to determine the posterior information of SWCC model parameters. It characterizes the uncertainty of the SWCC model and is of great significance for reliability analysis. The influences of the dataset size, prior distribution type, and time-varying monitoring data on the result of posterior distribution were studied. Finally, the uncertainty of the SWCC model parameters was characterized. Taking the water content monitoring data of the large-scale landslide model test as an example, the effectiveness of this method is explained. The main conclusions from this study are as follows:

1. This paper proposes a Bayesian-based analysis method that can take the parameters' uncertainty into account, and is different from conventional deterministic analysis methods. The feasibility of the Bayesian updating method in the monitoring data of a large landslide model was demonstrated. Through the Bayesian updating, the posterior value ranges of the SWCC parameters will be narrowed compared to the prior ones.
2. The Bayesian updating method has good application prospects in landslide monitoring. When the prior distribution is lognormal, the coefficient of variation of the posterior distribution is smaller than that of the normal distribution. The SWCC parameter a tends to increase when the posterior distribution changes with time.
3. In the process of Bayesian updating, a numerical simulation based in a FE model was used. Alternatively, a ANN surrogate model was also trained based on the FE model. This method can greatly improve the efficiency of Bayesian model updating—its accuracy being acceptable.

Author Contributions: Conceptualization, C.F. and M.B. (Michael Beer); data collection, C.F., X.L. and T.H.; methodology, C.F., M.B. (Matteo Broggi), and B.T.; software, C.F. and M.B. (Matteo Broggi); validation, C.F., B.X., and B.T.; data curation, C.F.; writing—original draft preparation, C.F.; writing—review and editing, S.B.; project administration, B.T.; funding acquisition, B.T. All authors have read and agreed to the published version of the manuscript.

Funding: This work is supported by the National Key Research and Development Program of China (2017YFC1501100), and the Research Fund for Excellent Dissertation of China Three Gorges University (2019SSPY005).

Acknowledgments: Thanks to the National Research Council and Three Gorges University for their support.

Conflicts of Interest: The authors declare no conflict of interest.

References

1. Tan, X.; Wang, X.; Khoshnevisan, S.; Hou, X.; Zha, F. Seepage analysis of earth dams considering spatial variability of hydraulic parameters. *Eng. Geol.* **2017**, *228*, 260–269. [CrossRef]
2. Zheng, D.; Huang, J.; Li, D.Q.; Kelly, R.; Sloan, S.W. Embankment prediction using testing data and monitored behaviour: A Bayesian updating approach. *Comput. Geotech.* **2018**, *93*, 150–162. [CrossRef]
3. Wang, B.; Li, J.; Jiang, Q.; Yang, Y.; Feng, X.T. An improved FE-Meshfree method for solving steady seepage problems. *Comput. Geotech.* **2020**, *119*, 103223. [CrossRef]
4. Guan, X.; Jha, R.; Liu, Y. Model selection, updating, and averaging for probabilistic fatigue damage prognosis. *Struct. Saf.* **2011**, *33*, 242–249. [CrossRef]
5. Mia, M.; Dhar, N.R. Response surface and neural network based predictive models of cutting temperature in hard turning. *J. Adv. Res.* **2016**, *7*, 1035–1044. [CrossRef]
6. Strauss, A.; Frangopol, D.M.; Kim, S. Use of monitoring extreme data for the performance prediction of structures: Bayesian updating. *Eng. Struct.* **2008**, *30*, 3654–3666. [CrossRef]
7. Hsein Juang, C.; Luo, Z.; Atamturktur, S.; Huang, H. Bayesian updating of soil parameters for braced excavations using field observations. *J. Geotech. Geoenviron. Eng.* **2013**, *139*, 395–406. [CrossRef]
8. Liu, W.; Luo, X.; Huang, F.; Fu, M. Uncertainty of the soil–water characteristic curve and its effects on slope seepage and stability analysis under conditions of rainfall using the Markov Chain Monte Carlo Method. *Water* **2017**, *9*, 758. [CrossRef]

9. Lam, H.F.; Yang, J.; Au, S.K. Bayesian model updating of a coupled-slab system using field test data utilizing an enhanced Markov chain Monte Carlo simulation algorithm. *Eng. Struct.* **2015**, *102*, 144–155. [CrossRef]
10. Rocchetta, R.; Broggi, M.; Huchet, Q.; Patelli, E. On-line Bayesian model updating for structural health monitoring. *Mech. Syst. Signal Process.* **2018**, *103*, 174–195. [CrossRef]
11. Bi, S.; Broggi, M.; Beer, M. The role of the Bhattacharyya distance in stochastic model updating. *Mech. Syst. Signal Process.* **2019**, *117*, 437–452. [CrossRef]
12. Van Genuchten, M.T. A closed-form equation for predicting the hydraulic conductivity of unsaturated soils. *Soil Sci. Soc. Am. J.* **1980**, *44*, 892–898. [CrossRef]
13. Neuman, S.P.; Witherspoon, P.A. Finite element method of analyzing steady seepage with a free surface. *Water Resour. Res.* **1970**, *6*, 889–897. [CrossRef]
14. Leong, E.C.; Rahardjo, H. Review of soil-water characteristic curve equations. *J. Geotech. Geoenviron. Eng.* **1997**, *123*, 1106–1117. [CrossRef]
15. Thu, T.M.; Rahardjo, H.; Leong, E.C. Elastoplastic model for unsaturated soil with incorporation of the soil-water characteristic curve. *Can. Geotech. J.* **2007**, *44*, 67–77. [CrossRef]
16. Fredlund, D.G.; Xing, A. Equations for the soil-water characteristic curve. *Can. Geotech. J.* **1994**, *31*, 521–532. [CrossRef]
17. Gardner, W. Some steady-state solutions of the unsaturated moisture flow equation with application to evaporation from a water table. *Soil Sci.* **1958**, *85*, 228–232. [CrossRef]
18. Li, D.Q.; Qi, X.H.; Phoon, K.K.; Zhang, L.M.; Zhou, C.B. Effect of spatially variable shear strength parameters with linearly increasing mean trend on reliability of infinite slopes. *Struct. Saf.* **2014**, *49*, 45–55. [CrossRef]
19. Li, D.Q.; Wang, L.; Cao, Z.J.; Qi, X.H. Reliability analysis of unsaturated slope stability considering SWCC model selection and parameter uncertainties. *Eng. Geol.* **2019**, *260*, 105207. [CrossRef]
20. Johari, A.; Talebi, A. Stochastic Analysis of Rainfall-Induced Slope Instability and Steady-State Seepage Flow Using Random Finite-Element Method. *Int. J. Geomech.* **2019**, *19*, 04019085. [CrossRef]
21. Han, Z.; Vanapalli, S.K. Model for predicting resilient modulus of unsaturated subgrade soil using soil-water characteristic curve. *Can. Geotech. J.* **2015**, *52*, 1605–1619. [CrossRef]
22. Brutsaert, W. Probability laws for pore-size distributions. *Soil Sci.* **1966**, *101*, 85–92. [CrossRef]
23. Chiu, C.; Yan, W.; Yuen, K.V. Reliability analysis of soil–water characteristics curve and its application to slope stability analysis. *Eng. Geol.* **2012**, *135*, 83–91. [CrossRef]
24. Ching, J.; Chen, Y.C. Transitional Markov chain Monte Carlo method for Bayesian model updating, model class selection, and model averaging. *J. Eng. Mech.* **2007**, *133*, 816–832. [CrossRef]
25. Patelli, E.; Govers, Y.; Broggi, M.; Gomes, H.M.; Link, M.; Mottershead, J.E. Sensitivity or Bayesian model updating: A comparison of techniques using the DLR AIRMOD test data. *Arch. Appl. Mech.* **2017**, *87*, 905–925. [CrossRef]
26. Wang, L.; Cao, Z.J.; Li, D.Q.; Phoon, K.K.; Au, S.K. Determination of site-specific soil-water characteristic curve from a limited number of test data—A Bayesian perspective. *Geosci. Front.* **2018**, *9*, 1665–1677. [CrossRef]
27. Kelly, R.; Huang, J. Bayesian updating for one-dimensional consolidation measurements. *Can. Geotech. J.* **2015**, *52*, 1318–1330. [CrossRef]
28. Papaioannou, I.; Straub, D. Reliability updating in geotechnical engineering including spatial variability of soil. *Comput. Geotech.* **2012**, *42*, 44–51. [CrossRef]
29. He, L.; Liu, Y.; Bi, S.; Wang, L.; Broggi, M.; Beer, M. Estimation of failure probability in braced excavation using Bayesian networks with integrated model updating. *Undergr. Space* **2019**. [CrossRef]
30. Tian, M.; Li, D.Q.; Cao, Z.J.; Phoon, K.K.; Wang, Y. Bayesian identification of random field model using indirect test data. *Eng. Geol.* **2016**, *210*, 197–211. [CrossRef]
31. Phoon, K.K.; Tang, C. Characterisation of geotechnical model uncertainty. *Georisk Assess. Manag. Risk Eng. Syst. Geohazards* **2019**, *13*, 101–130. [CrossRef]
32. Beck, J.L.; Katafygiotis, L.S. Updating models and their uncertainties. I: Bayesian statistical framework. *J. Eng. Mech.* **1998**, *124*, 455–461. [CrossRef]
33. Katafygiotis, L.S.; Beck, J.L. Updating models and their uncertainties. II: Model identifiability. *J. Eng. Mech.* **1998**, *124*, 463–467. [CrossRef]
34. Chib, S.; Greenberg, E. Understanding the metropolis-hastings algorithm. *Am. Stat.* **1995**, *49*, 327–335.

35. Arshad, I.; Baber, M. Finite element analysis of seepage through an earthen dam by using geo-slope (SEEP/W) software. *Int. J. Res* **2014**, *1*, 612–619.
36. Wang, L.; Tang, L.; Wang, Z.; Liu, H.; Zhang, W. Probabilistic characterization of the soil-water retention curve and hydraulic conductivity and its application to slope reliability analysis. *Comput. Geotech.* **2020**, *121*, 103460. [CrossRef]

applied sciences

MDPI

Article

3D Dilatometer Time-Series Analysis for a Better Understanding of the Dynamics of a Giant Slow-Moving Landslide

Jan Blahůt [1,*], Jan Balek [1], Michal Eliaš [2] and Stavros Meletlidis [3]

1 Institute of Rock Structure and Mechanics, The Czech Academy of Sciences, V Holešovičkách 94/41, 18200 Prague, Czechia; balek@irsm.cas.cz
2 Research Institute of Geodesy, Topography and Cartography, Ústecká 98, 25066 Zdiby, Czechia; michal.elias@pecny.cz
3 Centro Geofísico de Canarias, Instituto Geográfico Nacional, Calle la Marina 20, 38001 Santa Cruz de Tenerife, Spain; smeletlidis@mitma.es
* Correspondence: blahut@irsm.cas.cz; Tel.: +420-266-009-394

Received: 30 June 2020; Accepted: 3 August 2020; Published: 7 August 2020

Featured Application: Analysis of monitoring data from very slow-moving landslides.

Abstract: This paper presents a methodological approach to the time-series analysis of movement monitoring data of a large slow-moving landslide. It combines different methods of data manipulation to decrease the subjectivity of a researcher and provides a fully quantitative approach for analyzing large amounts of data. The methodology was applied to 3D dilatometric data acquired from the giant San Andrés Landslide on El Hierro in the Canary Islands in the period from October 2013 to April 2019. The landslide is a creeping volcanic flank collapse showing a decrease of speed of movement during the monitoring period. Despite the fact that clear and unambiguous geological interpretations cannot be made, the analysis is capable of showing correlations of the changes of the movement with increased seismicity and, to some point, with precipitation. We consider this methodology being the first step in automatizing and increasing the objectivity of analysis of slow-moving landslide monitoring data.

Keywords: slow-moving landslide; landslide monitoring; time-series analysis; San Andrés Landslide; El Hierro; Canary Islands

1. Introduction

Analysis of monitoring data from very slowly moving landslides faces very difficult issues considering the precision and accuracy of the used monitoring instruments, the influence of the different environmental factors on the measured data, the frequency of the measurements and the overall complexity of the phenomena [1–4]. In the case of large landslides (of more than several km^2) another problem arises with choosing the right placement of the monitoring devices on the ground, as remote sensing methods such as LiDAR or InSAR, despite their enormous applicability, still do not provide comparable sub-mm accuracy [5,6].

When analyzing the landslide movement monitoring results, the time-series analysis should be used. Time-series analysis comprises different statistical methods to understand the underlying context of data points or to make forecasts [7]. It can capture seasonal behaviors, trends and changes [8]. Time-series analysis is routinely applied in econometrics [9,10]. In geosciences, there is a constant use of time-series analysis within GNSS observations [11] or climatology [12]. Thus, it might be a bit surprising that in landslide science, time-series analysis studies are mostly related to remote sensing

(SAR) data analysis [13–15]. To some point, this is not surprising as SAR data are available in defined time-steps and form an undisrupted time series, which is the basic assumption of the analysis. In the case of landslides studies, time-series analysis is usually focused on trend behavior, sudden changes and seasonality analysis [16–18]. The big advantage of time-series analysis methods lies in the possibility to make predictions of future behavior based on previous activity. This is especially important in landslide disaster management. Complex non-linear time series are ubiquitous in geosciences [19]. The recently developed adaptive regression models (used in this work) appear to offer good solutions in multivariate non-stationary time-series analyses [20].

In the era of Big Data, it is more and more possible to harvest/gather a large amount of environmental and geophysical data related to slope stability using a variety of sensors and systems with high precision [21,22], which can be used for interpreting and monitoring the results and finding correlations to understand the landslide activity. However, a recent study [23] highlighted that the observations are prohibitively site-specific, highly subjective and fail to account for the spatial variability and correlations in slope movements and common trigger factors.

The landslide movement is usually neither smooth nor continuous. There are changes in the nature of the observed monitoring data at certain moments (changes of speed or direction of movement, jump events). We assume that these changes (events) during the development of landslide movements are caused by extreme states of factors (triggers) affecting the landslide movement (i.e., seismicity, precipitation) [24].

In this paper, we analyzed data from precise 3D dilatometers installed on the detachment planes of a giant complex landslide in a volcanic environment. The time-series analysis was performed on 1D, 2D and 3D data projections using adaptive regression. The main aim of this work was to find a suitable methodology for evaluating the time series of micro-movements to find the changes in landslide movement (events) in them and then compare them with the time moments of extreme external conditions including available climatic and seismic data to find correlations with the possible triggers of the landslide activity. The aim was fulfilled by applying known methods and combining them to understand better the landslide behavior and its responses to possible influencing factors.

2. Study Area

This study takes place on the island of El Hierro, Canary Islands, Spain. El Hierro is a composed oceanic volcano with the oldest subaerial rocks having an age of 1.12 Ma. Its youngest subaerially exposed rocks are represented by the Rift Series, with a maximum age of 0.16 Ma [25]. Over the past 33,000 years, onshore eruptions have reoccurred approximately once every 1000 years [26]. The latest, ongoing phase of volcanism began around 2.5 ka [27]. Recently a period of intense seismic activity spanned from July 2011 [28] to 2014 [29]. During this period, an offshore eruption started on 10 October 2011 and finished in March 2012 [30]. A more detailed description of the geology of the island has recently been presented elsewhere [31,32].

The characteristic three-point star morphology of El Hierro is a result of several enormous flank collapses (Figure 1). Until now, seven debris avalanches have been identified: Tiñor (<880 ka), Las Playas I (545–176 ka), Las Playas II (176–145 ka), El Julan (>158 ka), El Golfo A (176–133 ka), El Golfo B (87–39 ka), and Punta del Norte (unknown age) [27,33–42].

In addition to the previously mentioned flank collapses, a giant San Andrés Landslide (SAL) has been identified on the northeastern part of the island. The SAL is a deep-seated gravitational slope deformation [43] or a slump [44]. It has been supposed that the landslide underwent a single-event development at some point between 545 ka and 176 ka [45] and was inactive since then. Recent research shows, however, that there were at least two distinguishable separate slip events. The first one between 545 ka and 430 ka, and the second one between 183 ka and 52 ka [46]. Moreover, it has been proven that there is a continuous creep with a rate up to 0.5 mm·a^{-1}, suggesting that the landslide mass may be moving steadily to the east and southeast [32,47,48]. Numerical stability modelling [49] showed that creep movements are possible due to unconsolidated sediments at the sea bottom, and destabilization of the landslide might be possible during periods of intense seismic activity.

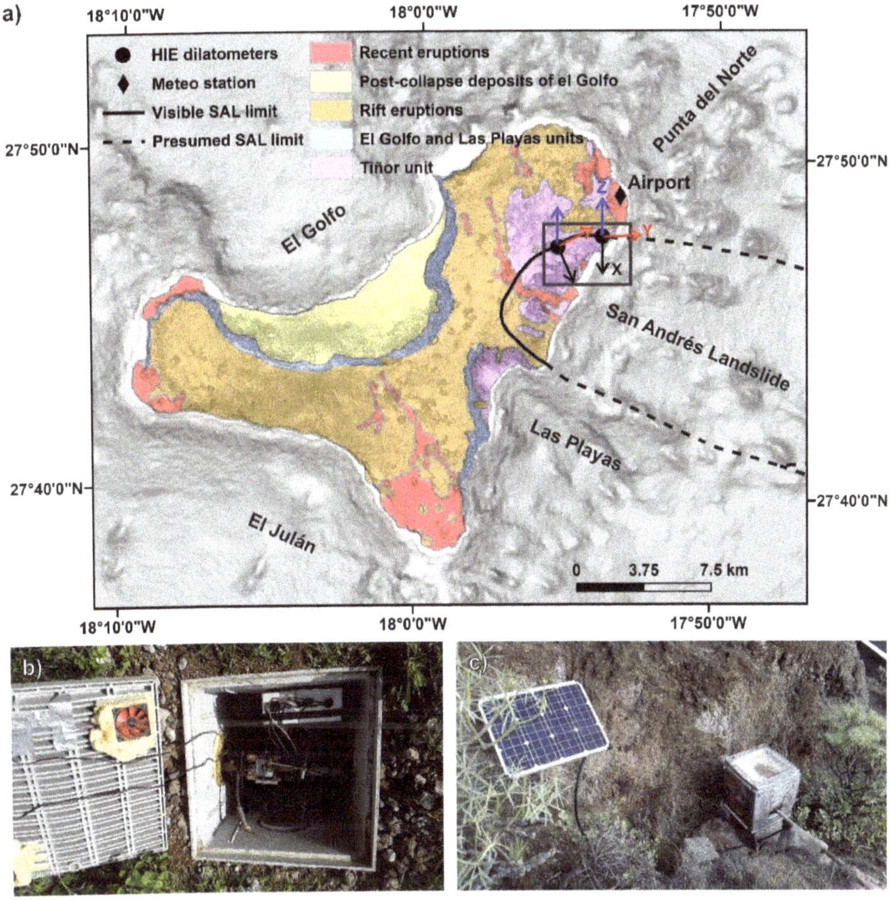

Figure 1. Location of the study area: (**a**) the El Hierro Island with a simplified geological map showing the location of the TM-71 3D dilatometers and showing the direction of the axis (X, Y, Z) of the measurement: grey rectangle shows the location of Figure 8c; (**b**) dilatometer HIE2 located underground on the upper scarp of the SAL; (**c**) dilatometer HIE3 located in the open air on the left side of the SAL.

This creep has been monitored by high-precision (±0.007 mm) 3D dilatometer gauges [3] with automatic data processing [50]. There are two devices installed on the landslide detachment plane in the upper scarp and on the left side of the landslide, respectively (Figure 1). The instrument records data based on optical–mechanical interferometry via the generation of moiré patterns, which result from the bending and interference of light rays as they pass through two specially designed optical grids [51].

3. Monitoring Data Collection and Preparation

The data used in this analysis can be split into three categories: (i) data from the 3D dilatometers placed on the SAL detachment plane; (ii) climatic variables and (iii) seismicity (Figure 2).

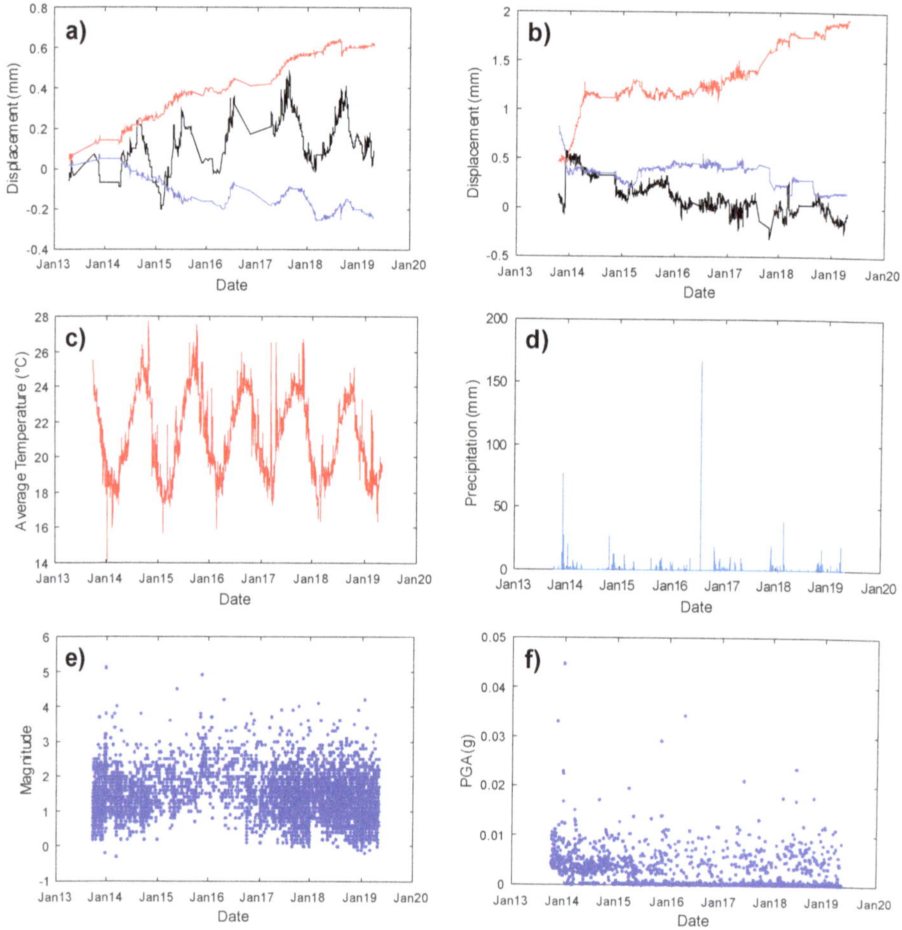

Figure 2. Data used in the analysis: (**a**) HIE2 dilatometer: *X*-axis—black line, *Y*-axis—red line, *Z*-axis—blue line; (**b**) HIE3 dilatometer: *X*-axis—black line, *Y*-axis—red line, *Z*-axis—blue line; (**c**) daily average air temperatures; (**d**) daily precipitation; (**e**) registered earthquakes; (**f**) calculated PGA.

3.1. 3D Dilatometer Data

Two 3D dilatometers named HIE2 and HIE3 are placed on the SAL's detachment plane. They are about 2.5 km apart (Figure 1) and the measurements are taken every 24 h at 00:00 UTC. The records (Figure 2) are analyzed in a Cartesian coordinate system with *X*-axis showing compression/extension perpendicular to the detachment plane; *Y*-axis showing strike-slip movement along the detachment plane and *Z*-axis showing vertical movement perpendicular to the detachment plane. As the records are taken at 00:00 UTC the records were correlated with climatic and geophysical variables from the previous day. Automated monitoring started on 16/10/2013 and covers a period of 2010 d till 17/04/2019. Due to technical difficulties, several instrument failures resulting in missing records happened. In the case of HIE2 device, there are 494 missing measurements (24.6%), in the case of HIE3 there are 358 missing records (17.8%).

3.2. Climatic Variables

Climatic variables are strongly influencing any landslide behavior. For that purpose, temperature, precipitation and insolation data from the El Hierro airport [52] were downloaded. The location of this station is not directly on the SAL. However, it is in its vicinity and provides daily undisturbed data for the whole monitoring period. Precipitation can play a principal role in landslide activation. For that purpose, the daily rainfall amounts were analyzed. Additionally, the 30-day antecedent precipitation index (API_{30}) [53,54] was calculated to better depict the influence of retrospective precipitation history and its influence on soil moisture [55]:

$$API_n = \sum_{i=0}^{n} \left(c^i \cdot P_{n-i}\right) \tag{1}$$

where n is the number of antecedent days; c is decay constant and it depends on watershed and seasonal parameters. The value of c ranges between 0.80 and 0.98 [56]. In this study, a c of 0.95 was used; i is the number of days counting backwards from the date on which the API is determined and P_n is the amount of precipitation i days before the causal rainfall (mm).

Mean and maximum daily temperatures can influence both the monitoring instrument and the rock on which it is attached by thermal dilation [57]. For that purpose, the temperature data were used to observe influence on the monitoring records and assess seasonal trends.

3.3. Seismic Data

Seismicity plays a major role in the stability of volcanic flanks. During the analyzed period, a total of 6537 seismic events were recorded around El Hierro Island by the Instituto Geográfico Nacional network [58]. We used this data to calculate the PGAs (Peak Ground Accelerations) of these events. The Canary Islands does not have developed local ground-motion attenuation relationship and there are no accelerometer network data available. For that reason, we applied an attenuation relationship developed for the Hawaiian Islands (to some point, a similar volcanic archipelago) by [59] and suggested to be used on the Canary Islands by [60]:

$$log_{10}PGA = 0.518 + 0.387\ (M - 6) - log_{10}r - 0.00256r + 0.335S \tag{2}$$

where PGA is measured in units of g, M is the magnitude, $r = (d^2 + 11.292)^{1/2}$ and S is 0 for lava sites and 1 at ash sites. The distance parameter d is the closest distance from the recording site to the surface projection of the fault rupture area.

For this analysis, we used the maximum recorded PGA for the particular day and also calculated the sums of daily PGAs (named PGA dose). The maximum PGAs and PGA doses were calculated for both HIE2 and HIE3 instruments considering their geographic location.

4. Methodological Approach

4.1. Input Data Pre-Processing

The relative motion on the detachment planes of the SAL is represented by time series which are not completely homogeneous. There are many missing values due to technical problems. This is not necessarily a problem as the reading frequency of the device (every 24 h) to the speed of the movement is very high. However, the methodology of time-series processing needs to be adapted to this fact.

The time series were decomposed into trend, residual and seasonal components (as is visible on Figure 3). The periodicity of the seasonal components reflects the effect of the thermal dilation of the anchoring elements holding the TM-71 device, but it can also be partially influenced by the volume changes of the monitored blocks. A Fourier model was fitted to estimate the characteristics of the periodic component, specifically the 1st harmonics. The period of the harmonic function was fixed at

365 d (Figure 3b). The applicability of the harmonic function 0.15 mm corresponds to the amplitude of the air temperature (4 °C) and the length of the steel anchoring elements (2 m) at the coefficient of thermal expansion $\alpha = 16.3 \times 10^{-6}$.

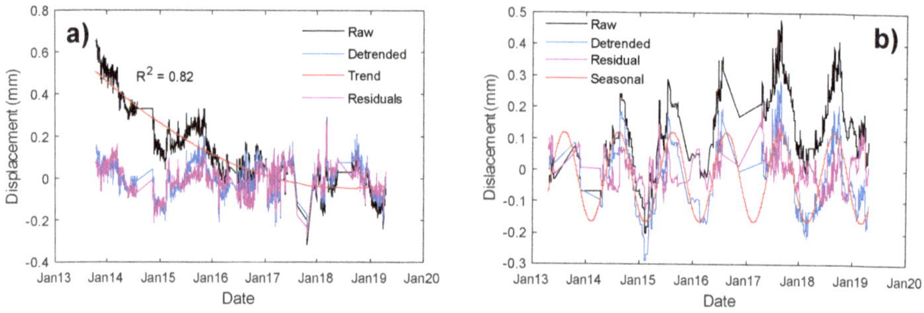

Figure 3. Example of the time-series decomposition: (**a**) close to a linear decrease of the velocity on HIE3 *Y*-axis; (**b**) significant seasonal periodicity visible on HIE2 *X*-axis.

The trend component is very well described by the approximation by a second-order polynomial. This indicates a linear change in velocity over time. This trend is clearly visible on the HIE3 *X*-axis (Figure 3).

4.2. Detection of Change-Points in Time Series of Individual Components

The time series of the measured movements are not stationary but often express sudden changes. We assume that these changes are caused by external factors. Over time, several methods have been developed to detect such changes such as CUSUM (Cumulative Sum) and its modifications (firstly introduced by [61]). Limiting factors of a wide range of methods are very often the computational demands in combination with a large volume of data or the requirement for time-series homogeneity. The problem of geotechnical measurements (field measurements in general) is the incompleteness of the time series or its inhomogeneity—different time spacing between recorded measurements. In this work, the PARCS (Paired Adaptive Regressors for Cumulative Sum) method was used, a method for detecting a change in the mean value of the multidimensional time series [20]. The time series of measurements are burdened with random errors. The mean value is then close to the value of the monitored variable in a large number of measurements.

The PARCS method looks for a predefined number of change-points in the time series. The input data to the algorithm are an array containing the time-series data and the scalar value indicating the number of searched change-points (N). The advantage over other methods is that the time series may not be completely consistent and may contain holes. The PARCS method was modified for data analysis so that the input parameter N was used only as an a priori (initial) value and the target mean error parameter was added. After the detection of change-points, the time series is divided into segments, which are approximated by linear regression. If the standard deviation on one of the sections exceeds the given limit value, the parameter N is increased (Figure 4).

The cut-off target value of the standard deviation was derived from the assumed accuracy of the dilatometer measurement as twice the standard deviation of the measurement found during laboratory testing [62].

A relatively large number of change-points was detected. Taking a closer look at the trajectory of motion, it is clear that some of these change-points do not lead to a change in the trajectory of motion but are the result of chaotical behavior. Therefore, we decided to directly analyze the trajectory of motion in the plane (2D) and space (3D) to find the key points of its change and their time stamps (Figure 5).

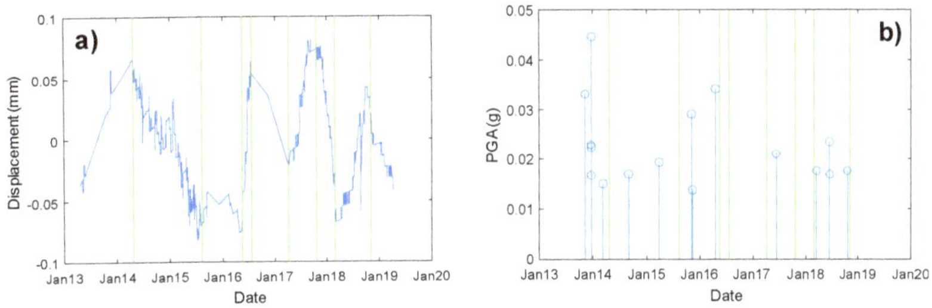

Figure 4. Modified PARCS method applied to the time series of residual values (HIE3 z-axis): (**a**) time stamps of detected change-points marked in green, target standard deviation of 0.05 mm; (**b**) comparison of detected time stamps and seismic events.

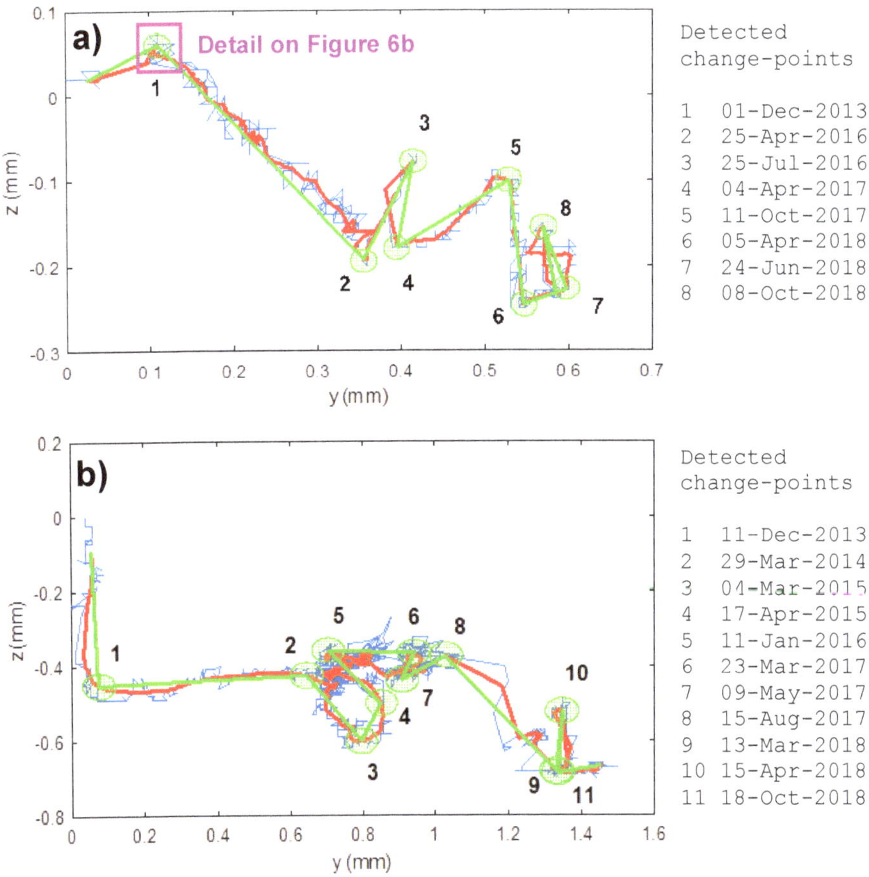

Figure 5. Example of the processing of the movement trajectory on a Z-Y plane on: (**a**) HIE2 device; (**b**) HIE3 device: raw trajectory (blue), smoothed trajectory using 30-day moving average (red), simplified trajectory (green).

The trajectory of the movement is represented by an array of coordinates with a removed periodic component. At first, the trajectory was smoothed using a 30-day moving average of the position to remove random noise. Then, the key breaking points of the motion curve were found, the curve was expressed by the minimum number of linear segments so that the accuracy criterion was preserved (Figure 5). For this purpose, the Ramer–Douglas–Peuckner (RDP) simplification algorithm was used [63]. The advantage of this approach is that all three components of motion can be analyzed together. Compared to other change-point detection methods applied to components, the RDP algorithm is simpler and enables very fast sampling of the curve and detection of change-points.

Importance of the change-points was rated to highlight the most significant changes in the pathway of movement of the monitored SAL detachment plane. We assume that the importance of change-point is increasing with its shortest distance from the connection of two adjacent change-points (Figure 6). The most significant changes in the trajectories were highlighted by red stars (Figures 10 and 11).

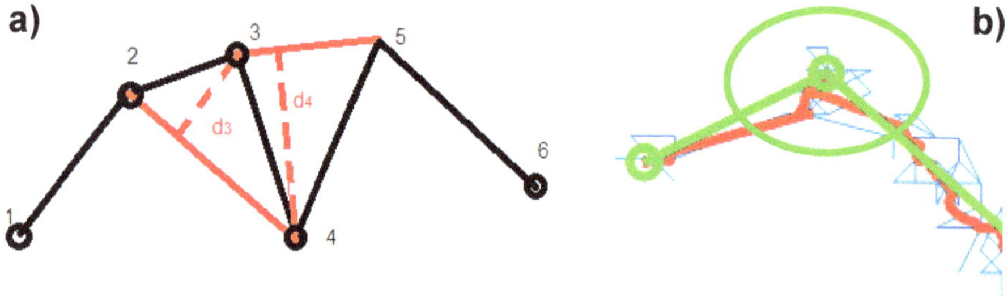

Figure 6. (**a**) Importance of change-points was quantified using minimal distance di; (**b**) change-point defined using error ellipse.

Note that in the figure above (Figure 6b), the change-point is displayed as one single moment, but a closer view to the trajectory of the movement shows, that in case of such slow movement (daily displacement is smaller than random noise) there should be some temporal buffer around the change-point where the trajectory is changing. This enabled to decide if the change of the trajectory is the reaction on a given event (for example seismic).

The time span of possible change-point location is in two- or three-dimensional space represented by error ellipse or ellipsoid, where its center is located in the change-point. The size of the major and minor axes of the ellipse/ellipsoid is derived from the magnitude of the standard deviations in the respective directions. All points of the trajectory located within the ellipse/ellipsoid are detected as possible change-points, and, therefore, each turn of the trajectory is represented by a set of relevant time stamps. As a consequence, one turn of the trajectory is now represented with various numbers of dates, which can be confronted with the time stamps of other events.

Both approaches: change-point detection in time series of components and trajectory simplification leads to detecting time stamps, which were then confronted with time stamps of above normal events in the time series of potential landslide triggers.

We assume that significant events in the movement are caused by significant events in the seismicity or precipitation. Statistics describe this relation as causality. We used Granger causality to quantify the relationship between sets of events with different extents, which allows determining if one set of values is caused by another on a given significance level [64].

There was a huge number of seismic events during the analyzed period—most of them with very low magnitude. To compare seismic events with significant events in the movement, it was needed to highlight above normal seismic events. For this reason, the average value and standard deviation of daily PGA maxima were computed. Then, only those seismic events were selected whose standard deviation of maximum PGA average was three times higher than the standard deviation. To highlight episodes with higher seismicity, five-day sums of PGA maxima were computed (Figure 7).

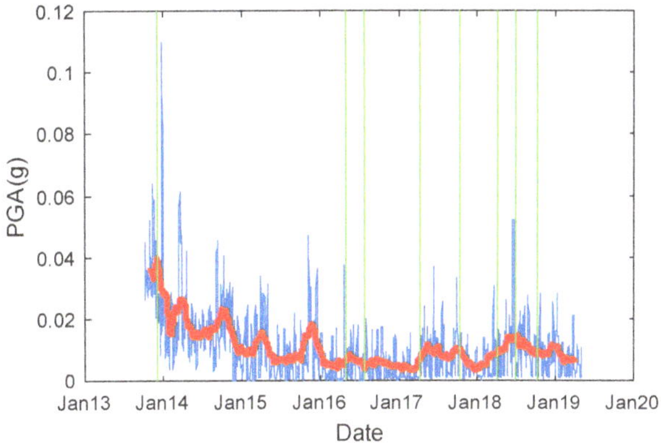

Figure 7. Cumulated PGA five-day sums and its moving 30-day average (red line).

5. Results and Discussion

Figure 8 shows the total displacement distance from the beginning of the monitoring. There is a visible decrease in movement speed on both devices. Although, in the case of HIE3 this is less pronounced.

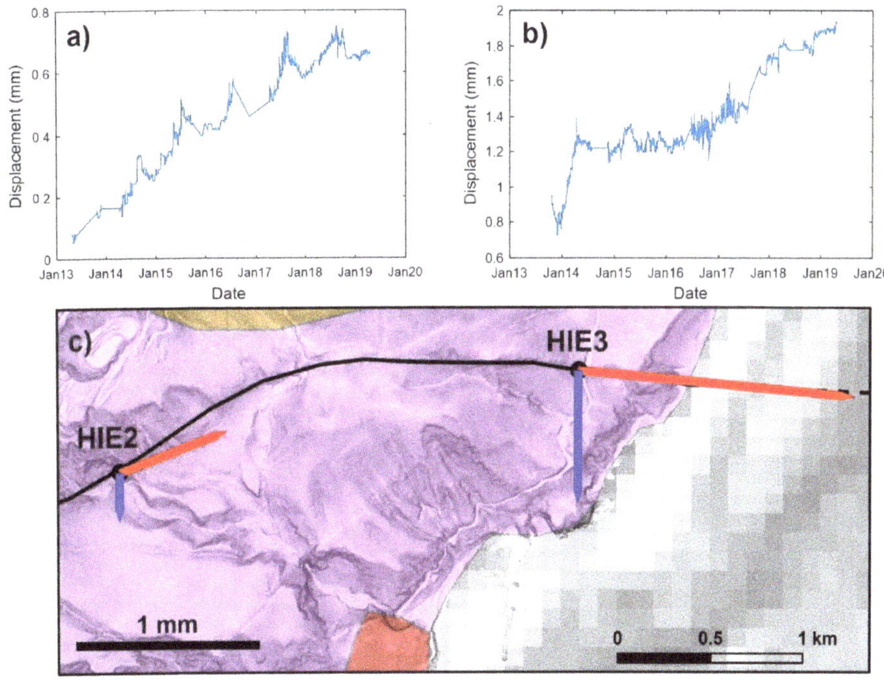

Figure 8. Cumulated spatial displacement since the beginning of the monitoring: (**a**) HIE2 dilatometer; (**b**) HIE3 dilatometer; (**c**) spatial interpretation: red line—horizontal plane, blue line—vertical plane.

Both 3D dilatometers HIE2 and HIE3 showed permanent very slow creep movement during the analyzed period, generally in the downslope sense. It can be noted from Figure 8c, that on HIE3 device the horizontal component of the movement is parallel to the general slope. On HIE2 device the horizontal component of the movement is almost perpendicular to the general slope (sinistral strike slip). This is probably a long-term behavior, which was already documented by [47] and it is recognizable on a step-like shape of the gully below the device. On both dilatometric devices the movement direction corresponds to the El Hierro earthquake and volcanic activity epicenters, which are concentrated to the west and southwest from the SAL landslide. The speed of the movement and its direction changed over time. Movement trend on the HIE2 device is well described by a second order polynomial (Figure 9a–c). The X and Y axes show a close-to-linear decrease in the speed of the movement.

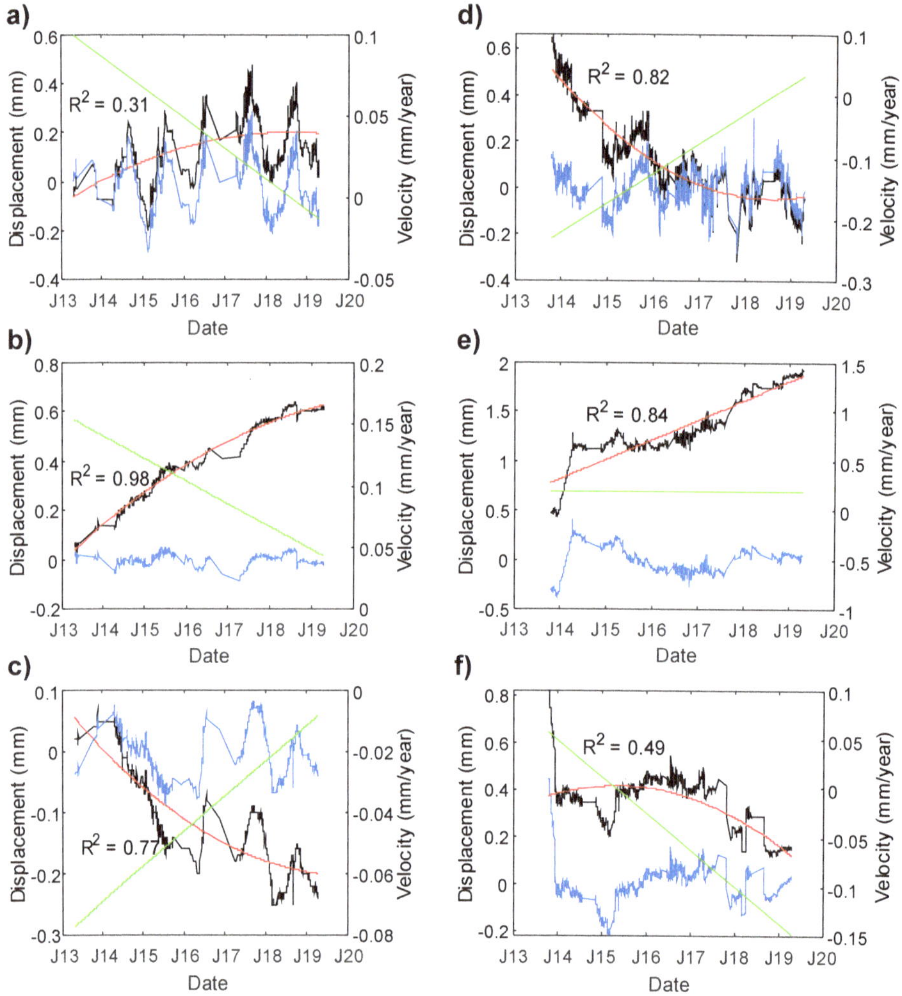

Figure 9. Decomposition of movements on the different axes showing the general decrease of the speed of the movement, black line—raw data; blue line—detrended data; red line—trend; green line—velocity.: (**a**) HIE2 *X*-axis; (**b**) HIE2 *Y*-axis; (**c**) HIE2 *Z*-axis; (**d**) HIE3 *X*-axis; (**e**) HIE3 *Y*-axis; (**f**) HIE3 *Z*-axis.

Analysis of GNSS and InSAR data from the beginning of the 3D dilatometer monitoring period [29] shows large absolute movements of the island in the period from July 2012 to March 2014. Although there are no permanent GNSS receivers on the SAL body, the nearest permanent GNSS stations show northeast shift and uplift of several centimeters. Similar results are shown from InSAR data, which are available for the island. The results, however, show largest surface movements in the western and southwestern parts of the island, while the SAL body on the northeast was much more stable at that time. It has to be noted, however, that HIE2 and HIE3 automatic devices are measuring relative movements on the detachment plane, while GNSS and InSAR are showing absolute movements of the island.

PS InSAR analysis performed for the period from October 2013 to October 2015 [65] shows rather different behavior on different parts of the SAL body, suggesting that the landslide moves as separate creeping blocks (as also shown in [32]). The speed of the movement during the two-year analyzed period reaches few mm.a^{-1} downslope.

It has to be noted that a principal road connecting the La Estaca port with the island's capital Valverde is crossing the SAL detachment plane just below the HIE3 dilatometer. This road has been damaged by cracks during 2012–2014 with general direction along the detachment plane, and the tarmac had to be repaired.

Movement trends on the HIE3 device are much more fluctuating (Figure 9d–f). This is not surprising considering the fact, that this device is installed in the open air and thus more prone to climatic influence (i.e., temperature changes, humidity). In the case of the *Z*-axis, the movement speed seems to slightly increase. This is most probably due to the fact, that this device is measuring a separated block within the entire landslide body, which exhibits slightly different behavior than the rest of the landslide [32].

It can be seen, that most significant changes in the direction of the movement happened in December 2013, at the very beginning of the monitoring. Comparison of the change-points in the trajectory on both HIE2 and HIE3 with the PGA and API are shown on Figures 10 and 11. There is a recognizable correlation of seismic activity and changes of movements on both devices. However, HIE2 shows more movement changes correlating with PGA than the HIE3 device. The most notable change-point probably influenced by the seismic activity occurred at the end of December 2013, when the strongest earthquake during the monitoring period (M = 5.1) also occurred. Other change-points presumably influenced by seismic activity happened in 14-Nov-2013; 23-Dec-2013; 24-Dec-2013; 25-Dec-2013; 28-Dec-2013; 16-Mar-2014; 04-Sep-2014; 30-Mar-2015; 08-Nov-2015; 13-Nov-2015; 21-Apr-2016; 08-Jun-2017; 14-Mar-2018; 14-Jun-2018; 15-Jun-2018 and 17-Oct-2018.

In case of API there is not a clearly visible correlation with change-points, except in December 2013; July 2016 and April 2018 affecting the HIE3 device and July 2016 affecting the HIE2 device. This suggests that the nature of the measured movement has, more probably, an endogenous cause. These observed results are in good accordance with the analysis performed by [32]. In their study, a period from October 2013 to June 2014 was analyzed on the HIE3 device. Three major impulses were recognized (rainfall event in December 2013 and seismic events in December 2013 and March 2014). All these three impulses were also identified using this proposed methodology.

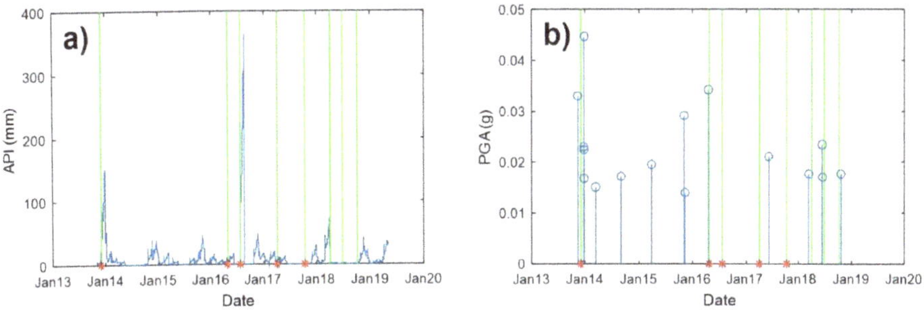

Figure 10. HIE2—Time stamps of change-points (green lines) detected using the Ramer–Douglas–Peuckner (RDP) algorithm vs. PGA (**a**) and vs. antecedent precipitation index (API) (**b**). Red stars show the time stamps of five most significant changes of the trajectory.

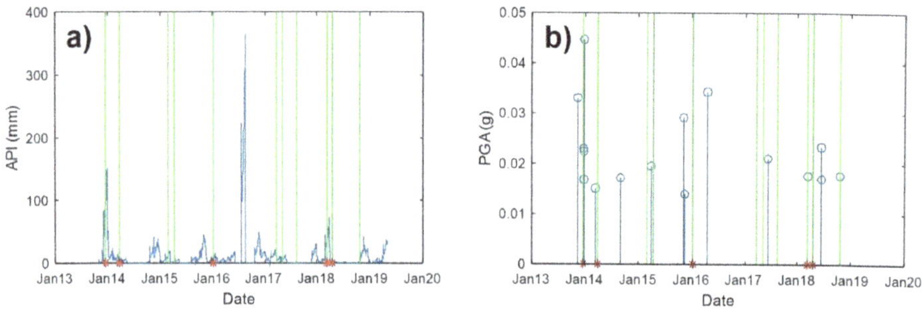

Figure 11. HIE3—Time stamps of change-points detected using the RDP algorithm vs. PGA (**a**) and vs. API (**b**). Red stars show the time stamps of five most significant changes of trajectory.

However, during most of the analyzed period, there is no clear correlation between change-points and seismic or rainfall activity. This can be caused by two reasons. Firstly, the seismic activity (PGA), which is presumed to be the main triggering factor of SAL movement [49], has ceased significantly since 2013, as is decreasing the average speed of movement of the SAL documented in this work; changes in the residuals are negligible. Secondly, the influence of precipitation (API) is probably negligible on the SAL behavior. Only three events were identified, mostly affecting the HIE3 device. This is probably caused by the presence of a shallower landslide block influenced by precipitation (see [32] for more details).

6. Conclusions

The proposed methodology allowed us to process a large amount of monitoring non-stationary data and to find critical time stamps—change-points, when the nature of the monitored landslide movement changed significantly. Moreover, it suppresses the influence of the scientist, who is always subjective and to some point tends to find what he is looking for.

The PARCS method detects change-points very well but is relatively hardware-demanding. It has to be noted, however, that the magnitude of the detected changes is close to the confidence limit of the measurements determined in laboratory conditions [62]. It is worth noting the significant change in the trajectory of motion detected on both measuring devices in December 2013 was influenced by both precipitation (at the beginning of December) and strong seismicity (24–26 December).

Despite not acquiring unambiguous results in terms of the geological interpretation, we believe that the proposed methodological approach can significantly contribute to the landslide monitoring data analysis of any slow-moving landslides in the future by both:

(i) decreasing the subjectivity of interpretation and
(ii) allowing fully quantitative analysis of the monitored data of a slow-moving landslide.

Future research should focus more on defining the influencing magnitude of the triggering factors to determine better the behavior of the SAL in the future.

Author Contributions: Conceptualization, J.B. (Jan Blahůt) and J.B. (Jan Balek); methodology, J.B. (Jan Blahůt) and J.B. (Jan Balek); formal analysis, J.B. (Jan Balek) and M.E.; investigation, J.B. (Jan Blahůt), S.M.; data curation, J.B. (Jan Blahůt), J.B. (Jan Balek) and M.A., writing—original draft preparation, J.B. (Jan Blahůt), J.B. (Jan Balek); visualization, J.B. (Jan Blahůt), J.B. (Jan Balek), writing—revision, J.B. (Jan Blahůt), J.B. (Jan Balek). All authors have read and agreed to the published version of the manuscript.

Funding: This research was funded by the Czech Science Foundation, grant number GJ16-12227Y and by the long-term conceptual development research organization RVO: 67985891.

Acknowledgments: The authors are indebted to María José Blanco and her staff at the Centro Geofísico de Canarias (IGN) for their help with installation and maintenance of the 3D dilatometer monitoring points on El Hierro and IGN for the open seismic data availability. OGIMET is acknowledged for providing daily meteorological data from the El Hierro Airport. Jana Šreinová and Monika Hladká are thanked for processing and analyzing the 3D dilatometer data. We also thank to the two anonymous reviewers who helped to improve the clarity of the manuscript.

Conflicts of Interest: The authors declare no conflict of interest. The funders had no role in the design of the study; in the collection, analyses or interpretation of data; in the writing of the manuscript, or in the decision to publish the results.

References

1. Corominas, J.; Moya, J.; Ledesma, A.; Lloret, A.; Gill, J.A. Prediction of ground displacements and velocities from groundwater level changes at the Vallcebre Landslide (Eastern Pyrenees, Spain). *Landslides* **2005**, *2*, 83–96. [CrossRef]
2. Van Asch, T.W.J.; Van Beek, L.P.H.; Bogaard, T.A. Problems in predicting the mobility of slow-moving landslides. *Eng. Geol.* **2007**, *91*, 46–55. [CrossRef]
3. Klimeš, J.; Rowberry, M.; Blahůt, J.; Briestenský, M.; Hartvich, F.; Košťák, B.; Rybář, J.; Stemberk, J.; Štěpančíková, P. The monitoring of slow-moving landslides and assessment of stabilisation measures using an optical–mechanical crack gauge. *Landslides* **2012**, *9*, 407–415. [CrossRef]
4. Carey, J.M.; Massey, C.I.; Lyndsell, B.; Petley, D.N. Displacement mechanisms of slow-moving landslides in response to changes in porewater pressure and dynamic stress. *Earth Surf. Dynam.* **2019**, *7*, 707–722. [CrossRef]
5. Singh, L.P.; Van Westen, C.J.; Champati Ray, P.K.; Pasquali, P. Accuracy assessment of InSAR derived input maps for landslide susceptibility analysis: A case study from the Swiss Alps. *Landslides* **2005**, *2*, 221–228. [CrossRef]
6. Schmidt, K.; Reimann, J.; Tous Ramon, N.; Schwerdt, M. Geometric accuracy of Sentinel-1A and 1B derived from SAR raw data with GPS surveyed corner reflector positions. *Remote Sens.* **2018**, *10*, 523. [CrossRef]
7. Sinharay, S. An Overview of Statistics in Education. In *International Encyclopedia of Education*, 3rd ed.; Peterson, P., Baker, E., McGaw, B., Eds.; Elsevier: Amsterdam, The Netherlands, 2010; pp. 1–11. [CrossRef]
8. Brillinger, D.R. Time Series: General. In *International Encyclopedia of the Social & Behavioral Sciences*, 1st ed.; Smelser, N.J., Baltes, P.B., Eds.; Elsevier: Amsterdam, The Netherlands, 2001; pp. 15724–15731. [CrossRef]
9. Lütkepohl, H. *New Introduction to Multiple Time Series Analysis*; Springer: Heidelberg, Germany, 2005; p. 764.
10. Neusser, K. *Time Series Econometrics*; Springer: Heidelberg, Germany, 2016; p. 409. [CrossRef]
11. Montillet, J.-P.; Bos, M.S. *Geodetic Time Series Analysis in Earth Sciences*; Springer: Heidelberg, Germany, 2020; p. 422. [CrossRef]
12. Mudelsee, M. *Climate Time Series Analysis*; Springer: Heidelberg, Germany, 2014; p. 454. [CrossRef]
13. Sun, L.; Muller, J.-P.; Chen, J. Time series analysis of very slow landslides in the Three Gorges Region through small baseline SAR offset tracking. *Remote Sens.* **2017**, *9*, 1314. [CrossRef]
14. Pfeiffer, J.; Zieher, T.; Rutzinger, M.; Bremer, M.; Wichmann, V. Comparison and time series analysis of landslide displacement mapped by airborne, terrestrial and unmanned aerial vehicle based platforms. *ISPRS Ann. Photogramm. Remote Sens. Spat. Inf. Sci.* **2019**, *IV-2/W5*, 421–428. [CrossRef]
15. Huang Lin, C.; Liu, D.; Liu, G. Landslide detection in La Paz City (Bolivia) based on time series analysis of InSAR data. *Int. J. Remote Sens.* **2019**, *40*, 6775–6795. [CrossRef]
16. Zvelebil, J.; Stemberk, J. Slope monitoring applied to rock fall management in NW Bohemia. In *Landslide Research, Theory and Practice: Proceedings of the 8th International Symposium on Landslides held in Cardiff on 26–30 June 2000*; Broomhead, E., Ed.; Thomas-Telford: London, UK, 2000; Volume 3, pp. 1659–1664.
17. Košťák, B. Deformation effects in rock massifs and their long-term monitoring. *Q. J. Eng. Geol. Hydrogeol.* **2006**, *39*, 249–258. [CrossRef]
18. Nikolakopoulos, K.; Kavoura, K.; Depountis, N.; Kyriou, A.; Argyropoulos, N.; Koukouvelas, I.; Sabatakakis, N. Preliminary results from active landslide monitoring using multidisciplinary surveys. *Eur. J. Remote Sens.* **2017**, *50*, 280–299. [CrossRef]

19. Guignard, F.; Laib, M.; Amato, F.; Kanevski, M. Advanced analysis of temporal data using Fisher-Shannon Information: Theoretical development and application in Geosciences. *Fron. Earth Sci.* **2020**, *8*, 255. [CrossRef]
20. Toutounji, H.; Durstewitz, D. Detecting multiple change points using adaptive regression splines with application to neural recordings. *Front. Neuroinform.* **2018**, *12*, 67. [CrossRef] [PubMed]
21. Alippi, C.; Camplani, R.; Galperti, C.; Marullo, A.; Roveri, M. An hybrid wireless-wired monitoring system for real-time rock collapse forecasting. In Proceedings of the 7th International Conference on Mobile Ad hoc and Sensor System (MASS), San Francisco, CA, USA, 8–12 November 2010; pp. 224–231.
22. Barile, G.; Leoni, A.; Pantoli, L.; Stornelli, V. Real-Time Autonomous System for Structural and Environmental Monitoring of Dynamic Events. *Electronics* **2018**, *7*, 420. [CrossRef]
23. Tordesillas, A.; Zhou, Z.; Batterham, R. A data-driven complex systems approach to early prediction of landslides. *Mech. Res. Commun.* **2018**, *92*, 137–141. [CrossRef]
24. Bontemps, N.; Lacroix, P.; Larose, E.; Jara, J.; Taipe, E. Rain and small earthquakes maintain a slow-moving landslide in a persistent critical state. *Nat. Commun.* **2020**, *11*, 780. [CrossRef]
25. Guillou, H.; Carracedo, J.-C.; Pérez Torrado, F.; Rodríguez Badiola, E. K-Ar ages and magnetic stratigraphy of a hotspot-induced, fast grown oceanic island: El Hierro, Canary Islands. *J. Volcanol. Geotherm. Res.* **1996**, *73*, 141–155. [CrossRef]
26. Becerril, L.; Ubide, T.; Sudo, M.; Martí, J.; Galindo, I.; Galé, C.; Morales, J.; Yepes, J.; Lago, M. Geochronological constraints on the evolution of El Hierro (Canary Islands). *J. Afr. Earth Sci.* **2016**, *113*, 88–94. [CrossRef]
27. Carracedo, J.-C.; Rodríguez Badiola, E.; Guillou, H.; de la Nuez, H.; Pérez Torrado, F. Geology and volcanology of the western Canaries: La Palma and El Hierro. *Estudios Geológicos* **2001**, *57*, 171–295. [CrossRef]
28. López, C.; Blanco, M.J.; Abella, R.; Brenes, B.; Cabrera Rodríguez, V.; Casas, B.; Domínguez Cerdeña, I.; Felpeto, A.; Fernández de Villalta, M.; del Fresno, C.; et al. Monitoring the volcanic unrest of El Hierro (Canary Islands) before the onset of the 2011–2012 submarine eruption. *Geophys. Res. Lett.* **2012**, *39*, L13303. [CrossRef]
29. Benito-Saz, M.; Parks, M.; Sigmundsson, F.; Hooper, A.; García-Cañada, L. Repeated magmatic intrusions at El Hierro Island following the 2011–2012 submarine eruption. *J. Volcanol. Geotherm. Res.* **2017**, *344*, 79–91. [CrossRef]
30. Meletlidis, S.; Di Roberto, A.; Domínguez Cerdeña, I.; Pompilio, M.; García-Cañada, L.; Bertagnini, A.; Benito-Saz, M.; Del Carlo, P.; Sainz-Maza Aparicio, S. New insight into the 2011–2012 unrest and eruption of El Hierro Island (Canary Islands) based on integrated geophysical, geodetical, and petrological data. *Ann. Geophys.* **2015**, *58*, S0546. [CrossRef]
31. Carracedo, J.-C.; Troll, V. *The Geology of Canary Islands*, 1st ed.; Elsevier: Amsterdam, The Netherlands, 2016; p. 622.
32. Blahůt, J.; Baroň, I.; Sokol, Ľ.; Meletlidis, S.; Klimeš, J.; Rowberry, M.; Melichar, R.; García-Cañada, L.; Martí, X. Large landslide stress states calculated during extreme climatic and tectonic events on El Hierro, Canary Islands. *Landslides* **2018**, *15*, 1801–1814. [CrossRef]
33. Masson, D. Catastrophic collapse of the volcanic island of Hierro 15 ka ago and the history of landslides in the Canary Islands. *Geology* **1996**, *24*, 231–234. [CrossRef]
34. Urgeles, R.; Canals, M.; Baraza, J.; Alonso, B. The submarine El Golfo debris avalanche and the Canary debris flow, west Hierro Island: The last major slides in the Canary Archipelago. *Geogaceta* **1996**, *20*, 390–393.
35. Urgeles, R.; Canals, M.; Baraza, J.; Alonso, B.; Masson, D. The most recent megalandslides on the Canary Islands: The El Golfo debris avalanche and the Canary debris flow, west El Hierro Island. *J. Geophys. Res. Solid Earth* **1997**, *102*, 20305–20323. [CrossRef]
36. Carracedo, J.-C.; Day, S.; Guillou, H.; Pérez Torrado, F. Giant quaternary landslides in the evolution of La Palma and El Hierro, Canary Islands. *J. Volcanol. Geotherm. Res.* **1999**, *94*, 169–190. [CrossRef]
37. Masson, D.; Watts, A.B.; Gee, M.J.; Urgeles, R.; Mitchell, N.; Le Bas, T.; Canals, M. Slope failures on the flanks of the western Canary Islands. *Earth Sci. Rev.* **2002**, *57*, 1–35. [CrossRef]
38. Longpré, M.; Chadwick, J.; Wijbrans, J.; Iping, R. Age of the El Golfo debris avalanche, El Hierro (Canary Islands): New constraints from laser and furnace 40Ar/39Ar dating. *J. Volcanol. Geotherm. Res.* **2011**, *203*, 76–80. [CrossRef]
39. Becerril, L.; Galve, J.; Morales, J.; Romero, C.; Sánchez, N.; Martí, J.; Galindo, I. Volcanostructure of El Hierro (Canary Islands). *J. Maps* **2016**, *12*, 43–52. [CrossRef]

40. León, R.; Somoza, L.; Urgeles, R.; Medialdea, T.; Ferrer, M.; Biain, A.; García-Crespo, J.; Mediato, J.; Galindo, I.; Yepes, J.; et al. Multi-event oceanic island landslides: New onshore-offshore insights from El Hierro Island, Canary archipelago. *Mar. Geol.* **2017**, *393*, 156–175. [CrossRef]
41. Blahůt, J.; Klimeš, J.; Rowberry, M.; Kusák, M. Database of giant landslides on volcanic islands—First results from the Atlantic Ocean. *Landslides* **2018**, *15*, 823–827. [CrossRef]
42. Blahůt, J.; Balek, J.; Klimeš, J.; Rowberry, M.; Kusák, M.; Kalina, J. A comprehensive global database of giant landslides on volcanic islands. *Landslides* **2019**, *16*, 2045–2052. [CrossRef]
43. Agliardi, F.; Crosta, G.; Zanchi, A. Structural constraints on deep seated slope deformation kinematics. *Eng. Geol.* **2001**, *59*, 83–102. [CrossRef]
44. Moscardelli, L.; Wood, L. New classification system for mass transport complexes in offshore Trinidad. *Basin Res.* **2008**, *20*, 73–98. [CrossRef]
45. Day, S.; Carracedo, J.-C.; Guillou, H. Age and geometry of an aborted rift flank collapse: The San Andres fault system, El Hierro, Canary Islands. *Geol. Mag.* **1997**, *134*, 523–537. [CrossRef]
46. Blahůt, J.; Mitrovic-Woodell, I.; Baroň, I.; René, M.; Rowberry, M.; Blard, P.-H.; Hartvich, F.; Balek, J.; Meletlidis, S. Volcanic edifice slip events recorded on the fault plane of the San Andrés Landslide, El Hierro, Canary Islands. *Tectonophysics* **2020**, *776*, 228317. [CrossRef]
47. Klimeš, J.; Yepes, J.; Becerril, L.; Kusák, M.; Galindo, I.; Blahůt, J. Development and recent activity of the San Andrés Landslide on El Hierro, Canary Islands, Spain. *Geomorphology* **2016**, *261*, 119–131. [CrossRef]
48. Blahůt, J.; Rowberry, M.; Balek, J.; Klimeš, J.; Baroň, I.; Meletlidis, S.; Martí, X. Monitoring giant landslide detachment planes in the era of big data analytics. In *Advancing Culture of Living with Landslides*; Mikoš, M., Arbanas, Ž., Yin, Y., Sassa, K., Eds.; Springer: Cham, Switzerland, 2017; Volume 3, pp. 333–340. [CrossRef]
49. Blahůt, J.; Olejár, F.; Rott, J.; Petružálek, M. Current stability modelling of an incipient San Andrés giant landslide on El Hierro Island, Canaries, Spain—First attempt using limited input data. *Acta Geodyn. Geomater.* **2020**, *17*, 89–99. [CrossRef]
50. Martí, X.; Rowberry, M.D.; Blahůt, J. A MATLAB® code for counting the moiré fringe patterns recorded on the optical-mechanical crack gauge TM-71. *Comput. Geosci.* **2013**, *52*, 164–167. [CrossRef]
51. Stemberk, J.; Briestenský, M.; Cacoń, S. The recognition of transient compressional fault slow-slip along the northern shore of Hornsund Fjord, SW Spitsbergen, Svalbard. *Pol. Polar Res.* **2015**, *36*, 109–123. [CrossRef]
52. Weather Information Service. Available online: http://www.ogimet.com/index.phtml.en (accessed on 6 June 2019).
53. Kohler, M.A.; Linsley, R.K., Jr. *Predicting Runoff from Storm Rainfall, Research Paper 34*; US Weather Bureau: Washington, DC, USA, 1951.
54. Mishra, S.K.; Singh, V.P. *Soil Conservation Service Number (SCS-CN) Methodology*; Kluwer Academic Publisher: Dordrecht, The Netherlands, 2003; p. 516.
55. Smolíková, J.; Blahůt, J.; Vilímek, V. Analysis of rainfall preceding debris flows on the Smědavská hora Mt., Jizerské hory Mts., Czech Republic. *Landslides* **2016**, *13*, 683–696. [CrossRef]
56. Viessman, W.; Lewis, G.L. *Introduction to Hydrology*, 4th ed.; Harper Collins: New York, NY, USA, 1996; p. 760.
57. Racek, O.; Blahůt, J.; Hartvich, F. Monitoring of thermoelastic wave within a rock mass coupling information from IR camera and crack meters: A 24-hour experiment on "Branická skála" Rock in Prague, Czechia. In *WLF5 Book—Volume 3 "Monitoring and Early Warning"*; Springer: Berlin/Heidelberg, Germany, 2019.
58. Instituto Geográfico Nacional. Available online: https://www.ign.es/web/ign/portal/sis-catalogo-terremotos (accessed on 6 June 2019).
59. Munson, C.G.; Thurber, C.H. Analysis of the attenuation of strong ground motion on the Island of Hawaii. *Bull. Seismol. Soc. Am.* **1997**, *87*, 945–960.
60. Gonzáles de Vallejo, L.I.; García-Mayordomo, J.; Insua, J.M. Probabilistic seismic-hazard assessment of the Canary Islands. *Bull. Seismol. Soc. Am.* **2006**, *96*, 2040–2049. [CrossRef]
61. Page, E.S. Continuous inspection schemes. *Biometrika* **1954**, *41*, 100–115. [CrossRef]
62. Balek, J.; Urban, R.; Štroner, M. Laboratory testing of the precision and accuracy of the TM-71 dilatometer. *Pap. SGEM* **2018**, *18*, 433–439. [CrossRef]
63. Visvalingam, M.; Whyatt, J.D. The Douglas-Peucker algorithm for line simplification: Re-evaluation through visualization. *Comput. Graph. Forum* **1990**, *9*, 213–228. [CrossRef]

64. Granger, C.W.J. Investigating causal relations by econometric models and cross-spectral methods. *Econometrica* **1969**, *37*, 424–438. [CrossRef]
65. Brasca Merlin, A.; Lanfri, M.; Carignano, C.; Pascual, I.; Schlögel, R.; Cuozzo, G. Sensado Remoto de procesos de remoción en masa: Pautas para el monitoreo operativo. In Proceedings of the 2018 IEEE Biennial Congress of Argentina (ARGENCON), San Miguel de Tucumán, Argentina, 6 June 2018; p. 8.

applied sciences

MDPI

Article

Integrated Field Surveying and Land Surface Quantitative Analysis to Assess Landslide Proneness in the Conero Promontory Rocky Coast (Italy)

Francesco Troiani [1], Salvatore Martino [1], Gian Marco Marmoni [1], Marco Menichetti [2], Davide Torre [1], Giulia Iacobucci [1] and Daniela Piacentini [2,*]

1 Department of Earth Sciences, Sapienza University of Rome, 00185 Rome, Italy; francesco.troiani@uniroma1.it (F.T.); salvatore.martino@uniroma1.it (S.M.); gianmarco.marmoni@uniroma1.it (G.M.M.); davide.torre1993@gmail.com (D.T.); giulia.iacobucci@uniroma1.it (G.I.)

2 Department of Pure and Applied Sciences, University of Urbino "Carlo Bo", 61029 Urbino PU, Italy; marco.menichetti@uniurb.it

* Correspondence: daniela.piacentini@uniurb.it

Received: 16 June 2020; Accepted: 8 July 2020; Published: 13 July 2020

Featured Application: Assessment of landslide proneness, with a particular emphasis on catastrophic landslide events.

Abstract: Rock slopes involved in extensive landslide processes are often characterized by complex morphodynamics acting at different scales of space and time, responsible for different evolutionary scenarios. Mass Rock Creep (MRC) is a critical process for long-term geomorphological evolution of slopes and can likewise characterize actively retreating coastal cliffs where, in addition, landslides of different typologies and size superimpose in space and time to marine processes. The rocky coast at the Conero promontory (central Adriatic Sea, Italy) offers a rare opportunity for better understanding the predisposing role of the morphostructural setting on coastal slope instability on a long-time scale. In fact, the area presents several landslides of different typologies and size and state of activity, together with a wide set of landforms and structural features effective for better comprehending the evolution mechanisms of slope instability processes. Different investigation methods were implemented; in particular, traditional geomorphological and structural field surveys were combined with land surface quantitative analysis based on a Digital Elevation Model (DEM) with ground-resolution of 2 m. The results obtained demonstrate that MRC involves the entire coastal slope, which can be zoned in two distinct sectors as a function of a different morphostructural setting responsible for highly differentiated landslide processes. Therefore, at the long-time scale, two different morphodynamic styles can be depicted along the coastal slopes that correspond to specific evolutionary scenarios. The first scenario is characterized by MRC-driven, time-dependent slope processes involving the entire slope, whereas the second one includes force-driven slope processes acting at smaller space–time scales. The Conero promontory case study highlights that the relationships between slope shape and structural setting of the deforming areas are crucial for reaching critical volumes to induce generalized slope collapse as the final stage of the MRC process. The results from this study stress the importance of understanding the role of morphostructures as predisposing conditions for generalized slope failures along rocky coasts involved in MRC. The findings discussed here suggest the importance of the assessment of the slope instability at the long time scale for a better comprehension of the present-day slope dynamics and its major implications for landslide monitoring strategies and the hazard mitigation strategies.

Keywords: coastal landslides; mass rock creep; coastal cliffs; land surface analysis; data analysis; Conero promontory

1. Introduction

Landslides of different typologies and size can occur along active retreating coastal cliffs, where the gravity-induced processes often superimpose, in space and time, to marine processes in shaping the coastal slopes. Rocky coasts represent about 80% of coasts worldwide [1], and they represent a great part of coasts around the Mediterranean basin [2]. Rocky coasts have been broadly studied so far, representing zones with high economic, social, cultural, and touristic values. In particular, many studies focused on the factors controlling erosional processes along both hard and soft rock coasts [3], whereas other studies focused on the present morphodynamics, with a particular emphasis on the role of active landslides on the present evolution of the coastline and the associated hazards [4,5]. Nonetheless, the understanding of the role of the slope shape and tectonic structures on the long-term morphoevolution of coastal slopes is crucial to enhance the awareness of slope instability on rocky coasts.

Complex morphodynamics, implicating multiscale morphoevolutionary scenarios, often characterize rock slopes affected by extensive landslide processes. Mass Rock Creep (MRC) *sensu* [6], has been demonstrated to represent a critical process for the long-term morphoevolution of rock slopes in different geological contexts [7–9]. Large disruptive landslides may represent the ultimate stage of the MRC process that can culminate in paroxysmal events (like rock avalanches); nevertheless, these ultimate events are anticipated by gravity-driven deformations which represent the strain effect of the rock mass viscous rheology. When these deformations develop into abrupt and generalized slope failures, it is caused by an increasing strain rate over time [10]. In particular, threshold conditions can be reached when the stationary creep stage evolves into an accelerating creep stage [11,12]. Generalized slope failures are typically associated to mass strength reduction due to rock mass damage, the latter occurring over time as an effect of viscous deformation [13,14]. Catastrophic failures that occur in the accelerating stage of MRC process can progress rapidly into rock avalanches and slides that are generally due to the fragmentation of the involved rock masses throughout their propagation and emplacement [15].

Coastal slope processes can be prone to evolve in paroxysmal events [16–18] or can display a continuous slow activity associated with MRC process [19–22]. Along rocky coasts, proneness to landslides needs to consider the relations between the cliff rheology and lithostructural setting, as well as external stresses [23,24], such as rainfall, earthquakes, or storm sea waves. Furthermore, considering the coastal slope instability at a longer time scale, changes in sea level, occurring over a wide range of space–time scales [25], should be taken into account.

The investigation of rocky coasts often represents a research challenge due to the accessibility of rock cliffs in all their portions. Remote sensing techniques from satellite imagery and aerial photos can be sometimes unhelpful due to cliff attitude and/or in the case of high-angle sloping coasts, as for example plunging cliffs. [26] proposes a methodological approach for the continuous acquisition of geological, geomorphological, hydrogeological, and biological datasets along rocky coasts based on a snorkeling approach. The latter is helpful for a detailed mapping of landforms and processes at the foot of the coastal slopes. [27] proposed a combined approach based on snorkeling-based and traditional fieldwork in the inland for a detailed mapping of the rocky coasts affected by gravity-induced slope processes.

The rocky coast at the Conero promontory (Central Adriatic Sea, Italy) presents several features that account for the predisposing role of morphostructural setting on coastal landslide processes involving extensive rock slopes [28,29]. Moreover, this area provides a wide set of geomorphic and tectonic elements that can be regarded as conditioning factors for the MRC process evolution [27].

The main goal of this study consists in unraveling the role of morphostructures as a controlling factor for the coastal morphodynamics and the implication of the latter on the gravity-induced slope processes that act at different spatial and time scales. Different long-term morphoevolutive scenarios are considered based on different geomorphological/structural settings and the possible triggering mechanisms, then their implications on the present-day slope instability are discussed.

2. Dataset and Methods

A combination of laboratory- and field-based techniques consisting of land surface quantitative analysis and geomorphological–structural surveys were performed. The topographic dataset, available for the study area in vector format at the scale of 1:10,000, was used as a basemap for the field surveys. These surveys were preliminarily supported by the interpretation of panchromatic orthophotos, available for the years 1988–1989 and 1994–1998 at www.pcn.minambiente.it as a web map service (WMS), together with panchromatic aerial photos available at the scale of 1:33,000 (year 1954) and 1:75,000 (year 1989). Land surface quantitative analysis was performed, starting from a Digital Elevation Model (DEM) with a ground resolution of 2 m and derived from the LiDAR (Light Detection and Ranging) dataset available for the Italian coastline at www.pcn.minambiente.it. A Geographic Information System (GIS) using the ESRI® ArcGIS 10.6 desktop software platform was used for storing geomorphological and geostructural data derived from the surveys, other than for managing the whole dataset and elaborating the quantitative analyses.

2.1. Geomorphological and Structural Survey

A series of geomorphological surveys, organized between the spring and summer of 2019, were carried out for corroborating previous surveys [27,30], which mainly focused on coastal processes and landforms, with particular attention on the detection of tidal notches and their morphoevolutive significance. The surveys performed here aimed at refining the recognition and mapping of coastal and gravity-induced processes and landforms, as well as at enlarging this detection along the entire coastal slope between the Portonovo and the Due Sorelle sites (Figure 1).

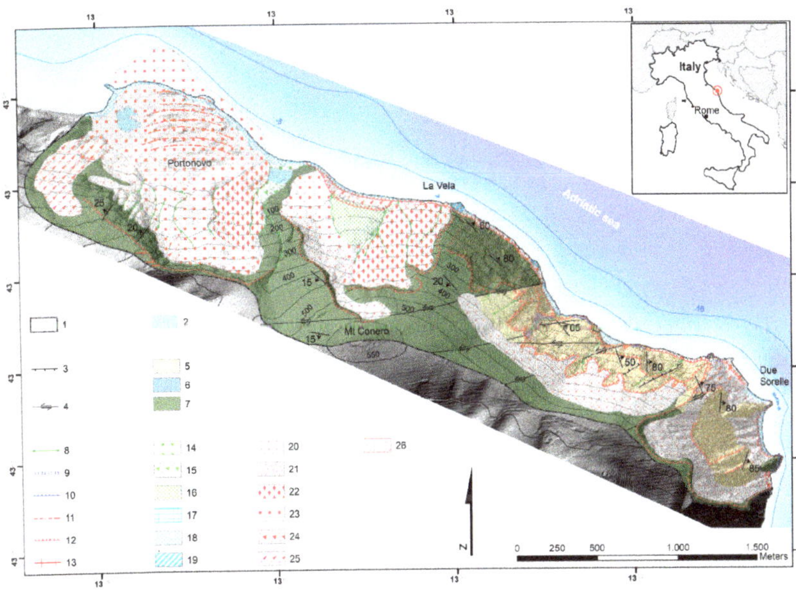

Figure 1. Geomorphological map of the Conero promontory showing the main geological and structural elements (map coordinates are in WGS84 cartographic reference system). Near-coast bathymetry (0–10 m below the sea level) is also indicated. 1: study area; 2: lake; 3: normal fault; 4: strike-slip fault; 5: Maiolica Fm.; 6: Marne a Fucoidi Fm.; 7: Scaglia Rossa Fm.; 8: gully erosion; 9: tidal notch; 10: plunging cliff; 11: landslide-toe compressional ridges; 12: rocky scarp; 13: landslide trench; 14: swamp; 15: alluvial fan; 16: stratified slope deposit; 17: sea stack; 18: beach; 19: shore platform; 20: slope debris and talus deposits; 21: debris flow; 22: debris avalanche; 23: rock avalanche; 24: rock fall; 25: rock slide; 26: diffuse gravity-driven slope deformations.

The detection and mapping of the processes and landforms were based on the approach proposed by the new Italian Geomorphological Mapping guidelines [31], which suggest mapping each landform based on its genesis, evolution, and present dynamics. The peculiar amount of landslide phenomena affecting the coastal slopes imposed special attention on the detection and mapping of the gravity-induced processes and landforms, and their complex evolutionary scenarios. Geomorphological fieldwork was preceded by the interpretations of the panchromatic orthophotos visualized in the GIS platform and by the interpretation of the historical aerial stereopair images, which were helpful for a preliminary delimitation of both large landslide bodies and debris deposits. Results of these preliminary tasks were also helpful for tracing the main structural lineaments. The optical satellite images available in Google Earth, together with the ones available in ArcGIS 10.6 (i.e., TerraColor dataset) were consulted for planning fieldwork and for supporting the mapping of the largest landslide features, other than for delimiting the main cliff escarpments. The multitemporal remote sensing analysis was also useful for defining the recent activity of the main landslides. After the geomorphological fieldwork, final refining of the geomorphological mapping was completed through the 3D visualization of the topography at high-resolution, thanks to a hillshade map derived from the 2 m-cell-sized DEM in the GIS platform.

Geostructural surveys, conducted together with the geomorphological ones, were carried out to detect and characterize the main tectonic structures that strongly control the rock mass behavior, as well as for refining the mapping of the lithological units composing the bedrock. The final aim was to unravel the connection between the local bedrock discontinuity frequency and distribution and the rock slope deformation and instability.

The fieldwork was conducted using digital survey techniques, based on mobile devices that allowed for collecting field data in digital form, and its direct storage through remote cloud-uploading and real-time updating.

2.2. Land Surface Quantitative Analysis

The objective classifications of terrain types, at the base of the quantitative geomorphic analysis, require the use of both local and statistical terrain parameters helpful for defining the topographic signature of different surface processes and landforms [32]. Nonetheless, given the strong variability of landforms in size and tridimensional shape, due to different morphogenetic factors, many terrain parameters have been proposed to provide a quantitative description of topography [32,33]. In this research, two main geomorphometric parameters were selected among numerous DEM-based terrain parameters, considering the specific research aim as well as the spatial scale of the analyses performed. The selected parameters were calculated starting from the available 2 m-cell-sized DEM.

The first parameter is the slope angle, derived by means of the surface slope algorithm running in the Spatial Analyst extension of the GIS platform. For each DEM cell, this algorithm computes the first derivative of the elevation surface, or the maximum change in elevation between a cell and its eight neighbors (i.e., the steepest downhill descent from the cell) [34].

The second parameter is the surface roughness that can be successfully used to define different landslide features and to examine the landslide activity [35,36]. Different methods have been proposed to compute the surface roughness [37–39], although one of the most used methods is simply based on the statistical dispersion of heights or slopes. In this work, the surface roughness was calculated as the standard deviation of the elevation values within a 4×4 moving window. This method, among many calculation methods revised by [36], was demonstrated as one of those useful for evidencing the topographic signature of large landslides. In order to obtain the surface roughness values, the focal statistics tool running in the ArcGIS 10.6 platform was adopted.

For supporting the spatial interpretation of both parameters, color-coded maps were produced and superimposed with a hillshade map, thus favoring the visualization of both terrain morphology and the spatial variation of parameters. Descriptive statistics for both parameters were calculated and reported as an inset for each map. Finally, for both parameters the number of pixels per bin (i.e., pixels in each parameter class) was reported using separate histograms for the southern and northern coastal sectors.

A composite altimetric profile parallel to the coastline was created, starting from the 2 m-cell-sized DEM. This composite profile reports the elevation distribution for progressive distance along three different tracks, thus allowing for a more accurate description of coastal morphology. The first track permits delineation of the cliff escarpment morphology (i.e., upslope zone); the second track favors the description of the morphology along the intermediate slope portion (i.e., midslope zone); and the third track portrays the base of the coastal slope, at about 50 m in from the present shoreline (i.e., footslope zone). The extraction and visualization of both slope angle and surface roughness values along two distinct transects support the study of the distribution of these parameters along different portions of the slope. This visualization increases the evidence of parameter value variation due to the presence of landslides of different type, size, and state of activity. The parameters were extracted along the midslope zone (i.e., at the +200 m a.s.l. contour line) and along the footslope (i.e., at the +20 m a.s.l. contour line).

3. Study Area

The Conero coastal sector stretches for about 5 km along the Adriatic Sea in the Central Italy, 10 km SE of the Ancona city, and represents its inflection point. Mt. Conero (43°33′04″ N, 13°36′18″ E) constitutes the highest relief along the Adriatic coast with an altitude of 572 m a.s.l. (Figure 1). The climate is of the Mediterranean type with mean annual precipitation of 780–790 mm and mean annual temperature of 14–15 °C. Summers are generally hot and dry, though a marked variability characterizes winters, mostly depending on the Atlantic cyclogenesis [40]. Prevailing winds approach from the SSE–SSW sectors [30] and account for the dominant SE main wave direction. During the winter months, sea-storm waves up to 3–3.5 m and with NE secondary wave direction can strike the coastline, during Bora wind conditions. The average tidal range in the Conero area is about 0.5 m and the action of tidal current is extremely low [27].

Concerning the geological setting, the Conero promontory consists of an asymmetric, NE–E verging anticline with NNW–SSE axial directions [29,41] and represents the easternmost outcropping part of the external stack of the Umbria–Marche foreland fold and thrust belt.

Limestones and marly calcareous terrains compose the bedrock of the study sector of the Conero promontory and are ascribable to the Umbria–Marche stratigraphic succession, from Cretaceous to Oligocene [29]. The complete stratigraphic sequence consists of the following formations, from oldest to youngest: Maiolica Formation (Fm.) (MA; Lower Cretaceous; prevalently composed of decimetric well-stratified limestones with chert and black shales), Marne a Fucoidi Fm. (MF; Lower Cretaceous; prevalently composed by centimetric marls and marly calcareous layers with black shales), and Scaglia Rossa Fm. (SR; Upper Cretaceous; composed of alternating and well-stratified centimetric to metric limestones, calcarenites, biocalcarenites and marly calcareous layers). In the area, this stratigraphic sequence presents some peculiarities with respect to the outcropping in the internal area of the Apennine chain. In particular, the thickness of the MF is about 15 m, whereas the regional thickness is generally up to 80 m. Furthermore, the paraconformity erosive contact between the MF and SR formations is likely caused by syn-sedimentary slumps [42]; in fact, in the area, the Scaglia Bianca Fm. together with a part of the basal portion of the Scaglia Rossa Fm. is totally absent [27]. The western side of the Mt. Conero anticline is gently dipping (around 25°) and affected in the inner part by N–S right-lateral transtensive strike-slip faults with an offset of hundreds of meters. The eastern, coastward flank of the Conero asymmetric anticline is strongly inclined and is intersected by several transversal left-transtensive faults with an average E–W direction and offset a few km (Figure 1). The shear zones are characterized by cataclastic deformation, often with calcite veins and a thickness up to a few meters. The joint pattern is well developed in the rock masses with three prevalent directions NE–SW, NW–SE, and E–W. A pervasive cleavage with subvertical SW dipping planes is observable especially in the Maiolica and in the calcareous Scaglia Rossa bedding. Subhorizontal diagenetic stylolites are quite common in the Maiolica Fm. Subvertical joints are developed mainly in the Scaglia Rossa Fm., with systems striking NW–SE and ENE–WSW, associated with breccias in related shear zones and systematic calcite veins.

The morphogenesis along the coastal slope is mainly conditioned by gravity-induced and marine processes, with the corresponding landforms and deposits (Figures 1 and 2). Landslides of different typologies, size, and state of activity are widespread in the area and principally characterize the entire coastal slope morphodynamics (Figure 1). Among the main mass movements, the landslides occurring in the northern coastal sector, at Portonovo and La Vela, are the largest ones (Figures 1 and 3A) [29]. Along the coastline, landslide deposits and plunging cliffs interrupt the poorly developed beaches and pocket beaches. The latter mainly consist of rocky blocks and gravels. Flat shore platforms, slightly inclined seaward, are also observable. Typical sea stacks at La Vela and Due Sorelle are located at short distances from the coastline (Figure 2). The current tidal notch occurs around these sea stacks [27]. The absence of either hanging or uplifted tidal notches, together with the absence of Tyrrhenian marine deposits and terrace remnants, testify of an active and intense morphodynamic in the entire coastal slope [27], mainly due to gravity-induced processes and the interrelated factors, such as geological factors or climate variations [43]. Several anthropic structures and man-made landforms, such as a local harbor, some touristic infrastructure, and parking areas, appear at the Portonovo landslide toe, close to the coastline.

Figure 2. Main landforms due to coastal processes at the foot of the cliff. (**A**) Shore platform along the central sector of the study area. Calcareous boulders due to recent rock falls are visible in the emerged portion of the platform. (**B**) Due Sorelle sea stacks and the adjacent pocket beach along the southern coastal sector.

Figure 3. Landforms and deposits due to gravity along the northern portion of the Conero rocky coast. (**A**) Panoramic view, towards the south, of the Portonovo and La Vela rocky coast (northern coastal sector). (**B**) Toe area of the Vela rockslide. In this sector, a late-glacial stratified slope deposit (i.e., èboulis ordonneè) overlays the landslide deposit. (**C**) Rock avalanche deposit at the toe of the Portonovo landslide exposed along a wave-cut scarp. The inset shows chaotic debris prevalently belonging to the Scaglia Rossa Fm. (SR) and incorporating a marly calcareous boulder belonging to the Marne a Fucoidi Fm. (MF). (**D**) Ridge-top depression in the Portonovo landslide backscarp area.

4. Results

The Conero promontory provided a set of data useful for better defining the morphoevolutive implications of the gravity-induced processes and landforms along the coastal zone. The entire coastal slope is affected by landslides of different typologies, which account for erosional landforms at the upslope zone and abundant landslide deposits along the mid- and footslope sites (Figure 1, Figure 2A, Figure 3, and Figure 4). Landslides are predisposed by low persistency of rock mass joints and by strata attitude whose dip is very close to the slope face direction.

The largest mass movements surveyed in the area occur in the northern coastal sector, between the Portonovo village and the La Vela sea stack (Figure 1), where two adjacent, large landslides masses can be observed and ascribable to rock-avalanche/slide mechanism [44], as they are associated with wide detachment scars that typify the sharp edge of the coastal cliff and delimit seaward the calcareous ridge top (Figures 1 and 3A). The northern landslide (Portonovo landslide) shows a runout of about 500 m that flowed seaward, accounting for the shaping of a bay where Portonovo village developed. A series of typical compressional ridges characterize the distal zone of this mass movement. The southern landslide (La Vela landslide), on the contrary, shows a truncated toe, though observing the trend and geometry of the contour lines down to −10 m b.s.l., the nearshore morphology of the seafloor indicates that a terminal landslide lobe is still detectable (Figure 1).

Figure 4. Landforms and deposits due to gravity along the southern portion of the Conero rocky coast. (**A**) Panoramic view of the Due Sorelle cliff. Falls and slides of rock blocks occur in the uppermost cliff portions, characterized by a sharp subvertical escarpment. Bedding attitude coincides with topography along the intermediate slope portion. At the cliff base, thick talus deposits occur, often interested by small debris flows and falls. (**B**) Evidence of a rock-fall detachment zone along the main cliff escarpment.

Patches of a thin layer of stratified slope-waste deposits form outcrops along the cliff base (Figures 1 and 2B) and unconformably overlay the chaotic landslide deposits. Stratified slope deposits are widespread within the Umbria–Marche Apennines, even along the coastline [45], principally where bedrock particularly prone to frost shattering, as Maiolica and Scaglia Rossa formations, is exposed. These deposits typify cold, periglacial environments [46] and represent in the Apennine area the termination of the last glacial slope–fluvial sedimentation cycle [47], thus their chronological collocation is at the end of Late Pleistocene [48].

A characteristic wave-cut scarp parallel to the present shoreline delimits the toe of both the landslide masses and provides suitable exposures of the deposits that here are those typical of rock avalanche/slide phenomena: disrupted rock masses, including rock blocks; chaotic calcareous debris with sandy clay; and a subordinately clay matrix (Figure 3B,C). Rock blocks and calcareous debris belong to the Scaglia Rossa Fm. and subordinately to the Marne a Fucoidi Fm., confirming that both bedrock formations are involved in the mass movement. At the toe of the Portonovo landslide, a deposit belonging to a shallow reactivation of the main landslide body uncomfortably overlays the rock avalanche/slide deposit (Figure 3C). Along the backscarp of both landslides, a series of trenches and ridge-top depressions mark the releasing zone (Figure 3D).

A plunging cliff characterizes the southern coastal sector, whose upslope and midslope zones are generally affected by rock-block slides and falls that strongly contribute to the cliff regression (Figure 4). A wide active talus deposit occurs at the footslope, often affected by secondary debris flows and fall phenomena. The main instability of the coastal cliff is strictly controlled by lithostructural factors, and just a few and limited zones along the cliff are subjected to the prevailing action of undercutting due to the sea actions.

Location and scale of gravity-induced instability processes appear related to the structural settings of the Mt. Conero anticline; in the southern sector, NE-dipping, high-angle bedding conditions the rock mass stability, providing the kinematic arrangement for planar and wedge-sliding, which seasonally affect the coastal slopes by mobilizing block volumes from tens up to hundreds of cubic meters. The repeated rock-block instabilities feed the slope system with a large amount of detritic material, which is mobilized by channelized flows after intense rainfalls.

The northern sector is instead featured by an E–ENE dip direction of bedding, resulting from fold-axis plunging and periclinal folding, as identifiable in the east escarpment of the Portonovo landslide. In this area, high-angle preorogenic transtensional faults [41] dissected the anticline forelimb, providing the inherited passive framework where slope-scale deformation evolved [49,50].

Evidence of slope-scale, gravity-driven processes occur along the entire coast where mesoscale features of MRC processes were surveyed (Figure 5).

Figure 5. Evidence of ongoing gravity-induced deformations along the coastal slope due to the MRC process. (**A**) Buckling folds affecting the calcareous layers at the Vela cliff. (**B**) Flexural toppling south of the Vela cliff. (**C**) Rock fall due to a buckling fold breakout. (**D**) Flexural block-toppling at the Due Sorelle cliff.

The structural observations indicate macroscopically ductile features of slope-scale flexural toppling along a steeply dipping and narrow-spaced primary discontinuity set, despite the presence of horizontal cross joints. The latter act as a breakout surface, driving the relative rotation of blocks with respect to a defined hinge surface [51]. These observations lead to a challenging distinction between flexural-toppling and brittle block-toppling [52], thus suggesting a transitional behavior strongly dependent on local lithological and joint density properties of the rock masses. Buckling folds also affect the calcareous bedding at the Vela cliff, causing tension and shearing fractures that can enhance the occurrence of rockfalls.

Along the analyzed coastal slope, the slope angle parameter shows a wide range of values (i.e., 0–88°) and its spatial distribution is good evidence for the subvertical geometry of the upslope zone of the cliff (Figure 6A). In particular, the slope map reveals a very good match between the maximum values and the main rock-block detachment zones and the large landslide scar areas, respectively, in the southern and northern coastal sector. In the latter sector, the lowest slope angle values also appear at the foot of the cliff, delimiting the Portonovo landslide toe. Statistics show the prevalence of the highest slope values in the southern sector (Figure 6B), while within the northern sector, the values are distributed heterogeneously (Figure 6C) with a prevalence of the lowest values at the foot of the cliff (Figure 6A,C).

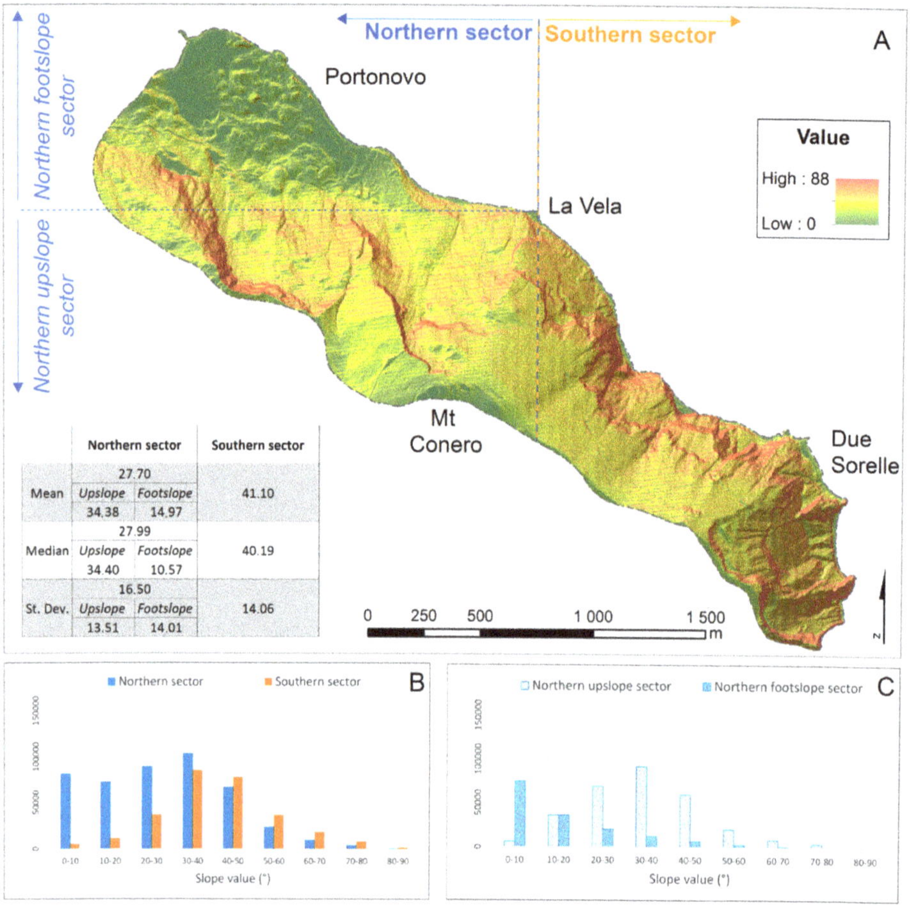

Figure 6. Spatial distribution of the slope values in the study area, based on a LiDAR-derived elevation dataset with ground resolution of 2 m. (**A**) Color-coded slope map overlain with a hillshade map, including a table with descriptive statistics. (**B**) Histogram with the distribution of slope values within the southern and northern sectors of the study area (y-axis reports the number of pixels per bin). (**C**) Histogram with the distribution of slope values within the footslope and upslope portions of the northern sectors of the study area (y-axis reports the number of pixels per bin).

The surface roughness parameter shows values up to 33 m (Figure 7). The highest values occur along the main rocky scarp, whereas the lowest ones occur in more homogeneous, flat or smoothed zones. The zones with the highest roughness coincide well with the ones with the highest slope values, along the main detachment areas for the rock-block falls. As for the slope parameter, even the surface roughness permits delineation of two almost homogenous zones along the coastal slope. The southern sector shows the largest values, while average small values are recorded in the northern sector (Figure 7B). More in detail, within the northern sector two distinct zones can be distinguished, based on surface roughness (Figure 7C). The first zone, involving the entire body of the La Vela landslide and the upper zone of the Portonovo landslide, rests within the base of the scar area downhill to the upper termination of the toe. The second zone coincides with the Portonovo landslide toe. This particular configuration is easily recognized along the transect at +200 m a.s.l. (Figure 8). This latter zone, along its northern portion shows similar values that range between 3 and 6 m; more in detail, the values are the same where the transect crosses both the La Vela and Portonovo landslide bodies (Figures 1, 3A and 9). Along the transect at +20 m a.s.l., on the contrary, the roughness values are dissimilar for the toe areas of the two adjacent landslides, with the La Vela foot zone showing the same values as the uppermost transect and the Portonovo footslope zone showing lesser values. The highest roughness values occur along the transect at +20 m a.s.l. along the southern coastal sector, marking well the topographic setting of the base of the active retreating cliff.

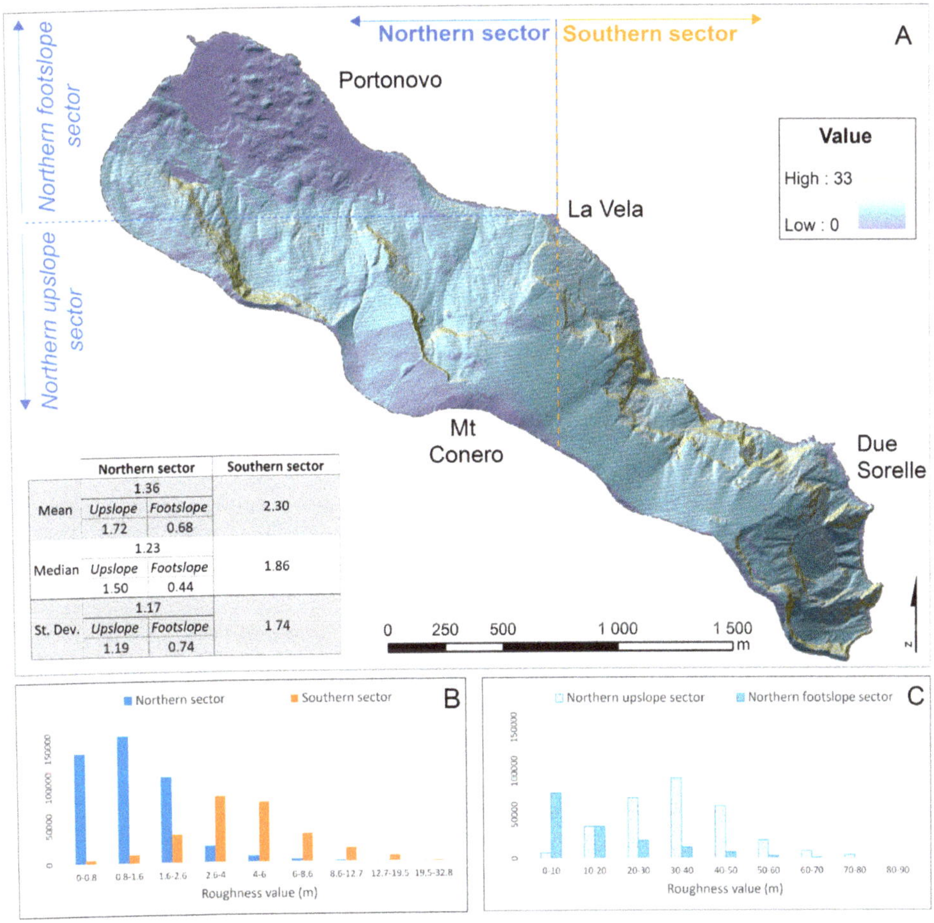

Figure 7. Spatial distribution of the surface roughness values in the study area. (**A**) Color-coded surface roughness map overlain with a hillshade map, including a table with descriptive statistics. (**B**) Histogram with the distribution of surface roughness values within the southern and northern sectors of the study area (y-axis reports the number of pixels per bin). (**C**) Histogram with the distribution of surface roughness values within the footslope and upslope portions of the northern sectors of the study area (y-axis reports the number of pixels per bin).

The quantitative analyses performed by means of altimetric profiles in the coastline direction improved the interpretation of the outcomes described above, as well as allowed the definition of the different topographic and geometric characteristics of the analyzed southern and northern sectors, and the ones of the upslope and midslope zones (at +200 m a.s.l.), and the footslope (at 50 m from the present shoreline).

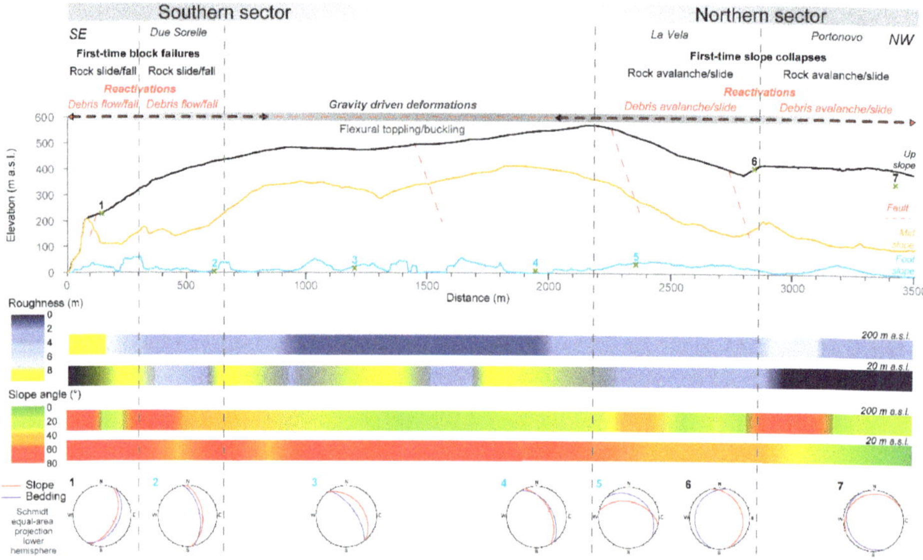

Figure 8. Composite altimetric profile of the entire coastal slope. The three superimposed profiles illustrate the morphology of the upslope, midslope, and footslope zones (the latter at 50 m in from the present shoreline). For the principal geomechanics measurement positions, the stereo plots illustrate the bedding attitude with respect to the slope geometry (topographic steepest descent). Below the profiles, the variations of both slope angle and surface roughness along the two distinct transects at the elevation of +20 and +200 m a.s.l. are reported.

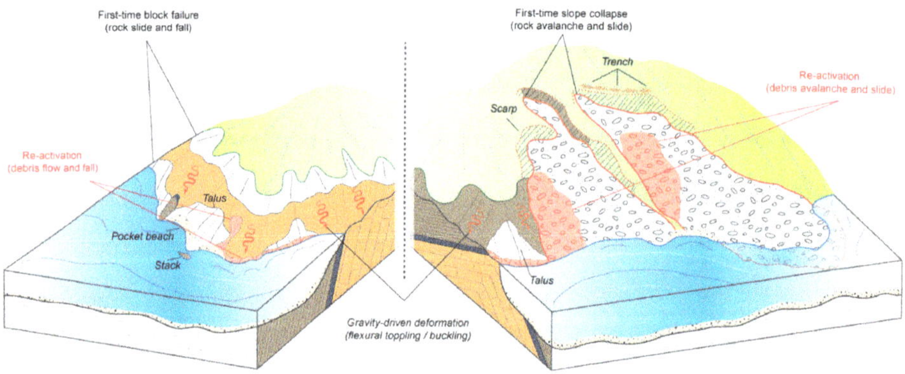

Figure 9. Conceptual landscape model. The 3D sketch illustrates the different morphodynamic styles characterizing the southern (left sketch) and northern (right sketch) coastal sectors. Morphostructural settings strongly influence the types and development of different gravity-induced processes, as well as diverse landform associations and their spatial organization and evolution at different scales. Slope-scale, MRC-driven processes dominate the long-term morphoevolution of the northern coastal sector, whereas the southern sector is generally characterized by cliff-scale, force-driven processes. See Figure 10 for further details.

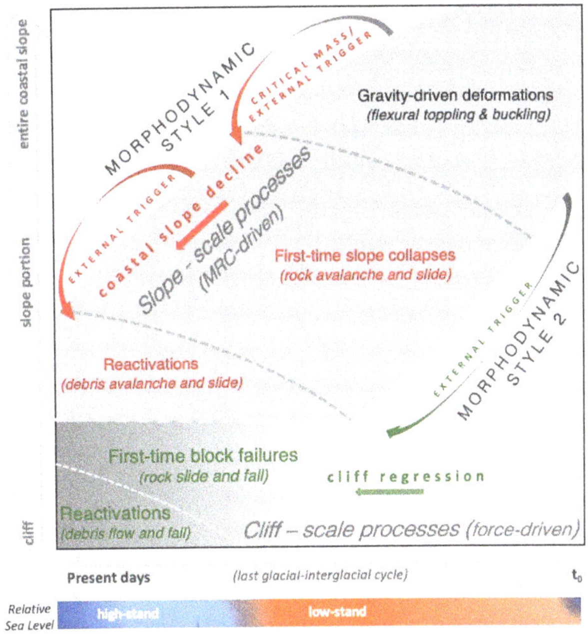

Figure 10. Conceptual space–time evolutive model for the active retreating coastal slope along the Conero promontory. Starting from the same process affecting the entire coastal slope, different long-term morphoevolutive scenarios are expected, depending on two different morphodynamic styles. The first scenario refers to slope-scale, time-dependent processes, while the second one refers to cliff-scale, time-independent (i.e., force-driven) processes. Both scenarios coexist along the analyzed coastal slope. The first one is typical of the northern coastal sector, where it favors active coastal slope downward erosion; the second scenario prevails along the southern coastal sector, where it accounts for active cliff regression.

The distribution of the slope parameter along the transect at +200 m a.s.l. (Figure 8) displays similar values, generally not exceeding 40–45° with two major exceptions at the Due Sorelle cliff, in the southern sector, and at the right flank of the Portonovo landslide, in the northern sector. Except for the Portonovo landslide toe zone (Figure 3A), where the average slope values are below 20°, along the transect at +20 m a.s.l. the average slope values are >40–45° (Figure 8), highlighting the presence of a high-angle or subvertical active retreating cliff (Figures 2A and 9). The +20 m transect crosses either the base of the rocky slope or the talus deposits at the footslope in those sectors where it occurs (Figure 4).

5. Discussion

Two distinct sectors can be recognized along the Conero slope, typified by convex, dip-slopes [53]. The first one includes the northern sector embracing the Portonovo and La Vela zones, whereas the second sector encompasses the whole southern portion of the coastal slope at the Due Sorelle sea stacks. The two slope portions, featured by the same geological formations, show many differences in both topographic (Figure 8) and structural settings (Figures 8 and 9). Consequently, these sectors experienced different morpho-evolutive styles due to gravity-induced processes acting at different space- and time scales (Figure 10). In the northern portion of the area, at the slope scale, massive rock slope failures and their related scar areas mainly characterize the landscape of the coastal area, whereas cliff-scale, rock fall, and rock slide typify the landforms along the southern sector.

It is reliable to assume that a first-time massive rock slope failure, representing the generalized slope collapse as the final stage of the MRC process, characterizes the entire slope encompassing both

Portonovo and La Vela localities with an estimated volume of about 40 × 10^6 m^3. The latter likely developed in at least two steps (i.e., multistep first-time event): the first one involving the whole coastal sector and the second one only the Portonovo coastal slope. Both events likely occurred during the last glacial time. The lack of remnants of the marine terrace representing the Middle Pleistocene highstand [27] and the presence of the periglacial deposits sealing the main landslide bodies both support this assumption. Moreover, the relative freshness of the landslide-related erosional and depositional landforms allows exclusion of the collocation of the mass movements during earlier glacial cycles. The spatial distribution of surface roughness values along the northern coastal sector seems to confirm that the two landslide bodies belong to the same first-time event, and that only the Portonovo area experienced a second massive rock slope failure event. In fact, the La Vela landslide body shows nearly the same roughness values as the upslope portion of the Portonovo landslide, indicating a similar postevent remodeling, hence the same timing. Both landslide bodies were partially reactivated during historical time (mainly debris avalanche/slide phenomena) and field evidence confirms this observation, especially at the Portonovo locality. At this location, a catastrophic large landslide event occurred during the Middle Ages, destroying the local harbor and an ancient Benedictine Abbey [29].

From an evolutive point of view, the Conero coastal slope can be considered a typical palimpsest landscape (*sensu* [54]), where landforms belonging to the same process that make up the landscape (the gravitational one in the present case) developed at different spatial scales, have different age and, at present, display dissimilar dynamics. In fact, the nested landslide system that characterize the Portonovo and La Vela coastal slope likely started to develop during the beginning of the last glacial time, under cold environmental conditions and relatively low sea level (only the surficial landslide reactivations, though catastrophic, developed under the present environmental conditions). This morphodynamic style accounts for the down-wearing evolution of the rocky coast, with the upslope portion eroding downward as the footslope receives the supply of upslope sediments and advances seaward (Figures 8 and 9). On the contrary, the southern sector is a typical active plunging cliff, whose morphodynamic style, though dominated by the gravity-induced processes, is in part actively related to the coastal processes due to continuous sea wave action, tidal range, and storm waves, in particular along the footslope portions. Rock-block slides and rockfalls dominate the upslope portion accounting for the upper escarpment retreat, whereas at the base of the coastal slope, in addition to rock-block falls and slides, debris flows and falls also account for the cliff retreat. Several earthquake-induced landslides were reported for this portion of the Conero promontory [55], often triggered by the low-magnitude, high-frequency seismic events that typically characterize the seismicity of the external portions of the Apennine chain.

It is assumed that the dominant role of the different morphoevolutive styles along the Conero coastal slope is strictly related to the different structural settings (Figures 8 and 9) that account for the rather different development of MRC processes.

A double scale morphodynamic conceptual scheme is proposed in Figure 10, where time is reported on the x-axis and space on the y-axis. According to the time scale, the processes considered evolve from gravity- to force-driven during recent times, as these processes involve more limited portions of slopes, causing smaller size landslide events. These last events can be triggered by time-independent local and transient forcing, such as sea storms and earthquakes, while the larger size rock mass failures (i.e., rock avalanches and slides), which can be observed in the northern sector of the Conero promontory, are substantially triggered by inertial forces as they result from MRC, causing viscous deformations over time (red arrows in the conceptual scheme, Figure 10). For a higher dip of strata attitude, the MRC deformations (occurring at the entire coastal slope scale) do not envelop critical volumes that can become prone to failure over a longer time [7]; in this last condition, the morphoevolution results in several small-size failures (occurring at the cliff scale), which are represented principally by the rock-falls and slides in the southern sector of the Conero promontory (green arrows in the conceptual scheme, Figure 10). Reactivations are also possible, after initial mass rock failures, due to debris avalanche/slide or debris flow occurrence. Concerning the climatic context and referring to the last glacial–interglacial cycle, the low-stand phase corresponded to the gravity-driven deformations evolving toward failures, while the high-stand phase mainly corresponds to the cliff-scale processes. The implications of this conceptual model in terms of risk are related to a higher probability of occurrence for the cliff-scale processes, acting at an entire slope scale, which represent the most hazardous ones with respect to the gravity-driven MRC deformations and failure.

The coexistence of large-scale MRC deformational processes and small-scale, time-independent failures is clearly visible in the Due Sorelle sector (southern portion of the Conero promontory) where observation of flexural toppling and buckling demonstrate that progressive rock mass deformations rapidly evolve toward failure (i.e., either falls or slides), thus avoiding larger volumes involving creep processes that reach the critical size for a generalized slope collapse.

Traditional field-based surveying and mapping with quantitative geomorphic analyses based on a high-resolution digital elevation dataset were integrated to represent a more suitable tool for better defining the morphostructural setting of the area under investigation, even in those sites where the field surveys were difficult to perform. DEM-based quantitative land surface analysis, thanks to different visualization techniques and the computation of two geomorphometric parameters, revealed a high-performance mapping methodology that allows for rapid completion of the time-consuming traditional techniques used in refining the survey and mapping slope processes and landforms due to gravity.

Results of this research can be valuable for coastal landslide analysis and propose a conceptual scheme for assessing and monitoring the current dynamics of slopes affected by mass movements of different typologies and size. In particular, this research reveals the importance of the evaluating the long-term morphoevolution of the entire slope for better comprehending the present activity and future development of the coastal landslides.

6. Conclusions

Based on field evidence, MRC processes involve the entire coastal slope along the Conero promontory. The coastal slope can be zoned into two distinct sectors, depending on different morphostructural settings and consequent rock slope instability processes. For the long timescale, two separate morphodynamic styles can be delineated along the coastal slope, which are each associated with specific evolutionary scenarios. The first scenario represents MRC-driven, time-dependent slope processes acting at the slope scale, whereas the second one includes force-driven, time-independent slope processes acting at smaller space–time scales. First-time rock slope failures (i.e., rock avalanche/slide mechanism) characterize the northern area of the promontory, which is interpreted as the result of extensive, multistep catastrophic failure that occurred at the final stage of the MRC process. Shallower landslides (i.e., debris avalanche/slide) can locally reactivate the main landslide masses. First-time rock-block slides and falls, occasionally triggered by earthquakes and/or storm sea waves, occur mainly along the southern sector of the coastal slope where they account for the active cliff retreat.

The present study highlights that slope shape and geostructural conditions of deforming slopes are concurrent in allowing the critical volumes for failures to be reached, thus inducing generalized slope collapse that can be regarded as the ultimate stage of MRC processes.

The findings discussed here also confirm the relevance in understanding the role of geomorphological vs. structural features as predisposing factors for slope collapse along rocky coasts. Moreover, the assessment of long-term slope instabilities allows for a better comprehension of the present-day, gravity-induced slope dynamics, suitably supporting landslide monitoring strategies and appropriate mitigation strategies even based on stress-strain time-dependent numerical analyses.

Author Contributions: Conceptualization, F.T. and S.M.; methodology, F.T. and S.M.; validation, M.M.; laboratory analysis, D.T. and G.M.M.; field investigation, F.T., D.P., M.M. and D.T.; data curation, G.I.; writing—original draft preparation, F.T., S.M., G.M.M. and D.P.; visualization, D.P. and G.I. All authors have read and agreed to the published version of the manuscript.

Funding: This research received no external funding.

Conflicts of Interest: The authors declare no conflict of interest.

References

1. Sunamura, T. *Geomorphology of Rocky Coasts*; Wiley: New York, NY, USA, 1992.
2. Furlani, S.; Pappalardo, M.; Gomez-Pujol, L.; Chelli, A. The rocky coasts of the Mediterranean and Black Sea. In *Rock Coast Geomorphology: A Global Synthesis*; Kennedy, D.M., Stephenson, W.J., Naylor, L.A., Eds.; Geological Society: London, UK, 2014; Memoirs 40; pp. 89–123.
3. Miccadei, E.; Mascioli, F.; Ricci, F.; Piacentini, T. Geomorphology of soft clastic rock coasts in the mid-western Adriatic Sea (Abruzzo, Italy). *Geomorphology* **2019**, *324*, 72–94. [CrossRef]
4. Soldati, M.; Maquaire, O.; Zezere, J.L.; Piacentini, D.; Lissak, C. Coastline at risk: Methods for multi-hazard assessment. *J. Coast. Res.* **2011**, *61*, 335–339. [CrossRef]
5. Gibson, A.D.; Culshaw, M.G.; Dashwood, C.; Pennington, C.V.L. Landslide management in the UK—The problem of managing hazards in a 'low-risk' environment. *Landslides* **2013**, *10*, 599–610. [CrossRef]
6. Chigira, M. Long-term gravitational deformation of rocks by mass rock creep. *Eng. Geol.* **1992**, *32*, 157–184. [CrossRef]
7. Esposito, C.; Martino, S.; Scarascia Mugnozza, G. Mountain slope deformations along thrust fronts in jointed limestone: An equivalent continuum modelling approach. *Geomorphology* **2007**, *90*, 55–72. [CrossRef]
8. Della Seta, M.; Esposito, C.; Marmoni, G.M.; Martino, S.; Scarascia Mugnozza, G.; Troiani, F. Morpho-structural evolution of the valley-slope systems and related implications on slope-scale gravitational processes: New results from the Mt. Genzana case history (Central Apennines, Italy). *Geomorphology* **2017**, *289*, 60–77. [CrossRef]
9. Delchiaro, M.; Della Seta, M.; Martino, S.; Dehbozorgi, M.; Nozaem, R. Reconstruction of river valley evolution before and after the emplacement of the giant Seymareh rock avalanche (Zagros Mts., Iran). *Earth Surf. Dynam.* **2019**, *7*, 829–947. [CrossRef]
10. Vick, L.M.; Böhme, M.; Rouyet, L.; Bergh, S.G.; Corner, G.D.; Lauknes, T.R. Structurally controlled rock slope deformation in northern Norway. *Landslides* **2020**. [CrossRef]
11. Saito, M. (1969) Forecasting time of slope failure by tertiary creep. In Proceedings of the 7th International Conference on Soil Mechanics and Foundation Engineering, Mexico City, Mexico, 29 August 1969; pp. 677–683.
12. Petley, D.N.; Allison, R.J. The mechanics of deep-seated landslides. *Earth Surf. Proc. Land.* **1997**, *22*, 747–758. [CrossRef]
13. Eberhardt, E.; Stead, D.; Coggan, J.S. Numerical analysis of initiation and progressive failure in natural rock slopes-the 1991 Randa rockslide. *Int. J. Rock Mech. Min.* **2004**, *41*, 69–87. [CrossRef]
14. Stead, D.; Eberhardt, E.; Coggan, J.S. Developments in the characterization of complex rock slope deformation and failure using numerical modelling techniques. *Eng. Geol.* **2006**, *83*, 217–235. [CrossRef]
15. Davies, T.R.; McSaveney, M.J. The role of rock fragmentation in the motion of large landslides. *Eng. Geol.* **2009**, *109*, 67–79. [CrossRef]
16. Qin, S.; Jiao, J.J.; Wang, S.A. Cusp Catastrophe Model of Instability of Slip-buckling Slope. *Rock Mech. Rock Eng.* **2001**, *34*, 119–134. [CrossRef]
17. Blikra, L.H.; Longva, O.; Harbitz, C.; Løvholt, F. Quantification of rock avalanche and tsunami hazard in Storfjorden, western Norway. In *Landslide and Avalanches*; Sennest, K., Flaate, K., Larsen, J.O., Eds.; Taylor and Francis Group: London, UK, 2005; pp. 7–63.
18. Bornhold, B.D.; Thomson, R.E. Tsunami hazard assessment related to slope failures in coastal waters. In *Landslides-Types, Mechanisms and Modeling*; Clague, J.J., Stead, D., Eds.; Cambridge University Press: Cambridge, UK, 2012; pp. 108–120.
19. Della Seta, M.; Martino, S.; Scarascia Mugnozza, G. Quaternary sea-level change and slope instability in coastal areas: Insights from the Vasto Landslide (Adriatic coast, central Italy). *Geomorphology* **2013**, *201*, 468–478. [CrossRef]
20. Chang, K.T.; Ge, L.; Lin, H.H. Slope creep behavior: Observations and simulations. *Environ. Earth Sci.* **2015**, *73*, 275–287. [CrossRef]

21. Piacentini, D.; Devoto, S.; Mantovani, M.; Pasuto, A.; Prampolini, M.; Soldati, M. Landslide susceptibility modelling assisted by Persistent Scatterers Interferometry (PSI): An example from the northwestern coast of Malta. *Nat. Hazards* **2015**, *78*, 681–697. [CrossRef]
22. Mantovani, M.; Devoto, S.; Piacentini, D.; Prampolini, M.; Soldati, M.; Pasuto, A. Advanced SAR interferometric analysis to support geomorphological interpretation of slow-moving coastal landslide (Malta, Mediterranean Sea). *Remote Sens.* **2016**, *8*, 443. [CrossRef]
23. Bezerra, F.H.R.; Barreto, A.M.F.; Suguio, K. Holocene sea-level history on the Rio Grande do Norte State coast, Brazil. *Mar. Geol.* **2003**, *196*, 73–89. [CrossRef]
24. Pellicani, R.; Miccoli, D.; Spilotro, G.; Gallipoli, M.R.; Mucciarelli, M.; Bianca, M. Dynamic response of a rocky cliff under the sea wave pulse: A study along the Adriatic coast of Polignano (Apulia, Italy). *Environ. Earth Sci.* **2015**, *73*, 6243–6257. [CrossRef]
25. Lambeck, K.; Antonioli, F.; Purcell, A.; Silenzi, S. Sea level change along the Italian coast for the past 10,000 yrs. *Quat. Sci. Rev.* **2004**, *23*, 1567–1598. [CrossRef]
26. Furlani, S. Integrating observational targets and instrumental data on rock coasts through snorkel surveys: A methodological approach. *Mar. Geol.* **2020**, *425*. [CrossRef]
27. Furlani, S.; Piacentini, D.; Troiani, F.; Biolchi, S.; Roccheggiani, M.; Tamburini, A.; Tirincanti, E.; Vaccher, V.; Antonioli, F.; Devoto, S.; et al. Tidal notches (Tn) along the western Adriatic coast as markers of local coastal stability during late Holocene. *Geogr. Fis. Din. Quat.* **2018**, *41*, 33–46.
28. Aringoli, D.; Gentili, B.; Materazzi, M.; Pambianchi, G.; Farabollini, P. Il ruolo della gravità nell'evoluzione geomorfologica di un'area di falesia: Il caso del Monte Conero (mare Adriatico, Italia Centrale). *Studi Costieri* **2014**, *22*, 19–32.
29. Montanari, A.; Mainiero, M.; Coccioni, R.; Pignocchi, G. Catastrophic landslide of medieval Portonovo (Ancona, Italy). *Geol. Soc. Am. Bull.* **2016**, *128*, 1660–1678. [CrossRef]
30. Mastronuzzi, G.; Aringoli, D.; Aucelli, P.; Baldassarre, M.A.; Bellotti, P.; Bini, M.; Biolchi, S.; Bontempi, S.; Brandolini, P.; Chelli, A.; et al. The geomorphological map of the Italian coast: From a descriptive to a morphodynamic approach. *Geogr. Fis. Din. Quat.* **2017**, *40*, 161–196.
31. Campobasso, C.; Carton, A.; Chelli, A.; D'Orefice, M.; Dramis, F.; Graciotti, R.; Guida, D.; Pambianchi, G.; Peduto, F.; Pellegrini, L. *Aggiornamento ed Integrazioni Delle Linee Guida Della Carta Geomorfologica d'Italia Alla Scala 1:50.000*; Quaderni serie III; ISPRA, Servizio Geologico d'Italia: Roma, Italy, 2018; Volume 13.
32. Florinsky, I.V. An illustrated introduction to general geomorphometry. *Progr. Phys. Geogr.* **2017**, *41*, 723–752. [CrossRef]
33. Hengel, T.; Reuter, H.I. *Geomorphometry: Concepts, Software, Applications*; Elsevier: Amsterdam, The Netherlands, 2009.
34. Burrough, P.A.; McDonell, R.A. *Principles of Geographical Information Systems*; Oxford University Press: New York, NY, USA, 1998.
35. McKean, J.; Roering, J. Objective landslide detection and surface morphology mapping using high-resolution airborne laser altimetry. *Geomorphology* **2004**, *57*, 331–351. [CrossRef]
36. Berti, M.; Corsini, A.; Daehne, A. Comparative analysis of surface roughness algorithms for the identification of active landslides. *Geomorphology* **2012**, *182*, 1–18. [CrossRef]
37. Hobson, R.D. Surface roughness in topography: Quantitative approach. In *Spatial Analysis in Geomorphology*; Chorley, R.J., Ed.; Harper & Row: New York, NY, USA, 1972; pp. 221–245.
38. Shepard, M.K.; Campbell, B.A.; Bulmer, M.H.; Farr, T.G.; Gaddis, L.R.; Plaut, J.J. The roughness of natural terrain: A planetary and remote sensing perspective. *J. Geophys. Res.* **2001**, *106*, 777–795. [CrossRef]
39. Guth, P.L. Terrain organization calculated from digital elevation models. In *Concepts and Modelling in Geomorphology: International Perspectives*; Evans, I.S., Dikau, R., Tokunaga, E., Ohmori, H., Hirano, M., Eds.; Terrapub: Tokyo, Japan, 2003; pp. 199–220.
40. Savelli, D.; Troiani, F.; Cavitolo, P.; Nesci, O. Rocky cliffs joining velvet beaches: The northern Marche coast. In *Landscapes and Landforms of Italy*; Marchetti, M., Soldati, M., Eds.; World Geomorphological Landscapes, Springer International Publishing AG: Cham, Switzerland, 2017; pp. 271–280.
41. Cello, G.; Coppola, L. Assetto geologico-strutturale dell'area anconetana e sua evoluzione Plio-Quaternaria. *Boll. Soc. Geol. Ital.* **1984**, *103*, 97–109.
42. Coltorti, M.; Nanni, T.; Rainone, M.L. Il contributo delle scienze della terra nell'elaborazione di un piano paesistico: L'esempio del Monte Conero (Marche). *Mem. Soc. Geol. Ital.* **1987**, *37*, 629–647.

43. Mortimore, R.N.; Lawrence, J.; Pope, D.; Duperret, A.; Genter, A. Coastal cliff geo hazards in weak rock: The UK chalk cliffs of Sussex. In *Coastal Chalk Cliff Instability*; Mortimore, R.N., Duperret, A., Eds.; Engineering Geology Special Publ. 20/1; Geological Society: London, UK, 2004; pp. 3–31.
44. Hungr, O.; Leroueil, S.; Picarelli, L. The Varnes classification of landslide types, an update. *Landslides* **2013**, *11*, 167–194. [CrossRef]
45. Coltorti, M.; Dramis, F. The chronology of Upper Pleistocene stratified slope-waste deposits in central Italy. *Permafr. Periglac.* **1995**, *6*, 235–242. [CrossRef]
46. Van Steijn, H. Stratified slope deposits: Periglacial and other processes involved. In *Ice-Marginal and Periglacial Processes and Sediments*; Martini, I.P., French, H.M., Perez Alberti, A., Eds.; Special Publications 354; Geological Society: London, UK, 2011; pp. 213–226.
47. Nesci, O.; Savelli, D. Cicli continentali tardo-quaternari lungo i tratti vallivi mediani delle Marche settentrionali. *Geogr. Fis. Din. Quat.* **1986**, *9*, 192–211.
48. Savelli, D.; Troiani, F.; Bruciapaglia, E.; Calderoni, G.; Cavitolo, P.; Dignani, A.; Ortu, E.; Teodori, S.; Veneri, F.; Nesci, O. The landslide-dammed paleolake of Montelago (North Marche Apennines, Italy): Geomorphological evolution and paleoenvironmental outlines. *Geogr. Fis. Din. Quat.* **2013**, *36*, 267–287.
49. Di Luzio, E.; Saroli, M.; Esposito, C.; Bianchi Fasani, G.; Cavinato, G.P.; Scarascia Mugnozza, G. Influence of structural framework on mountain slope deformation in the Maiella anticline (Central Apennines, Italy). *Geomorphology* **2004**, *60*, 417–432. [CrossRef]
50. Alfaro, P.; Delgado, J.; Esposito, C.; Tortosa, F.G.; Marmoni, G.M.; Martino, S. Time-dependent modelling of a mountain front retreat due to a fold-to-fault controlled lateral spreading. *Tectonophysics* **2019**, *773*, 228–233. [CrossRef]
51. McAffee, R.P.; Cruden, D.M. Landslides at Rock Glacier Site, Highwood Pass, Alberta. *Can. Geotech. J.* **1996**, *33*, 685–695. [CrossRef]
52. Nichol, S.L.; Hungr, O.; Evans, S.G. Large scale brittle and ductile toppling of rock slopes. *Can. Geotech. J.* **2002**, *39*, 773–788. [CrossRef]
53. Scarascia Mugnozza, G.; Bianchi Fasani, G.; Esposito, C.; Martino, S.; Saroli, M.; Di Luzio, E.; Evans, S.G. Rock avalanche and mountain slope deformation in a convex, dip-slope: The case of the Majella Massif (Central Italy). In *Landslide from Massive Rock Slope Failure*; Evans, S.G., Scarascia Mugnozza, G., Strom, A., Hermanns, R.L., Eds.; NATO Science Series 49; Springer: Dordrecht, The Netherlands, 2006.
54. Bloom, A.L. Teaching about relict, no-analog landscapes. *Geomorphology* **2002**, *47*, 303–311. [CrossRef]
55. Martino, S.; Prestininzi, A.; Romeo, R.W. Earthquake-induced ground failures in Italy from a reviewed database. *Nat. Hazards Earth Syst. Sci.* **2014**, *14*, 799–814. [CrossRef]

applied sciences

MDPI

Article

Physical Model Experiments on Water Infiltration and Failure Modes in Multi-Layered Slopes under Heavy Rainfall

Junfeng Tang [1], Uchimura Taro [1], Dong Huang [2], Jiren Xie [3] and Shangning Tao [1,*]

1 Department of Civil & Environmental Engineering, Saitama University, Saitama 338-8570, Japan; tanger@imde.ac.cn (J.T.); uchimurataro@mail.saitama-u.ac.jp (U.T.)
2 Institute of Mountain Hazards and Environment, Chinese Academy of Science, Chengdu 610041, China; dhuang@imde.ac.cn
3 Department of Civil Engineering, Central South University, Changsha 410012, China; jirenxie198911@csu.edu.cn
* Correspondence: to.s.670@ms.saitama-u.ac.jp; Tel.: +81-048-858-9002

Received: 30 March 2020; Accepted: 11 May 2020; Published: 17 May 2020

Featured Application: Authors are encouraged to provide a concise description of the specific application or a potential application of the work. This section is not mandatory.

Abstract: To assess the influence of an intermediate coarse layer on the slope stability during heavy rainfall, knowledge about water movement and how slope failure occurs is important. To clarify the characteristics of water infiltration in a multi-layered slope and assess its influence on the slope failure modes, eight groups of physical slope models were investigated. It was found that the unsaturated hydraulic conductivity in the coarse layer (5.54×10^{-6} cm/s) was much lower than that of the fine layer (1.08×10^{-4} cm/s), which resulted in the capillary barrier working at a lower water content. Intermediate coarse layers embedded between finer ones may initially confine the infiltration within the overlying finer layers, delaying the infiltration and eventually inducing a lateral flow diversion in the inclined slope. Two different failure modes occurred in the model experiments: surface sliding occurred at the toe in the single-layer slope group and piping occurred at the toe in the multi-layered slope as the rainfall water accumulated, was diverted along the interface, and then broke through in the downslope direction of the intermediate coarse layer. The lateral flow diversion caused by the capillary barrier and the tilt angle may be the major factors influencing the difference of the failure modes. The result also revealed that the coarser layers may have negative effects on the slope stability.

Keywords: unsaturated soil; capillary barrier; multi-layer slope; slope failure

1. Introduction

Rainfall-induced slope failure is one of the most destructive natural disasters that occur in shallow natural slopes. The impacts of such catastrophic events are known, but these recurring natural hazards still result in many significant casualties and economic losses [1,2].

This study deals with a slope consisting of a fine layer and an intermediate coarse layer. The presence of soil layers with different unsaturated hydraulic conductivities in a shallow depth of soil affects the process of water infiltration, distribution, and pore water pressure in the slope, which results in different failure modes [3]. Rainfall infiltration water is due to the build-up of capillary barriers [4], which accumulate at the interface between fine and coarser soil layers. The capillary barrier effect has been widely studied in terms of its use in cover systems in waste disposal sites [5,6], whilst more recent research has focused on slope stabilization [7].

Capillary barriers can maintain a high degree of saturation in the soil above them, which results in different failure parts in a multi-layer slope [8]. These phenomena are related to the capillary tension, which limits the downward movement of the wetting front from a finer soil into underlying coarser soil. In an inclined interface, under continuous water infiltration from the slope surface, the accumulated water above the interface between the fine and the coarse material leads to the formation of a gravity-driven sub-surface water flow along the interface. This depends on the geometrical or boundary variations, at a certain distance downslope, often reported as the "diversion length", which can be estimated using a model proposed by Ross [9]. In natural slopes, characterized by a length in the order of hundreds of meters and an irregular layer geometry, the conditions leading to the penetration of infiltrating water into the underlying coarse layer are mainly governed by the materials, slope angle, and infiltration rate [10].

In some field investigations, a multi-layered slope with different hydraulic conductivities is a common situation in layered hillside slopes. Such a slope consists of fine sand deposits and medium-coarse sand deposits that are a few meters thick [11]. Additionally, the inclined angles range from 10 to 35 degrees. The coarse sand deposits involved have been found to have large pores and a higher hydraulic conductivity in a saturated condition, which have an important influence on the subsurface water flow and water content distribution during the infiltration, steady percolation, and drainage [12,13]. A schematic of a three-layer distribution in a slope is presented in Figure 1.

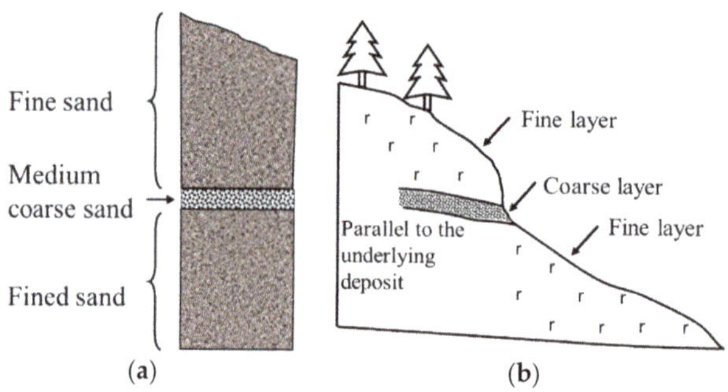

Figure 1. (a) Schematic profile of a multi-layered slope; (b) a description of the geometry of a multi-layered slope.

Many previous field investigations have been conducted and have found that layered soil influences the process of rainfall water movement and the distribution of water content, which determines the slope failure. However, clarifying the influence of an intermediate coarse layer on slope failure is still a complex issue, since many uncontrollable factors exist in a natural slope, such as the slope angle, rainfall intensity, etc. Hence, slope model experiments replacing field monitoring have been conducted, as these allow the same slope in different conditions to be investigated, reduce the cost by reducing the duration, and consider variable slope types [14].

To investigate the infiltration process in a layered soil profile in simplified and known geometrical and boundary conditions, physical model experiments have been performed by many researchers. Infiltration experiments have been conducted and have indicated that the wetting front temporarily stops above the interface of the fine layer and coarse layer, and the infiltration rate slows down [15,16]. Additionally, the pore water pressure head of the finer layer above the soil interface could not increase when the water started to infiltrate into the coarse layer [17]. The presence of a coarser lower-most layer may confine the infiltration within the upper finer layer up to a high degree of saturation. This capillary barrier effect occurred in a slope and was considered to be the cause of landslide initiation [7]. A capillary barrier at the upper interface of a coarse layer could have developed, favoring the accumulation and a lateral distribution of infiltrating rainfall and a possible diversion of flow down the slope, thus leading to a localized increase in the water content and loss of strength [18].

A diversion of the flow from the vertical direction towards the slope direction occurs and part of the water crosses the coarse layer infiltrate into the deepest layer. The settlement of soil is accompanied by progressive saturation of the soil, but general slope failure does not take place in the whole process [19].

Although some model experiments have been conducted to analyze the water movements and post-failure evolution due to the capillary barrier effect, few model experiments have focused on the slope failure time and modes in single- and multi-layer slope conditions or analyzed how an intermediate coarser layer may affect the slope stability.

Therefore, the aims of this study were as follows: (1) to experimentally clarify the influence of intermediate layers on water movement in a horizontal and inclined slope, respectively, especially after the breakthrough of the capillary barrier and (2) to investigate its influence on slope failure modes and the failure time in a single-layer and multi-layer slope by monitoring the water content, pore water pressure, and failure modes at different tilt angles. Soil-water characteristic curves (SWCC) [20] of fine sand and coarse sand were also measured to explain how the capillary barrier works. The proposed physical model experiments can provide a perspective on the failure time and modes of a single- or multi-layered slope during a rainfall event.

2. Materials and Methods

2.1. Basic Material Properties

Silica sands S1 and S7 were chosen for the physical model experiments as the fine material and coarse material, respectively, by the difference of D_{50} and size distribution. Figure 2 shows the size distribution of S1 (fine material) and S7(coarse material) used in model experiment. Table 1 summarizes the specific properties of S1 and S7 used in the model experiments. The sieve tests were conducted using the JGS Geotechnical Society standard test methods (JGS0131-2009).

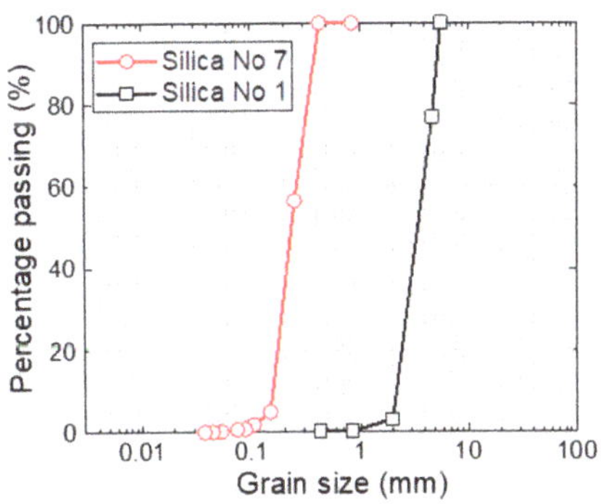

Figure 2. Grain size distribution of S1 and S7 in physical model experiments.

Table 1. Basic properties of sand materials.

Description	Silica No 7	Silica No 1
Specific gravity Gs	2.63	2.62
Gravel content (>4.75 mm; %)	0	97.13
Sand content (%)	87.31	2.86
Fines content (<0.075 mm; %)	11.64	0
D_{10} (mm)	0.043	2.26
D_{50} (mm)	0.152	3.52
D_{60} (mm)	0.165	4.21
Maximum dry density (g/cm^3)	1.556	1.52
Minimum dry density (g/cm^3)	1.271	1.3

The unsaturated hydraulic properties were measured by the variable head method (ASTM 2006) D2434-68. Table 2 and Figure 3a show the fitting parameters of the VG (van Genuchten–Mualem) model [21] and soil-water characteristic curves of silica No 1 and 7 in both the drying process and wetting process. The unsaturated hydraulic conductivities of sands were obtained by a pressure plate apparatus in the lab. As shown in Figure 3a, the drying (desorption) process and the wetting (absorption) process of the SWCCs of sand cause hysteretic behavior [15] for the same suction value, and the sand can retain more water in the drying process than in the wetting process. In the wetting process, the air entry value (AEV) of the fine layer (silica No 7) is about 1.5 kPa. The AEV of sand and gravel is about 0.45 kPa, being lower than the AEV of the fine layer. The capillary barrier effect occurs at the fine-coarse sand interface during rainfall infiltration. In Figure 3b, the unsaturated hydraulic conductivities of two types of sand are reported. The obtained values range between 10^{-2} and 10^{-7} cm/s under different suction conditions. The results show that the hydraulic conductivity of the coarse sand is higher than that of the fine sand at almost saturation, while it is significantly lower when the soil is unsaturated.

The soil-water characteristic curves of the soil were modeled with the van Genuchten–Mualem model [5], as follows:

$$S_e = \{1 + (-\alpha h)^n\}^{-m}, S_e = \frac{\theta - \theta_r}{\theta_s - \theta_r}, m = 1 - \frac{1}{n}. \tag{1}$$

In the above equation, the water retention curve is expressed in terms of the effective degree of saturation. θ is the volumetric water content; θ_r and θ_s indicate the residual and saturated values of the water content, respectively; a, m, and n are the fitting parameters; h is the matric suction; a is a scaling parameter (units of m^{-1}); and the exponents n and m are parameters that determine the shape of the retention curve. The hydraulic parameters are given in Table 2.

Table 2. Hydraulic properties of sand materials.

Description	Symbol (unit)	Soil	
		No 7	No 1
Drying curve			
Saturated volume water content	θ_s	0.44	0.42
Air-entry value	ψa (kPa)	2.44	0.62
Residual volume water content	θ_r	0.12	0.05
van Genuchten model	a	0.41	1.59
Fitting parameter	n	4.07	3.11
	m	1.42	0.67
Wetting curve			
van Genuchten model	a	0.42	0.03
Fitting parameter	n	4.51	1.42
	m	0.78	0.29

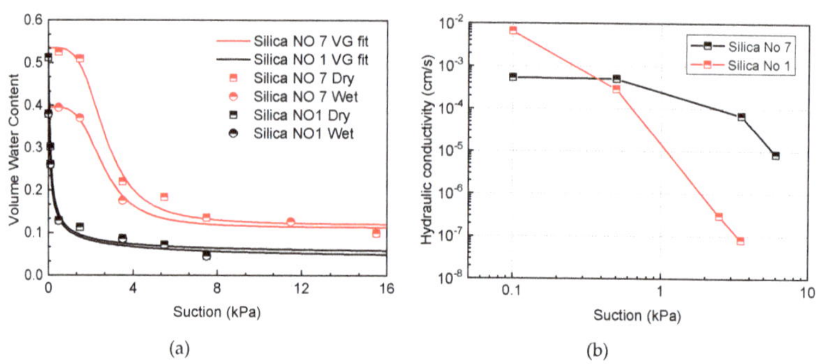

Figure 3. (**a**) Soil-water characteristic curves (SWCCs) of silica No 1 and No 7 in the drying and wetting process; (**b**) unsaturated hydraulic conductivity of silica No 1 and No 7 in the wetting process.

2.2. *Water Flow and Slope Failure in Multi-Layer Slope Models*

The multi-layer models were set up to evaluate the effect of the capillary barrier acting on a slope. This model system was built to show the advancement of the wetting front, and monitor the volumetric water content in the soil, pore water pressure, and slope failure process. In addition, the failure process was directly recorded by cameras to obtain a better understanding of failure modes and the effect of the capillary barrier. Table 3 summarizes the physical experimental conditions.

Table 3. General information for the slope model experiment.

Experiment	Sediment Type	No. of Layers	Tilt Angle (°)	Rainfall Intensity	Depth of Layers
Case I	S1, S7	3	0	75 mm/h	0.2 m, 0.05 m, 0.2 m
Case II	S1, S7	3	7	75 mm/h	0.2 m, 0.05 m, 0.2 m
Case III	S1, S7	3	15	75 mm/h	0.2 m, 0.05 m, 0.2 m
Case IV	S1, S7	3	21	75 mm/h	0.2 m, 0.05 m, 0.2 m
Case V	S7	1	0	75 mm/h	0.45 m
Case VI	S7	1	7	75 mm/h	0.45 m
Case VII	S7	1	15	75 mm/h	0.45 m
Case VIII	S7	1	21	75 mm/h	0.45 m

2.2.1. Flume Model System

Figure 4a shows the apparatus used for the physical model experiments, which consisted of an inclined steel box, a rainfall simulation system, and a set of pore water pressure and VWC (volumetric water content) sensors. The details of each subsystem are as follows: (i) the inclined steel box had dimensions of 1.0 m (length) × 0.3 m (width) × 0.5 m (height); (ii) the sidewalls of the box were made of an acrylic plate to observe the advancement of the wetting front and failure process during rainfall. The gap between the steel plates and the acrylic plate was sealed with epoxy adhesive.

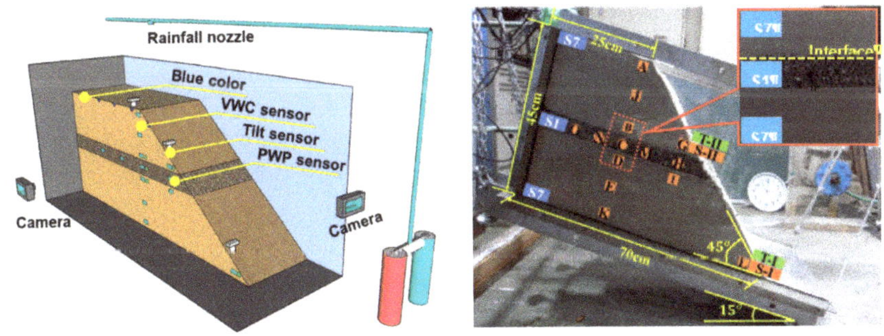

(a) (b)

Figure 4. (**a**) Schematic diagram of the experimental apparatus used for a multi-layered slope under rainfall: side-view of the multi-layered slope; (**b**) side view of the experimental apparatus used for the slope. The yellow dashed line shows the fine–coarse interface.

Furthermore, the inclined angle of the model box could be lifted by a crane to simulate the inclined cover from 0 to 60 degrees. Experiment pictures were recorded by cameras in different locations around the model. Table 3 summarizes the physical model experimental conditions.

2.2.2. Rainfall Simulation System

A rainfall nozzle was placed 60 cm above the model box to simulate rainfall with a constant intensity. The intensity and duration of rainfall were controlled by a control value and air pressure gauge. The rainfall intensity was kept at a constant intensity by air pressure, and ranged from 35 to 100 mm/h. The sensors ECH2O EC-5 were used to determine the volumetric water content. Soil-specific calibration is recommended for obtaining the best possible accuracy in volumetric water content measurements [22,23]. Calibration of the EC-5 sensors has been shown to result in an increased accuracy of 1–2% for all soils with soil-specific calibration [24,25]. The calibration of EC-5 sensors employed in silica No 1 and No 7 is shown in Figure 5b.

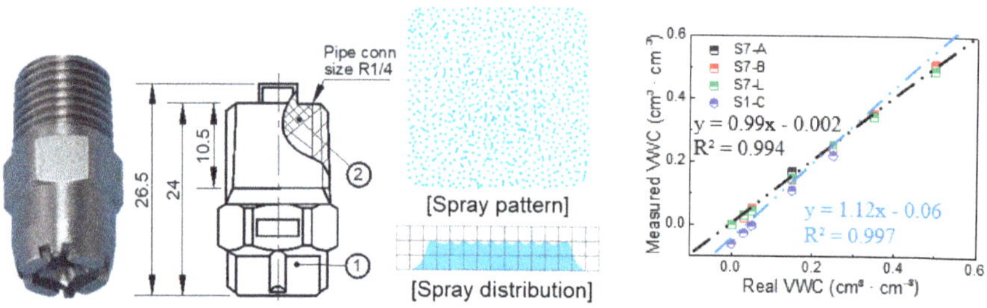

(**a**) (**b**)

Figure 5. (**a**) Nozzle body and spray pattern and distribution for rainfall simulation; (**b**) calibration of the volumetric water content sensor in silica No 1 and No 7.

2.2.3. Theory of Measurement Devices

The instruments used in the model experiments were calibrated before installation, including the pore water pressure, tilt sensors (Figure 6a), VWC sensors (Figure 5b), and rainfall simulators (Figure 5a). The intensity and uniformity of artificial rainfall created by the simulator were calculated based on the weight of the sample at a certain time.

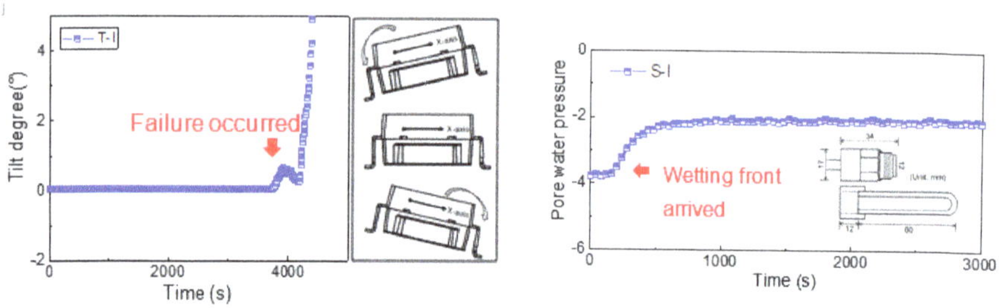

(**a**) (**b**)

Figure 6. (**a**) MEMS (Micro-Electro-Mechanical System) tilt sensor and its working theory; (**b**) typical pore water pressure response in soil and the structure and size of the transducer.

2.2.4. Testing Procedure

Soil Preparation

S1 and S7, which were used to make the slope, were dried in an oven for 48 h. Then, an amount of water was added to the soil to achieve an initial water content of around 6%.

Compaction of Soils

The prepared soil was compacted and placed in a series of horizontal layers. Silica No 7 and No 1 were placed in the model box in layers and compacted to achieve a dry density of 1.33 and 1.43 g/cm^3, respectively. Each layer was tamped equally with a rod to a thickness of 5 cm and these procedures were repeated until the height of the slope was achieved.

Positions of Sensors and Cameras

During the soil placement, VWC sensors, tensiometers, and tilt sensors were placed at specific locations in the three different layers and the time of recording the quantity of water content was 10 s.

Table 3 summarizes all of the experimental conditions.

3. Results

3.1. Failure Situations in All Cases

Figure 7 shows the failure situations for different cases during rainfall. No runoff on the surface of the slope was observed in all cases, as the infiltration capacity of the soil layer was higher than the rainfall intensities used in this study. Two failure modes were observed in this study. One was soil piping, which occurred in the multi-layered slope. Slight soil piping occurred at the toe of the slope, finer materials mixed with amounts of infiltrate water flowed out from the piping as the seepage surface, and the seepage surface grew gradually and increased until cracks appeared (Figure 7c). As Figure 7a shows, the piping phenomenon occurred slightly, and as the tilt angle increased, the soil pipe size developed rapidly and more seriously (see Figure 7b,c).

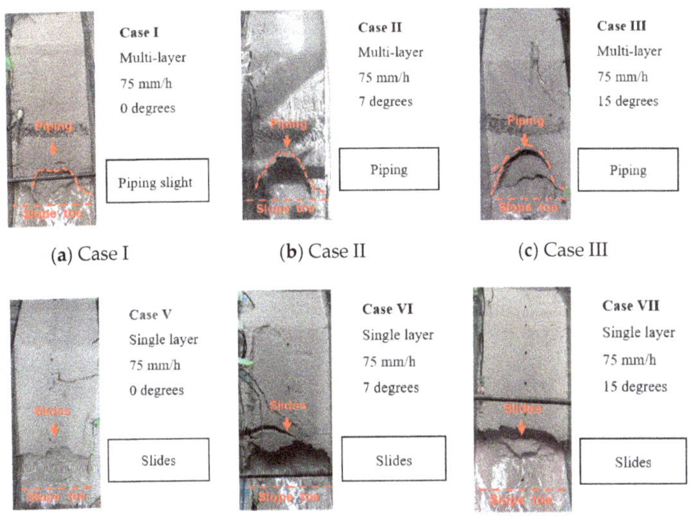

(**a**) Case I (**b**) Case II (**c**) Case III

(**d**) Case V (**e**) Case VI (**f**) Case VII

Figure 7. Comparison of failure situations of the different model experiment cases (each case was performed three times to ensure the repeatability of the results, and typical results of individual experiments were chosen to show the failure situation in each case). (**a**) case I (tilt angle $\alpha = 0°$); (**b**) case II (tilt angle $\alpha = 7°$); (**c**) case III (tilt angle $\alpha = 15°$); (**d**) case V (tilt angle $\alpha = 0°$); (**e**) case VI (tilt angle $\alpha = 7°$); (**f**) case VII (tilt angle $\alpha = 15°$).

Another failure mode was surface sliding from the toe of the slope in a single-layer slope. Small slide failure occurred at the toe at first, and then relatively larger slide failure followed. This type of failure mode was clearly observed for cases IV, V, and VI, as shown in Figure 7d.

The failure modes and initial failure time t for all eight cases are summarized in Table 4. The initial slope failure occurred at the toe of the slope in case I and case IV at about 1.18 and 0.95 h after the rainfall has been applied, respectively. This means that failure in a multi-layer slope occurs later than that in a single layer under the same condition in the horizontal group. In Figure 7b,e, in case II and case V, failure occurred at around 0.84 and 0.89 h, respectively. In case III and case VI, more rapid movement of slope failure occurred in an almost fully saturated condition and they thus had a lower strength. The failure times of the 15 degrees inclined group were 0.78 and 0.85 h. It should be noted that failure occurred earlier in the multi-layer slope than in the single-layer slope in the inclined group, which is different from the result obtained for the horizontal group.

Table 4. Failure conditions of physical model experiments.

Experiment	Rainfall Intensity *I*	Tilt Angle α (°)	Failure Modes	Initial Failure Time *t* (h)
Case I	75 mm/h	0	Piping slightly	1.18 h
Case II	75 mm/h	7	Piping	0.84 h
Case III	75 mm/h	15	Piping	0.78 h
Case IV	75 mm/h	21	Piping	0.62 h
Case V	75 mm/h	0	Surface slides	0.95 h
Case VI	75 mm/h	7	Surface slides	0.89 h
Case VII	75 mm/h	15	Surface slides	0.85 h
Case VIII	75 mm/h	21	Surface slides	0.76 h

3.2. *Profile of the Volumetric Water Content in a Slope*

During the tests, the hydrological response was monitored by means of VWC (volumetric water content) sensors located at different locations within the slope and crossing the entire soil thickness, allowing the retrieval of volumetric water content profiles at different depths. The VWC sensors in section I in different cases were shown to explain how the capillary barrier effects influence the water infiltration and distribution during the rainfall and drying process, as is shown in Figure 8 (the sensor locations are shown in Figure 4b). Throughout the experiment, the VWC of the toe of slopes increased slowly with time toward a saturated value in response to the saturation process, until failures occurred. It should be noted that the capillary barrier clearly controlled the rate of changes of the VWC in a multi-layer slope, which determined the failure time of slope.

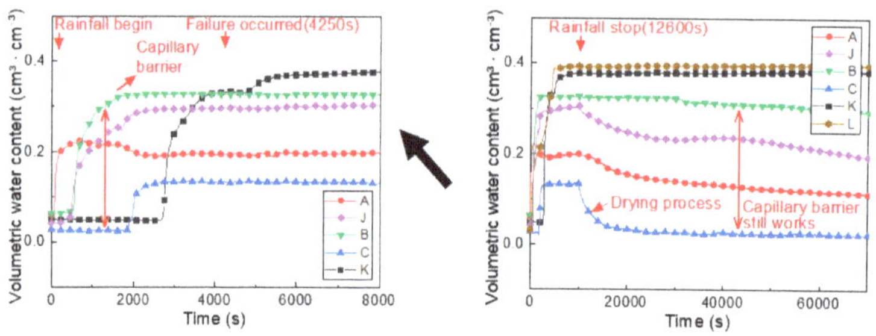

(a)

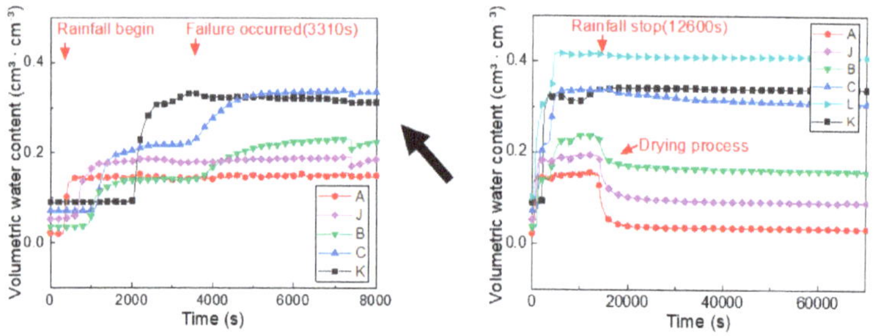

(b)

Figure 8. *Cont.*

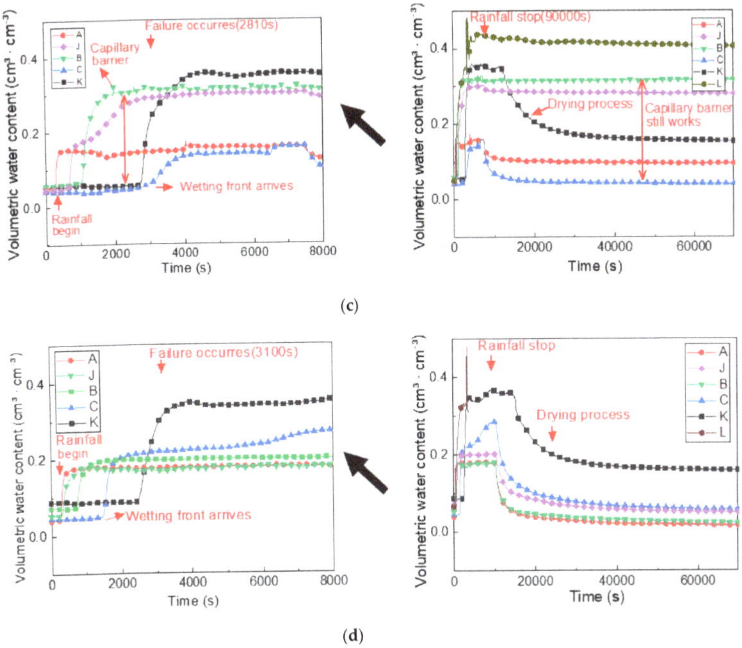

Figure 8. Time histories of VWC during rainfall and the drying process in case I, III, V, and VII (the left figures show the time from 0 to 8000 s, while the right figures show the whole wetting and dry process in the rainfall event). (**a**) Case I. VWC trends in a multi-layer slope in the flat group (tilt angle $\alpha = 0°$); (**b**) Case V. VWC trends in a single-layer slope in the flat group (tilt angle $\alpha = 0°$); (**c**) Case III. VWC trends in a multi-layer slope in the inclined group (tilt angle $\alpha = 15°$); (**d**) Case VII. VWC trends in a multi-layer slope in the inclined group (tilt angle $\alpha = 15°$).

In case I and case V (see Figure 8a,b), the advancement of the wetting front is evident from the time history of the VWC sensors. It shows that the rise of the VWC was rapid, and then became gradual, as the toe of the slope approached a fully saturated condition. When the VWC at point L reached around 0.4, failure occurred at the toe of the slope. In addition, compared with the VWC at point L in the flat group, point L reached the saturated condition later in the multi-layer model, since the capillary barrier prevented the water from infiltrating into the bottom. This made the slope more stable and caused a delay in the failure time, which was 1.2 and 0.92 h in case I and case III, respectively. This capillary barrier effect can also be explained by the VWC histories obtained during the experiments. For example, it took around 26 min for the wetting front to pass through the interface from point B to point C [11].

Figure 8c,d show that when rainfall was applied, the VWC increased quickly above the interface (point B), and when rainfall was stopped, the soil above the coarse layer (point B) remained wetter than the same location in a single-layer slope. Additionally, for the VWC at the toe of the slope (point L) in the inclined group, the failure occurred earlier (0.78 h), while it was 0.84 h in a single-layer slope, which was contrary to the flat group experimental results.

3.2.1. Beginning of Rainfall (t = 0 h)

In case I and V, the VWC in the bottom of the upper fine layer (point B) increased from 0.05 to 0.33 with the depth in the upper fine layers (height was from 25 to 45 cm), while that in the coarse layer was about 0.03 before capillary barrier breakthrough at the beginning of rainfall in case I (t = 0.5 h).

This indicates that the wetting front above the interface was stopped for a while and rainfall water was stored in the upper fine layer due to the influence of the capillary barrier. In the lower section I (height was from 0 to 20 cm), the VWC maintained a constant condition while the capillary barrier was present.

3.2.2. The Breakthrough of the Capillary Barrier (Case I, II, and III)

During the first 0.8 h of rainfall, as shown in case I and case II in Figure 9 a,b, the presence of the intermediate coarse layer caused a significant time delay in the infiltration process: the VWC in section I above the interface increased to 0.34 in case I, whereas the VWC in the coarse layer changed from 0.03 to 0.05 in case I at the depth of 22 cm, while the VWC was from 0.04 to 0.22 at the same depth in case II (t = 0.6 h). In this stage, the interface between the fine and coarse layer acted as a capillary barrier and gradually broke when the bottom of the upper fine layer was almost at saturation. In case II, the VWC in section I above the interface increased from 0.06 to 0.32, while that of the coarse layer was from 0.03 to 0.04. A lateral diversion flow may have occurred along the inclined interface, resulting in capillary barrier breakthrough occurring later in section I compared with case I.

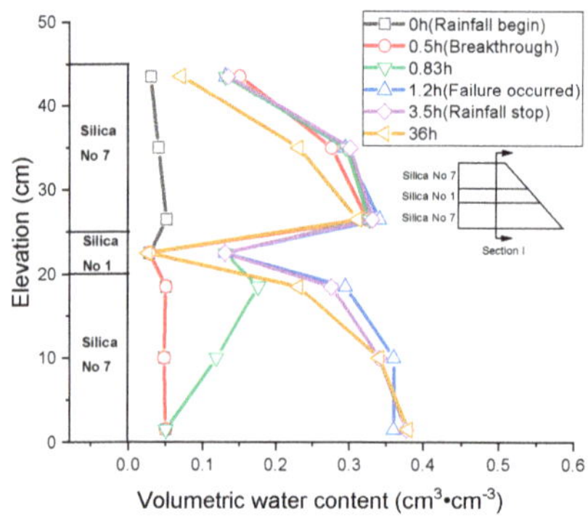

(a)

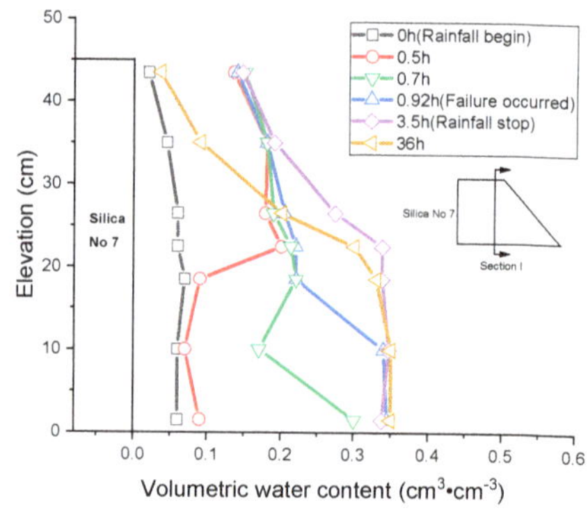

(d)

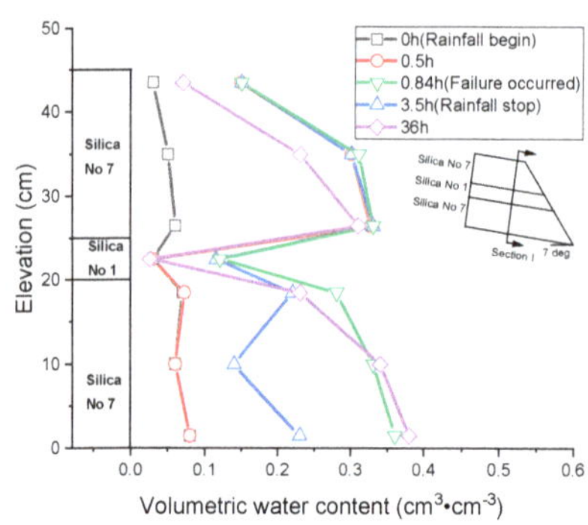

(b)

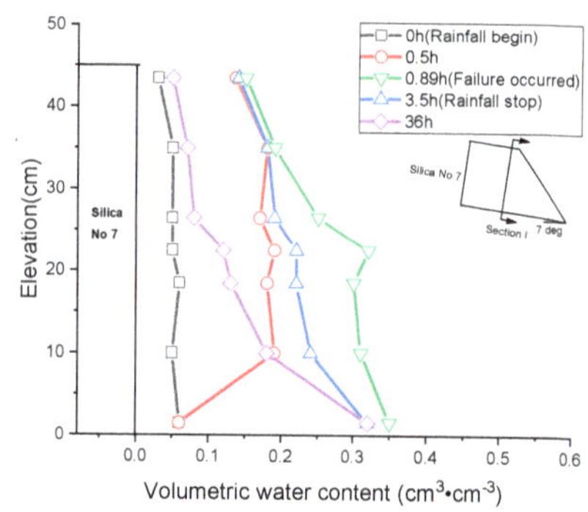

(e)

Figure 9. *Cont.*

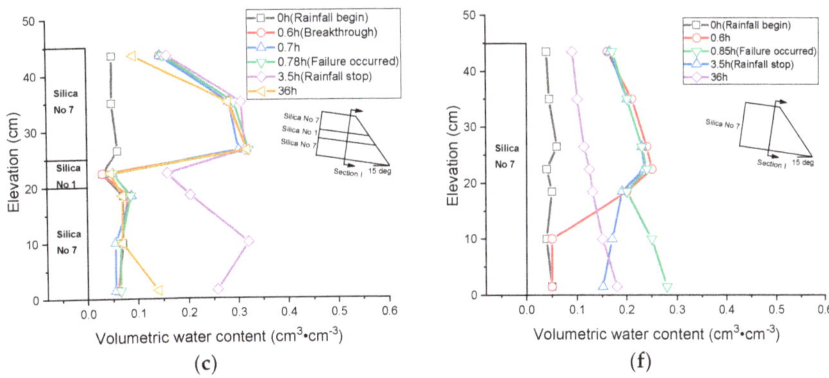

Figure 9. Profile of the VWC in section I in different cases (the VWC distribution after slope failure was verified by obtaining soil samples from various locations at the end of the experiments). (**a**) Case I; (**b**) Case II; (**c**) Case III; (**d**) Case V; (**e**) Case VI; (**f**) Case VII.

3.2.3. Post-Breakthrough of the Capillary Barrier (t = 0.6–0.78 h in Case I, II, and III)

After 0.6 h of rainfall, the VWC of the upper fine layer maintained a constant value in section I in case I and III, since it reached steady state infiltration in the soil. Additionally, rainfall water infiltrates into the next layer following capillary barrier breakthrough. For example, the VWC of the coarse slayer increased from 0.05 to 0.12 at 0.83 h and 0.13 at 1.2 h (see Figure 9a) in case I and the VWC of the coarse layer increased from 0.05 to 0.07 at 0.78 h in case III. It should be noted that the VWC of the bottom of the lower finer layer increased to 0.36 at 0.8 h in case I and to 0.08 at 0.78 h in case III, respectively. This means that rainfall water could not arrive at the lower fine layer along section I because of the lateral water flow along the inclined interface in case III.

3.2.4. Failure Occurred (t = 0.78–1.2 h)

In this part, the failure process is analyzed using the response of the tilt angle, pore water pressure, and VWC measurements in different locations in the slope. In case III and VII, the experimental flume was tilted to 15° and subjected to the same rainfall intensity. This section aims to evaluate the effects of an inclined angle on the water movement and slope stability.

The main results of this part are reported in Figures 10 and 11. The initial pore water pressure values were negative at the locations of S-I and S-II. The pore water pressure and VWC increased during the infiltration and then stabilized when the infiltration reached a steady state condition. During the rainfall, the tilt angle of point G and point A almost maintained a constant value, while the soil in these locations maintained a negative value. A sudden change occurred when the pore water pressure approached 0 kPa at point L in all cases.

Comparing the difference of the pore water pressure measured at S-II in case I and case V, presented in Figure 10a,b), the pressure grew from −4 kPa at the beginning of the rainfall to around −1 kPa after around 1.2 h above the coarse interface, while it reached −3.5 kPa in case V. This suggests that there was a higher water content and pressure head above the coarse layer than at the same location in a single-layer slope, as the capillary barrier effect led to the storage of water and high water pressure head. In this respect, it is worth noticing that in case I, the pore water pressure at point L (S-I) increased suddenly after capillary barrier breakthrough at the interface.

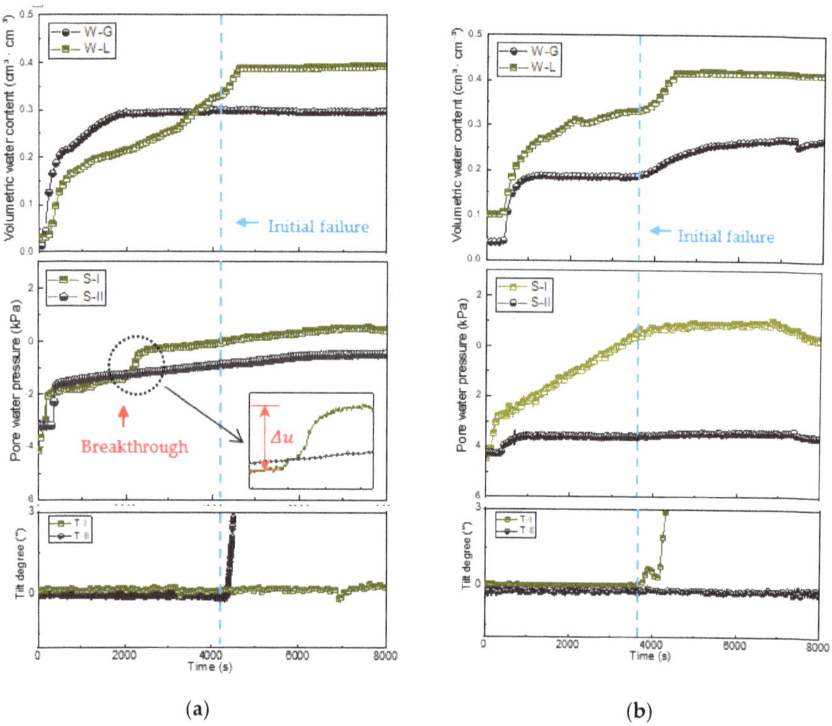

Figure 10. Volumetric water content, pore water pressure, and tilt angle change trends at different locations during infiltration. (**a**) case I (tilt angle $\alpha = 0°$, multi-layer), where the pore water pressure increased after capillary barrier breakthrough; (**b**) case V (tilt angle $\alpha = 0°$, single layer). For the positions of sensors, see Figure 3b.

The same behavior, in terms of the pore water pressure and VWC, was observed in tests with the tilt angle $\alpha = 15°$ (Figure 11). In these two cases, the greater slope angle induced the most sudden variations of water content and pore water pressure, and also resulted in an earlier response of tilt sensors located at the toe of the slope. For case III with the tilt angle $\alpha = 15°$, failure occurred in about 0.78 h, and the increase of VWC was more rapid than that of case VII, which was contrary to case VII. In the inclined group, the later failure occurred at 0.84 h in a single-layer slope, which proved that an inclined multi-layer slope is more dangerous than a single-layer one under this condition, which is contrary to the findings of the flat group.

It is evident in Figures 10a and 11a that the increase of the pore water pressure can be divided into two periods. A negative pore pressure was measured at the beginning of rainfall (S-I), and then, around 30 min later, a sudden increase of the pore water pressure value Δu could be measured after capillary barrier breakthrough at point L (Figure 3b). Moreover, a few minutes after breakthrough, piping occurred for a pore pressure of 0 kPa at the toe, which seems to indicate a negative influence for the slope instability. Similar to case I, pore pressure sensor S-I in case III also showed the same trends after the breakthrough of the capillary barrier. It is clear from both data sets that the pore water pressure value increased from −1.5 to −0.3 kPa in case I and from −1.1 to −0.21 kPa in case III, respectively.

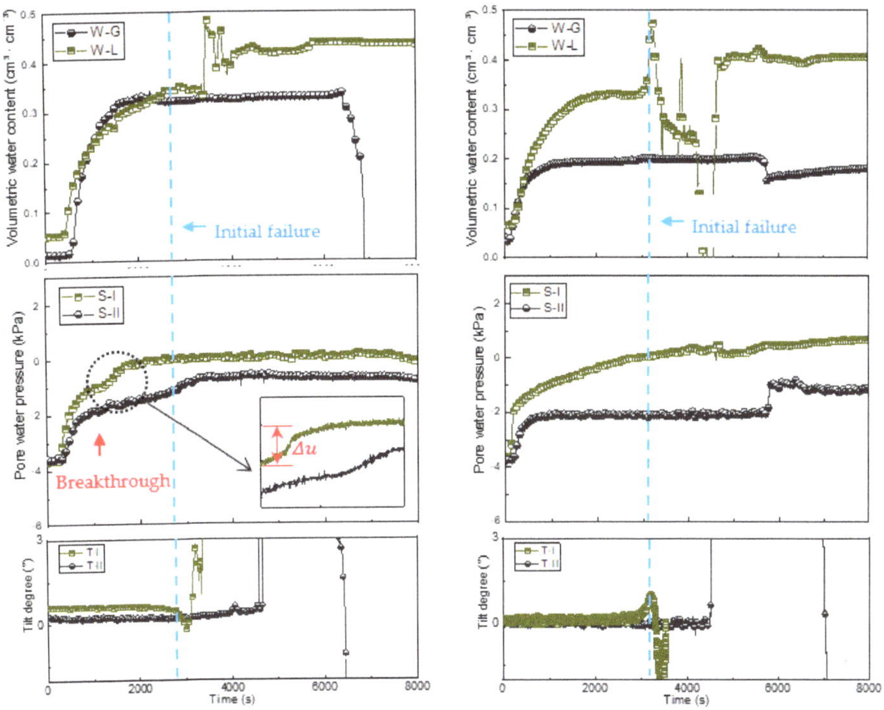

(a) (b)

Figure 11. Volumetric water content, pore water pressure, and tilt angle change trends at different locations during infiltration. (**a**) Case III (tilt angle $\alpha = 15°$, multi-layer), where the pore water pressure increased after capillary barrier breakthrough; (**b**) case VII (tilt angle $\alpha = 15°$, single layer). For the positions of sensors, see Figure 3b.

3.2.5. Capillary Barrier Restoration in the Drying Process (t = 3.5–36 h)

The VWC in soil decreased once the rainfall stopped in case I (Figure 9a) and V ((Figure 9c). However, the bottom of the upper fine layer maintained a higher water content condition at 0.33 compared with the single-layer slope, which was only 0.19 in case I and V, respectively. The capillary barrier broken through could be restored to its pre-breakthrough condition once rainfall had stopped. The VWC of the finer soil above the interface decreased as the rainfall water continued to drain out. As the VWC at the interface decreased, the unsaturated hydraulic conductivity of the coarser lower layer also decreased, and eventually approached zero. The capillary barrier was completely restored when the intermediate coarser layer could not accept any more rainfall water from the overlying fine layer.

3.3. *Influence of the Tilt Angle*

3.3.1. Influence of the Tilt Angle on Water Movement

To show the water movement and its lateral diversion length in a multi-layer slope, the slope surfaces were marked with blue colors before the rainfall. With the aid of the backlighting and the dye traces, it was simple to visualize the dyed streamlines with the water movement once a steady state was achieved. The lateral diversion length and breakthrough zone could also be measured directly. Figure 11 shows photographs of the dyed trace streamlines from single-layer and multi-layer slopes with different tilt angles. The coarse layer is shown in deep gray in the pictures. For all of the inclined experimental cases, the dye traces were diverted downslope (referred to as capillary diversion, see Figure 12) and, in most cases, penetrated the coarse layer at different points (breakthrough). The slope at 7° with a breakthrough zone maintained a constant value when infiltration reached a steady state. In these three cases, a clear lateral flow region without infiltrate water passing through the coarse layer formed near the toe of the coarse layer. An amount of water entered the bottom layer through the breakthrough zone. Additionally, the breakthrough region was measured as the total length along the interface through which breakthrough was observed. This will be referred to as the piping in the subsequent discussion.

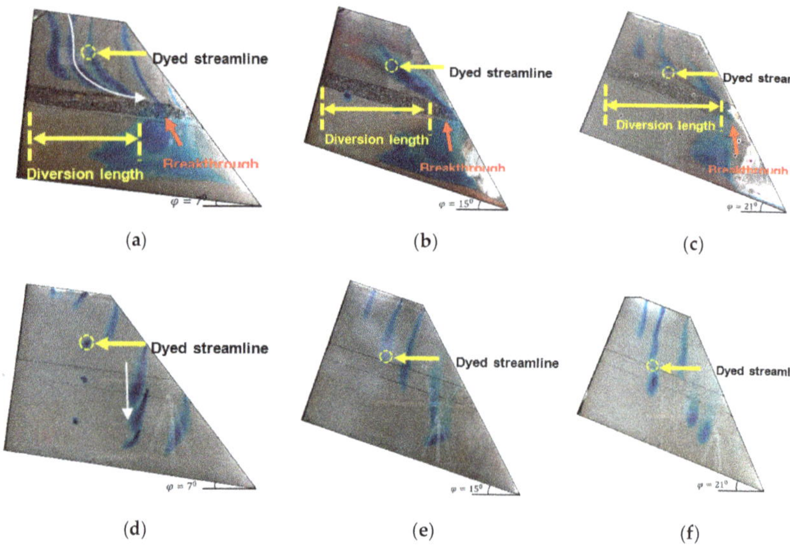

Figure 12. Side view of model experiments with blue dye traces during rainfall for the slopes in the inclined group. (**a**) case II (tilt angle $\alpha = 7°$); (**b**) case III (tilt angle $\alpha = 15°$); (**c**) case VI (tilt angle $\alpha = 21°$); (**d**) case VI (tilt angle $\alpha = 21°$); (**e**) case VII (tilt angle $\alpha = 21°$); (**f**) case VIII (tilt angle $\alpha = 21°$). The capillary barrier diversion in the multi-layer slope is shown by the movement of blue dye traces. The white arrow shows the approximate flow direction (case II and case VI).

Figure 13 shows the lengths of the three cases observed, and the capillary diversion length occurring upslope on the fine-coarse interface. The length of capillary diversion was measured from the initial point of the blue dye trace closest to the upper interface to the point where the dye first penetrated into the coarse layer. Figure 14 shows the VWC contour maps at 0.5 h, when failure occurred, rainfall stopped (t = 3.5 h), and the drying process stopped (t = 36 h) for case I, II, III, and VII. The initial VWC values were about 0.06 in the slopes.

In Figure 14a,b, comparing the VWC in the flat group at t = 0.5 h, the area above the interface reached a higher degree of saturation in the multi-layer slope than in the single-layer slope, since the capillary barrier prevented the water from infiltrating into the coarse layer, which made the slope more stable and caused a delay in the failure time. Comparing Figure 13, it was proven that lateral diversion occurred along the interface in the inclined slope, which resulted in a higher water content near the toe of the slope, and water could not infiltrate into the lower finer layer in Figure 13. This result may be associated with the sloping of the cover system [24]. The lateral diversion in the coarse layer was around 0.4 m, possibly because of the inclined angle, material properties, and rainfall intensity used in the test. Afterward, water could infiltrate into the deeper finer layer at the end of lateral diversion [26].

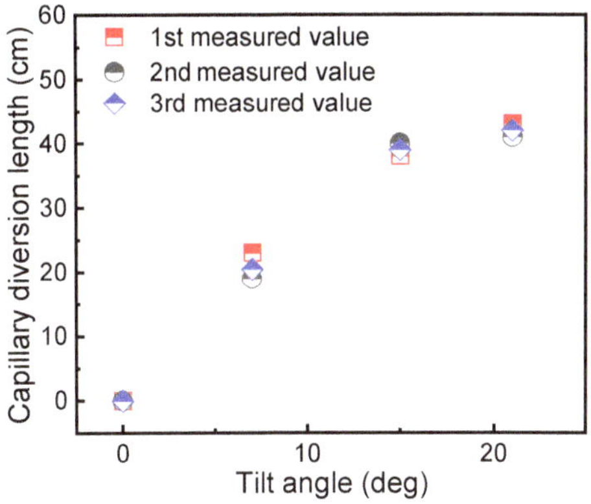

Figure 13. Effect of the tile angle on the capillary barrier diversion length in a multi-layer slope in case I, II, III, and IV (each case was performed three times for the same experimental conditions).

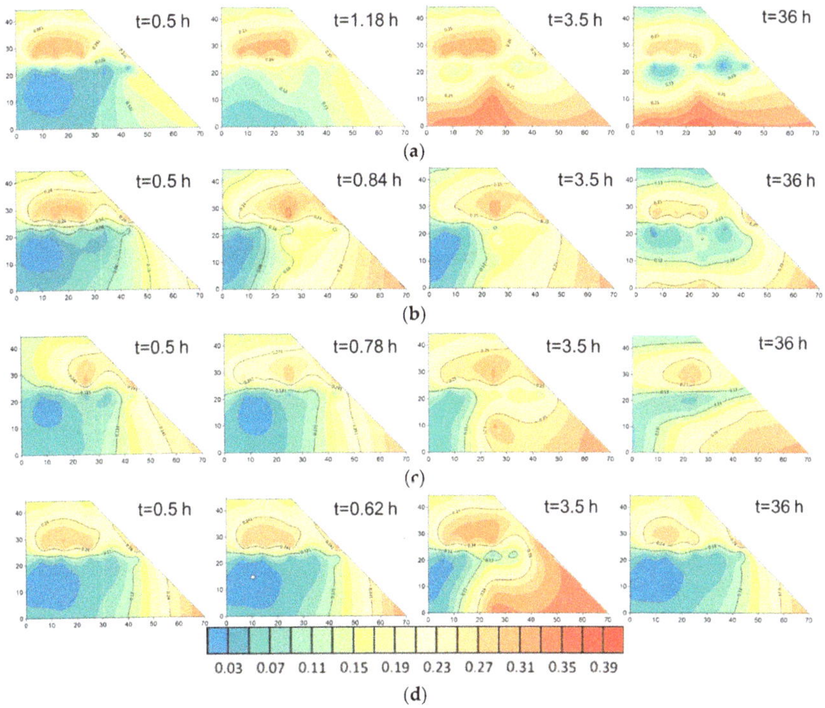

Figure 14. The VWC changes with elapsed time in case I, II, III, and IV (the VWC distribution after slope failure was verified by obtaining soil samples from various locations at the end of the test). (**a**) Case I. Multi-layer slope, 0 deg, 75 mm/h; (**b**) Case II. Multi-layer slope, 7 deg, 75 mm/h; (**c**) Case III. Multi-layer slope, 15 deg, 75 mm/h; (**d**) Case VII. Multi-layer slope, 21 deg, 75 mm/h.

3.3.2. Influence of the Tilt Angle on the Pore Pressure and VWC

To investigate the effects of the tilt angle, the pore pressure sensor (S-I) and VWC sensor (W-L) at the bottom were considered for comparison in different cases, as these are most crucial to the knowledge on slope failure. Detailed in Figure 14 is the progressive build-up of VWC and pore water pressure throughout the experiments I, III, and IV, until failure occurred.

For Figure 15a,b, it is clear that an increased tilt angle has a drastic effect on the build-up of volumetric water content and pore water at the bottom of the slope, progressively resulting in quicker failure times as the tilt angle increases.

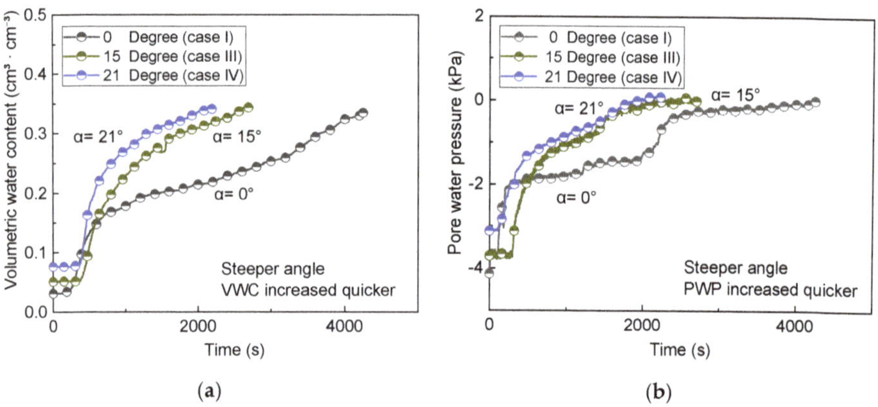

Figure 15. Time series of the VWC and pore pressure at the bottom of the slope for case I, III, and IV, with α ranging from 0° to 21° (multi-layer slope). (**a**) Time series of the VWC from initiation to failure; (**b**) time series of the pore pressure from initiation to failure.

Once the wetting front arrived, sensor spikes in measurements occurred. It took around 1.18, 0.78, and 0.62 h for the pressure head and VWC to approach the maximum value in case I, III, and IV, respectively. The peak value of VWC and pore pressure values were similar for each case ranging between 0.348 and 0.351 and −0.021 and −0.016 kPa, respectively, at times of failure. Due to the rapid progression of the wetting front at an increased tilt angle, the pore water pressure and VWC increased at a faster rate, as is evident in results showing that the α = 21° (case V) failed 47% sooner than α = 0° (case I) and 34% sooner than α = 15° (case III). These similar trends were exhibited in all experiments where the tilt angle was increased [27].

4. Discussion

4.1. Mechanism of the Capillary Barrier

A capillary barrier (Figure 16a) forms in unsaturated conditions when the hydraulic conductivity of finer soil is higher than that of coarse soil, which limits the downward infiltration of water due to the difference of the hydraulic conductivity. Figure 15b shows the relations between the hydraulic conductivity and suction of the two sands measured in the lab. The intersect of two hydraulic conductivity curves is 0.4 kPa. From the intersect to the higher suction (suction > 0.4 kPa), the hydraulic conductivity in the coarse layer is much lower than that in the fine layer, which caused the capillary barrier to form at the fine-coarse interface. Otherwise, suction at the interface decreased gradually after the wetting front arrived and was located to the left of the intersection, and the hydraulic conductivity in the coarse layer was higher than that of the fine one, which allowed the rainfall water to infiltrate into the next layer after capillary barrier breakthrough.

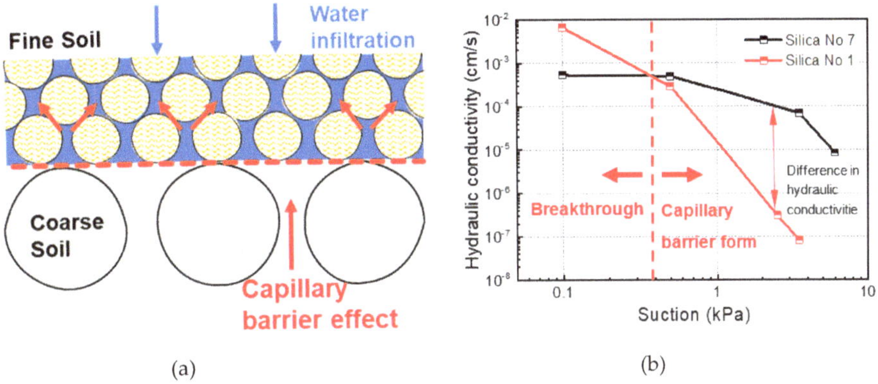

Figure 16. (**a**) Capillary barrier at the interface between different layers with different conductivities. (**b**) The relationship between the hydraulic conductivity and suction of silica No 1 and No 7.

4.2. D Flow and Multi-Layer Slope Stability

As is shown in Figure 17a, the blue color was used to stain the surface of the slope and to show the movement of rainfall water, and it was applied under the coarse layer to check that the capillary barrier still worked. At the beginning of case III, the water flow showed a vertical movement, mainly controlled by gravity, before the wetting front arrived at the interface. Secondly, the water flow could not go across the interface directly upward of the slope and showed a significant velocity component parallel to the inclined interface after the wetting front arrived [28].

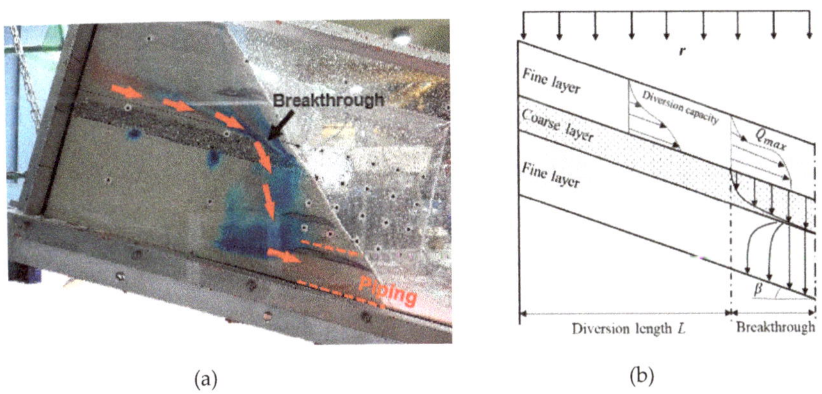

Figure 17. (**a**) Side view of the diversion length and breakthrough area of the capillary barrier (the movement and the breakthrough zone are shown by the blue dye traces); (**b**) calculation of the capillary diversion length L.

Lateral diversion is essentially gravity-driven unsaturated drainage within the finer layer of a sloped capillary barrier [29,30]. The unsaturated hydraulic conductivity is the maximum as the VWC of the upper finer soil increases with depth, where lateral diversion is concentrated at the fine-coarse interface. Besides, the diverting water will increase the water content in the downslope direction with the influence of the inclined angle, which may result in failure of the barrier and then infiltrate into the next layer.

In Figure 17b, the maximum length of the capillary diversion can be calculated by [31]:

$$K(\psi) = \frac{\left\{1-(\alpha\psi)^{mn}\left[1+(\alpha\psi)^{n}\right]^{-m}\right\}^2}{\left[1+(\alpha\psi)^{n}\right]^{m/2}} * K_s \ , \tag{2}$$

$$L \leq \tan(\beta)\frac{K(\psi)}{q}\left[\frac{1}{\alpha} + \left(|\psi_a| - |\psi_w^*|\right)\right] \tag{3}$$

where, ψ is the matric suction in the soil; ψ_w^* is the water entry value of the coarse layer; ψ_a is the air entry value of the fine layer; β is the tilt angle of the slope (°); r is the infiltration rate (mm/h); $K(\psi)$ is the hydraulic conductivity of the fine layer; and a, m, and n are the fitting parameters.

Physical model experiments with a tilt angle α = 0°, 7°, 15°, and 21° were performed three times in order to ensure the repeatability of the results. An important point to be noted is that every physical model was assumed to be identical, but there were slight differences in the model construction and preparation process, which could have resulted in some dissimilarity between experiments. Taking this fact into account, the result of the capillary diversion length at different tilt angles (Figure 18a), which produced similar results, demonstrated the repeatability of the experiment. The pictures of failure situations in case III (tilt angle α = 7°) are described in detail in Figure 18b. Although the details of soil piping of each experiment were unique, the results of different experiments were consistent.

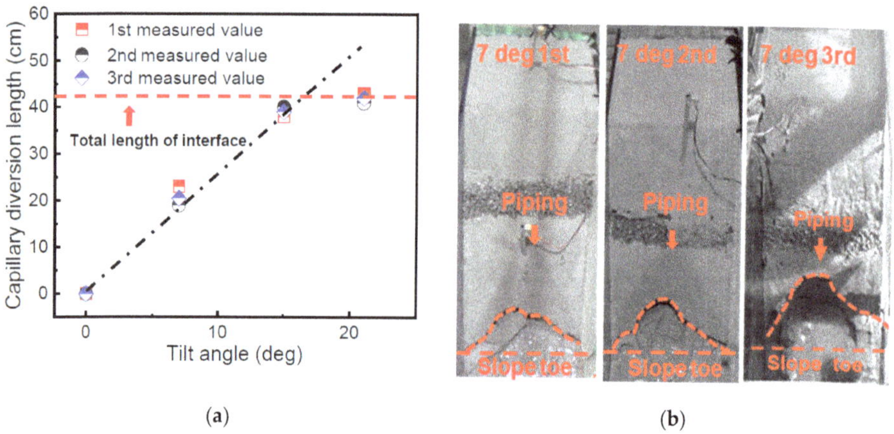

Figure 18. (**a**) Repeatability of the capillary diversion length (tilt angle α = 0°, 7°, 15°, and 21°, where each case was performed three times to ensure the repeatability of the results); (**b**) experiments of failure situations were conducted in case III under the same experimental conditions three times.

The capillary barrier effects caused by the inter-mediate coarser layer initially confines the rainwater infiltration in the uppermost soil. Depending on the characteristics of rainfall and the inclined angle condition, top finer soils are almost permanently unsaturated [32]. In some slope steeper than 40° at higher elevations, this may lead to instability of the top fine soil before capillary barrier breakthrough, when the coarse layer and bottom layer are still far from the point of saturation [33]. Depending on the slope angle and shear strength of soil, failure can occur at the bottom before complete saturation, while the pore water pressure is still negative. These phenomenon in natural slope seem to disprove the possibility of failure mechanism due to piping in this study.

In our experiment, the piping failure occurred above the impermeable layer after the loss of the capillary barrier. This is consistent with the field investigation of the development of soil pipe in the base of the fine soil [34]. However, the actual mechanism of soil piping in these slope that still remains unclear. Some mechanisms have been proposed by previous researches to explain this phenomenon such as internal erosion, flowslides and grain coarsening [34,35]. Internal erosion of the finer soil fraction driven by the seepage forces is thought to have played a significant influence on the slope failure [36,37]. It seems reasonable that the rainfall water caused the build-up of excess pore water pressure inside slope and the water inflow occurred during the rainfall condition, as the sharp increase of pore water pressure Δu measured at base of multi-layered slope (see Figure 10a). Further measurement such as the pore water pressure inside slope, internal displacements are still needed to improve the understanding of mechanism of failure of the slope [38].

5. Conclusions

Four groups of laboratory model experiments were performed to investigate the water movement, failure time, and modes in multi-layer and single-layer slopes caused by rainfall infiltration. In addition, the unsaturated hydraulic conductivity and SWCCs were also measured in the lab to clarify how the capillary barrier works under different conditions.

The results of hydraulic conductivity show that the unsaturated hydraulic conductivity K in the coarse layer is lower than that in the fine layer in a lower suction condition (0.4 kPa), which results in the development of the capillary barrier at the interface of the fine-coarse layer.

Different failure modes occurred in an inclined multi-layer slope and single-layer slope: sliding and piping failure. In the flat group, the capillary barrier was presented, which prevented the rainwater from infiltrating into the coarse layer for a while and caused a delay of the failure time. However, in the inclined group, the inclined intermediate coarser layer formed a capillary barrier, resulting in a significant amount of water being diverted to the downward slope and causing piping failure at the toe of the slope that resulted in earlier failure, which has a negative influence on the slope stability.

An increased tilt angle has a drastic effect on the capillary diversion length, in which more infiltrate water will be diverted to the downslope side and then infiltrate into the bottom of slope, resulting in quicker failure times.

The present study does not provide a model to be used in a specific site problem. Instead, the model is suitable for studies on hypothetical multi-layer hillsides to assess water movement patterns and general failure mechanisms. The results from such studies can prove useful in the development of an appropriate strategy for resolving problems in individual, site-specific multi-layer slopes.

In the present manuscript, we have mainly focused on the influence of the tilt angle and the effect of the capillary barrier on water movement and slope failure modes compared with a single-layer slope. It should be noted that in multi-layer slopes, different geometric characteristics, such as different heights, slope ratios, and fronts, also have an influence on the water infiltration and water content distribution in the slopes, significantly affecting slope failure initiation. Model experiments with different geometries of the shallow soil cover will be conducted in further research [39].

Author Contributions: Conceptualization, J.T. and U.T.; investigation, D.H.; methodology, S.T.; resources, J.X.; supervision, U.T.; Writing—original draft, J.T. All authors have read and agreed to the published version of the manuscript.

Funding: This research was supported by a Japanese Government Scholarship for PhD studies of the first author, the Sichuan Science and Technology Program (2018JY0613), Grants-in-Aid for Scientific Research of Japan Society for the Promotion of Science (JSPS), and Research Fellowships of Japan Society for the Promotion of Science for Young Scientists.

Acknowledgments: The authors would like to thank the reviewers for constructive comments that improved the paper.

Conflicts of Interest: The authors declare no conflict of interest.

References

1. Santi, P.M. Debris-flow hazards and related phenomena: (Edited by matthias jakob and oldrich hungr). *Environ. Eng. Geosci.* **2007**, *13*, 75. [CrossRef]

2. Sidle, R.C.; Ochiai, H. Landslides: Processes, prediction, and land use. In *Water Resources Monograph*; American Geophysical Union: Washington, DC, USA, 2006.
3. Mancarella, D.; Simeone, V. Capillary barrier effects in unsaturated layered soils, with special reference to the pyroclastic veneer of the Pizzo d'Alvano, Campania, Italy. *Bull. Eng. Geol. Environ.* **2012**, *71*, 791–801. [CrossRef]
4. Webb, R.W.; Fassnacht, S.R.; Gooseff, M.N.; Webb, S.W. The presence of hydraulic barriers in layered snowpacks: TOUGH2 simulations and estimated diversion lengths. *Transp. Porous Media* **2018**, *123*, 457–476. [CrossRef]
5. Khire, M.V.; Benson, C.H.; Bosscher, P.J. Capillary Barriers: Design Variables and Water Balance. *J. Geotech. Geoenviron. Eng.* **2000**, *126*, 695–708. [CrossRef]
6. Abdolahzadeh, A.M.; Vachon, B.L.; Cabral, A.R. Evaluation of the effectiveness of a cover with capillary barrier effect to control percolation into a waste disposal facility. *Can. Geotech. J.* **2011**, *48*, 996–1009. [CrossRef]
7. Rahardjo, H.; Santoso, V.A.; Leong, E.C.; Ng, Y.S.; Tam, C.P.H.; Satyanaga, A. Use of recycled crushed concrete and Secudrain in capillary barriers for slope stabilization. *Can. Geotech. J.* **2013**, *50*, 662–673. [CrossRef]
8. Garcia, E.F.; Gallage, C.P.K.; Uchimura, T. Function of permeable geosynthetics in unsaturated embankments subjected to rainfall infiltration. *Geosynth. Int.* **2007**, *14*, 89–99. [CrossRef]
9. Morel-Seytoux, H.J.; Nimmo, J.R. Soil water retention and maximum capillary drive from saturation to oven dryness. *Water Resour. Res.* **1999**, *35*, 2031–2041. [CrossRef]
10. Kämpf, M.; Holfelder, T.; Montenegro, H. Identification and parameterization of flow processes in artificial capillary barriers. *Water Resour. Res.* **2003**, *39*, 1–9. [CrossRef]
11. Rulon, J.J.; Freeze, R.A. Multiple seepage faces on layered slopes and their implications for slope-stability analysis. *Can. Geotech. J.* **1985**, *22*, 347–356. [CrossRef]
12. Hideki, I. Relationship between slope failure and the active fault based on kumamoto earthquake in 2016 and following heavy rainfall induced disaster. *J. Jpn. Soc. Eng. Geol.* **2017**, *58*, 188–196.
13. Matsushi, Y.; Hattanji, T.; Matsukura, Y. Mechanisms of shallow landslides on soil-mantled hillslopes with permeable and impermeable bedrocks in the Boso Peninsula, Japan. *Geomorphology* **2006**, *76*, 92–108. [CrossRef]
14. Hamrouni, F.; Trabelsi, H.; Jamei, M.; Olivella, S. Numerical analysis of landslides caused by rainfall in a reduced physical slope model. *Eur. J. Environ. Civ. Eng.* **2019**, *0*, 1–22. [CrossRef]
15. Hillel, D. *Environmental Soil Physics: Fundamentals, Applications, and Environmental Considerations*; Elsevier: Amsterdam, The Netherlands, 1998; ISBN 0080544150.
16. Wang, C.; Mao, X.; Hatano, R. Modeling ponded infiltration in fine textured soils with coarse interlayer. *Soil Sci. Soc. Am. J.* **2014**, *78*, 745–753. [CrossRef]
17. Yang, H.; Rahardjo, H.; Leong, E.C.; Fredlund, D.G. A study of infiltration on three sand capillary barriers. *Can. Geotech. J.* **2004**, *41*, 629–643. [CrossRef]
18. Mancarella, D.; Doglioni, A.; Simeone, V. On capillary barrier effects and debris slide triggering in unsaturated layered covers. *Eng. Geol.* **2012**, *147–148*, 14–27. [CrossRef]
19. Damiano, E.; Greco, R.; Guida, A.; Olivares, L.; Picarelli, L. Investigation on rainwater infiltration into layered shallow covers in pyroclastic soils and its effect on slope stability. *Eng. Geol.* **2017**, *220*, 208–218. [CrossRef]
20. Hillel, D. *Introduction to Soil Physics*; Academic Press: Cambridge, MA, USA, 1982.
21. Van Genuchten, M.T. A closed-form equation for predicting the hydraulic conductivity of unsaturated soils1. *Soil Sci. Soc. Am. J.* **1980**, *44*, 892–898. [CrossRef]
22. Evett, S.R.; Tolk, J.A.; Howell, T.A. Soil profile water content determination: Sensor accuracy, axial response, calibration, temperature dependence, and precision. *Vadose Zone J.* **2006**, *5*, 894–907. [CrossRef]
23. Di Matteo, L.; Pauselli, C.; Valigi, D.; Ercoli, M.; Rossi, M.; Guerra, G.; Cambi, C.; Ricco, R.; Vinti, G. Reliability of water content estimation by profile probe and its effect on slope stability. *Landslides* **2018**, *15*, 173–180. [CrossRef]
24. Zhan, L.T.; Li, G.Y.; Jiao, W.G.; Wu, T.; Lan, J.W.; Chen, Y.M. Field measurements of water storage capacity in a loess-gravel capillary barrier cover using rainfall simulation tests. *Can. Geotech. J.* **2017**, *54*, 1523–1536. [CrossRef]
25. Sinai, G.; Zaslavsky, D.; Golany, P. The effect of soil surface curvature on moisture and yield—Beer sheba observation. *Soil Sci.* **1981**, *132*, 367–375. [CrossRef]

26. Yeh, W.W.-G. Reservoir Management and operations models: A state-of-the-art review. *Water Resour. Res.* **1985**, *21*, 1797–1818. [CrossRef]
27. Xie, J.; Uchimura, T.; Chen, P.; Liu, J.; Xie, C.; Shen, Q. A relationship between displacement and tilting angle of the slope surface in shallow landslides. *Landslides* **2019**, *16*, 1243–1251. [CrossRef]
28. Junfeng, T.; Taro, U.; Shangning, T.; Dong, H.; Jiren, X. Water movement and deformation in unsaturated multi-layered slope under heavy rainfall. *Int. J. Geomate* **2020**, *19*, 174–181. [CrossRef]
29. Morris, C.E.; John, C. Stormont parametric study of unsaturated drainage layers in a capillary barrier. *J. Geotech. Geoenviron. Eng.* **1999**, *125*, 1057–1065.
30. Morris, C.E.; Stormont, J.C. Capillary barriers and subtitle D covers: Estimating equivalency. *J. Environ. Eng.* **1997**, *123*, 3–10. [CrossRef]
31. Walter, M.T.; Kim, J.S.; Steenhuis, T.S.; Parlange, J.Y.; Heilig, A.; Braddock, R.D.; Selker, J.S.; Boll, J. Funneled flow mechanisms in a sloping layered soil: Laboratory investigation. *Water Resour. Res.* **2000**, *36*, 841–849. [CrossRef]
32. Calcaterra, D.; Parise, M.; Palma, B.; Pelella, L. The influence of meteoric events in triggering shallow landslides in pyroclastic deposits of Campania, Italy. *Hazards Mitig.* **2000**, *11*, 99–107.
33. Del Prete, M.; Guadagno, F.M.; Hawkins, A.B. Preliminary report on the landslides of 5 May 1998, Campania, southern Italy. *Bull. Eng. Geol. Environ.* **1998**, *57*, 113–129. [CrossRef]
34. Hu, W.; Dong, X.J.; Xu, Q.; Wang, G.H.; van Asch, T.W.J.; Hicher, P.Y. Initiation processes for run-off generated debris flows in the Wenchuan earthquake area of China. *Geomorphology* **2016**, *253*, 468–477. [CrossRef]
35. Wang, G.; Sassa, K. Factors affecting rainfall-induced flowslides in laboratory flume tests. *Géotechnique* **2001**, *51*, 587–599. [CrossRef]
36. Hu, W.; Scaringi, G.; Xu, Q.; Huang, R. Internal erosion controls failure and runout of loose granular deposits: Evidence from flume tests and implications for postseismic slope healing. *Geophys. Res. Lett.* **2018**, *45*, 5518–5527. [CrossRef]
37. Hu, W.; Scaringi, G.; Xu, Q.; Pei, Z.; Van Asch, T.W.J.; Hicher, P.Y. Sensitivity of the initiation and runout of flowslides in loose granular deposits to the content of small particles: An insight from flume tests. *Eng. Geol.* **2017**, *231*, 34–44. [CrossRef]
38. Huang, A.B.; Lee, J.T.; Ho, Y.T.; Chiu, Y.F.; Cheng, S.Y. Stability monitoring of rainfall-induced deep landslides through pore pressure profile measurements. *Soils Found.* **2012**, *52*, 737–747. [CrossRef]
39. Tao, S.; Uchimura, T.; Fukuhara, M.; Tang, J.; Chen, Y.; Huang, D. Evaluation of soil moisture and shear deformation based on compression wave velocities in a shallow slope surface layer. *Sensors* **2019**, *19*, 3406. [CrossRef]

applied sciences

MDPI

Article

Exploring the Impact of Multitemporal DEM Data on the Susceptibility Mapping of Landslides

Jiaying Li [1,2,3], Weidong Wang [1,2], Zheng Han [1,*], Yange Li [1] and Guangqi Chen [3]

1 School of Civil Engineering, Central South University, Changsha 410075, Hunan, China; 184801058@csu.edu.cn (J.L.); 701139@csu.edu.cn (W.W.); liyange@csu.edu.cn (Y.L.)
2 Key Laboratory of Heavy-haul Railway, Ministry of Education, Changsha 410075, Hunan, China
3 Department of Civil and Structural Engineering, Kyushu University, Fukuoka 819-0395, Japan; chen@civil.kyushu-u.ac.jp
* Correspondence: zheng_han@csu.edu.cn; Tel.: +86-18874163071

Received: 5 March 2020; Accepted: 2 April 2020; Published: 6 April 2020

Featured Application: Landslide susceptibility assessment and other geological disaster assessment issues.

Abstract: Digital elevation models (DEMs) are fundamental data models used for susceptibility assessment of landslides. Due to landscape change and reshaping processes, a DEM can show obvious temporal variation and has a significant influence on assessment results. To explore the impact of DEM temporal variation on hazard susceptibility, the southern area of Sichuan province in China is selected as a study area. Multitemporal DEM data spanning over 17 years are collected and the topographic variation of the landscape in this area is investigated. Multitemporal susceptibility maps of landslides are subsequently generated using the widely accepted logistic regression model (LRM). A positive correlation between the topographic variation and landslide susceptibility that was supported by previous studies is quantitatively verified. The ratio of the number of landslides to the susceptibility level areas (RNA) in which the hazards occur is introduced. The RNA demonstrates a general decrease in the susceptibility level from 2000 to 2009, while the ratio of the decreased level is more than fifteen times greater than that of the ratio of the increased level. The impact of the multitemporal DEM on susceptibility mapping is demonstrated to be significant. As such, susceptibility assessments should use DEM data at the time of study.

Keywords: multitemporal DEM; control factors; susceptibility assessment; LRM; historical landslide events

1. Introduction

Geohazards are some of the most uncontrolled impacts on local and global economies, as well as on people's livelihoods. Almost nowhere on the planet is free from the damage of geohazards [1]. An essential component of predicting possible geohazard zones is the identification of an area that is vulnerable to future landslides [2,3]. Landslides are common geohazards that destroy local resources and environments. Therefore, susceptibility assessments of landslides have been widely investigated to improve their capability for use with these hazards [4,5].

Current studies commonly focus on two major issues regarding the susceptibility assessments of landslides—selection of evaluation indices and establishment of a rational assessment model. Relevant reviews [6–9] demonstrate that exiting studies have proposed various remarkable models for susceptibility assessment of landslides, such as logistic regression models (LRMs) [10]. Meanwhile, in recent decades, the rapid development of machine learning algorithms in susceptibility assessments has been seen over time. LRMs are some of the most classic and commonly used methods, which have the advantages of minimal computation, high detection speed, and good adaptability [11]. LRMs have extensive applications in forecasting [12,13] and susceptibility assessment [14,15]. These studies provide a solid foundation for the susceptibility assessment of landslides, and advance the knowledge of susceptibility assessment with machine learning algorithms.

In addition to an assessment model, the rational selection of indices is quite important for susceptibility assessment of landslides. Generally, indices such as lithology, elevation, and topography are introduced as the controlling factors, which are strongly related to the susceptibility assessment of landslides [16–20]. The rational selection of these indices remains a subject of scientific debate in many studies because the importance of different indices may vary from case to case. A review of previous studies [21] provides a guideline for addressing this issue. However, the generalizability of these results is subject to certain limitations, while another essential problem regarding the temporal variation of indices and the influence of this on susceptibility is still confusing and requires more attention. The susceptibility result will be questionable if the data for one index is out of date owing to the absence of a current data source, which is the case in digital elevation models (DEMs). So far, however, few researchers have highlighted the impact of outdated data sources on the reliability of susceptibility results, and few previous studies have investigated the mechanisms involved.

A DEM is the digital representation of a terrain surface. These models have been widely applied in geohazard planning, terrain surface analysis, and other fields [22]. Generating a DEM generally involves data from different sources [23], including global multiresolution topography (GMRT) models, shuttle radar topography mission digital elevation models, and advanced space-borne thermal emission and reflection radiometer global digital elevation models (ASTER GDEMs).

Among the abovementioned indices, elevation, topography, and slope are closely related to the DEM data [24]. In the majority of the previous studies, it has been suggested that DEM data should be consistent with the study date [25–29]. However, because of landscape changes and reshaping processes (e.g., earthquakes and engineering construction), the DEM data usually show obvious temporal variation. An example of significant temporal variation of DEM data was shown by Cucchiaro et al., who substantiated that the DEM difference of the eastern Italian Alps within one month is up to 0.14 m [30]. Pineux et al. demonstrated that the changes experienced by a DEM over time are obvious and unpredictable [31]. DEM data have significant influence on susceptibility assessments of landslides, mainly through influencing factors such as elevation, topography, and slope. Elevation is one of the most influential factors controlling landslide occurrences in a study area, while topography and slope are also important in susceptibility assessments [18,29,32]. However, studies have not considered the obvious temporal variation in DEMs.

In the present paper, multitemporal DEMs area are obtained for the study and the elevation changes demonstrated in the DEMs are analyzed. The DEM data for three years (1992, 2000, and 2009) are used to analyze factors affected by a DEM. The susceptibility results for the three years are evaluated using the different DEM-dependent factors and the same DEM-independent factors. Quantifying the index, the susceptibility assessments are then obtained using a generally accepted LRM model. Additionally, the maps of susceptibility levels in the study area are obtained within the geographic information system (GIS). The influence of DEM data at different times on a susceptibility assessment and the influence of the specific values of the elevation differences on the assessment level are obtained. Meanwhile, the impact of historical landslide events on the susceptibility assessment is analyzed. The study aims to achieve a better evaluation of the impact of multitemporal factors on the susceptibility assessment of landslides.

2. Study Area

2.1. Study Area and Landslide Data

The southern area of Sichuan province in China is a region with frequently occurring landslides. The study area covers the three districts of Panzhihua, Liangshan, and Ya'an (Figure 1). The area is located at a latitude of 26°03′–30°56′ north and a longitude of 100°03′–103°52′ east. The area also belongs to the Sichuan Basin, adjacent to the Qinghai–Tibet Plateau in the west. It has a complex topography mainly composed of plains, hills, mountains, and plateaus; furthermore, the relative elevation is more than 3200 m.

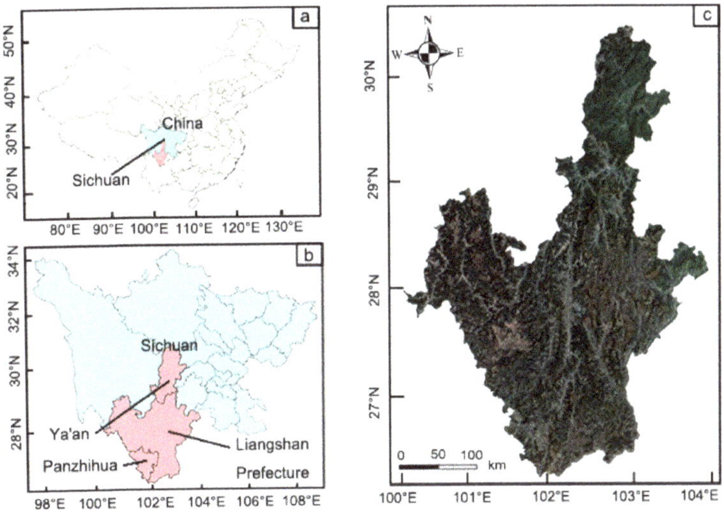

Figure 1. Location of the study area: (**a**) Sichuan province in China; (**b**) the study area; (**c**) an aerial photograph of the study area.

The climate in the study area is a typically humid subtropical monsoonal climate that is hot in summer and mild and humid in winter, with plentiful rainfall and mist. The vertical climate differs greatly, and the annual average temperature distribution gradually increases from northwest to southeast. The average temperature ranges from 5.3 to 15.7 °C, with annual average rainfall ranging from 1500 to 1800 mm. The study area is, therefore, extremely rainy. The rainfall is mainly concentrated from June to October, which accounts for approximately 70%–75% of the annual rainfall. The maximum daily precipitation reaches 300–500 mm. Meanwhile, part of the study area is located in the Himalaya earthquake zone, which suffers from permafrost, avalanches, and landslides. The stratum of the study area is complete (from Archean to Quaternary). The dominant lithology is sedimentary rock, majorly consisting of dolomite, limestone, siliceous rock, shale, and sandstone [33]. The surface water system is developed, and there are many tributaries of the Yangtze River. The groundwater is distributed widely and buried shallowly, which is affected significantly by the rainfall and landform [34].

The Chinese Geological Environment Monitoring Institute collected the landslides information in this region through the China Geological Survey, including locations and occurrence times, and released it to public in its Bulletin of National Geological Hazards [35]. The annual total number of historical landslide events is shown in Figure 2. More than 85% of the historical landslide events were induced by natural environmental factors, such as rainfall and earthquakes, while only a few events were induced by human factors, such as mining and slope cutting. As the source data in the Bulletin of National Geological Hazards are not distinguished by landslide type, the general landslides analyzed in the study are of various types.

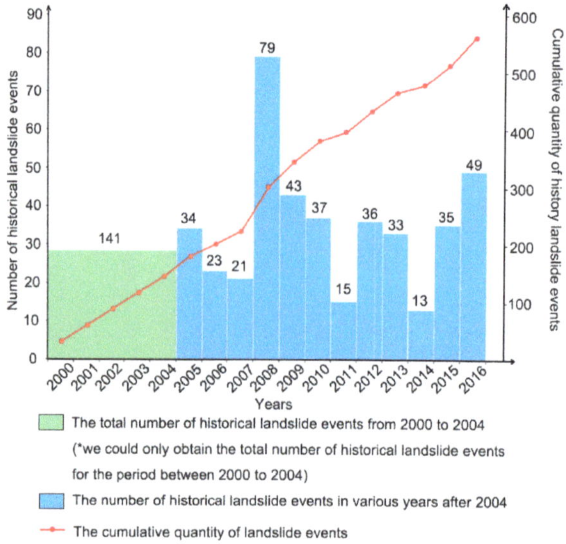

Figure 2. Number of historical landslide events in various years.

In this study, the total number of historical landslide events from 2000 to 2016 and the number of those hazards in various years after 2004 are obtained. However, we can only obtain the total number of historical landslide events for the period between 2000 and 2004. Thus, to better analyze the impact of a DEM on susceptibility assessment, we divide the historical landslide events into two broad types: the landslide events during the period ranging from 2000 to 2009, and the period ranging from 2010 to 2016. The historical landslide events in the study area from 2000 to 2009 and from 2010 to 2016 (Figure 3) can be obtained from the China Geological Environment Information database (http://www.cigem.cn). A total of 341 historical landslide events from 2000 to 2009 and 218 historical landslide events from 2010 to 2016 were recorded.

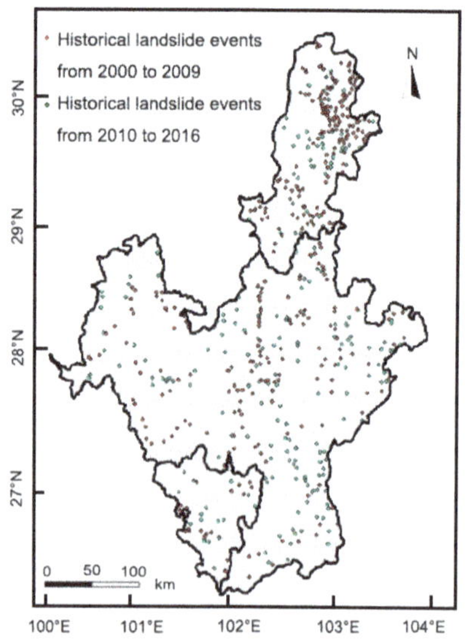

Figure 3. Locations of historical landslide events.

2.2. *Influencing Factors*

Selecting the influencing factors is a key step in a landslide susceptibility assessment [36,37]; however, the causes of landslides are complex. To date, hundreds of influencing factors have been identified that are potentially important to susceptibility assessments [38]. For example, areas with high elevation and steep slopes are highly prone to landslides [39]. Lithology is also related to landslide occurrences, because different lithologies can withstand different levels of triggering factors. The distances to structure lines, rivers, and roads have important impacts on the spread and size of landslides in the study area [40]. Pourghasemi et al. investigated global susceptibility during the period of 2005–2016 [28]. Nearly 100 factors were summarized to reveal the commonly used factors. In accordance with the relevant studies [29,41,42] and the study area, we select eight influencing factors: elevation, topography, slope, lithology, distance to a structure line, distance to a river, average annual rainfall, and distances to roads.

Table 1 lists all the DEM data for the study area and their sources. The Open Topography Facility provided the GMRT data, which is hosted at the San Diego Supercomputer Center, University of California San Diego. This facility has built a strong cyberinfrastructure framework for managing and processing high-resolution topography data from light detection and ranging (LiDAR) (http://opentopo.sdsc.edu/datasets?listAll=true). Meanwhile, the Shuttle Radar Topography Mission (SRTM) DEM and Global Digital Elevation Model (GDEM) DEM were provided by the Geospatial Data Cloud site of the Computer Network Information Center in the Chinese Academy of Sciences (http://www.gscloud.cn/). The DEM data at the same resolution for 1992, 2000, and 2009 were then obtained using a resampling tool in the GIS environment (Figure 4). There are three resampling methods in the resampling tool, namely nearest neighbor, bilinear interpolation, and cubic convolution methods. Nearest neighbor is selected in the present study because the method is simple, fast, and applicable.

Table 1. Digital elevation model (DEM) data used in this study and their sources.

DEM Data	Survey Date	Data Type	Resolution	Coordinates	Source
GMRT * Data Synthesis	1992	KML	90 m	WGS84	Open Topography
SRTM ** DEM	2000	IMG	90 m	UTM/WGS84	Geospatial Data Cloud site
GDEM *** DEM	2009	IMG	30 m	UTM/WGS84	Geospatial Data Cloud site

* GMRT is short for Global Multi-Resolution Topography. ** SRTM is short for Shuttle Radar Topography Mission. *** GDEM is short for Global Digital Elevation Model.

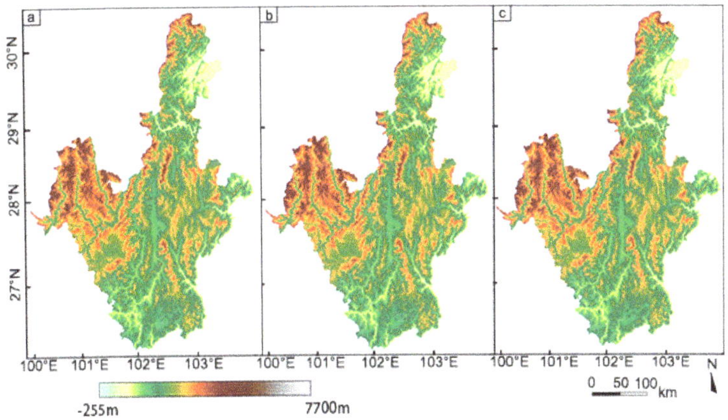

Figure 4. DEM data for the study area in (**a**) 1992, (**b**) 2000, and (**c**) 2009.

The DEM data for 1992, 2000, and 2009 were used to analyze factors affected by the DEM. Meanwhile, the data of other factors, namely lithology, distance to the structure line, distance to the river, and average annual rainfall, remained the same and were provided by China Railway Number 4 Engineering Group Co., Ltd. (745 Heping Road, Wuhan, Hubei, China), and the Roads and Traffic Authority of China. Because of the diversity of lithological layers in the study area, the lithology is divided into four groups in Table 2 [43]. Thus, the zoning maps of the four factors were obtained (Figure 5).

Table 2. Lithology and structure in the study area.

Groups	Structure	Lithology
Group 1	Loose structure	Clay, gravelly soil, clay rock, thin layer siltstone
Group 2	Cataclastic structure	Siltstone, shale, phyllite, thin layer slate
Group 3	Stratified structure	Thick layer sandstone, conglomerate with argillaceous rocks, siliceous rock with argillaceous shale
Group 4	Block structure	Limestone, siliceous rock, thick layer conglomerate, dolomite, phosphate rock

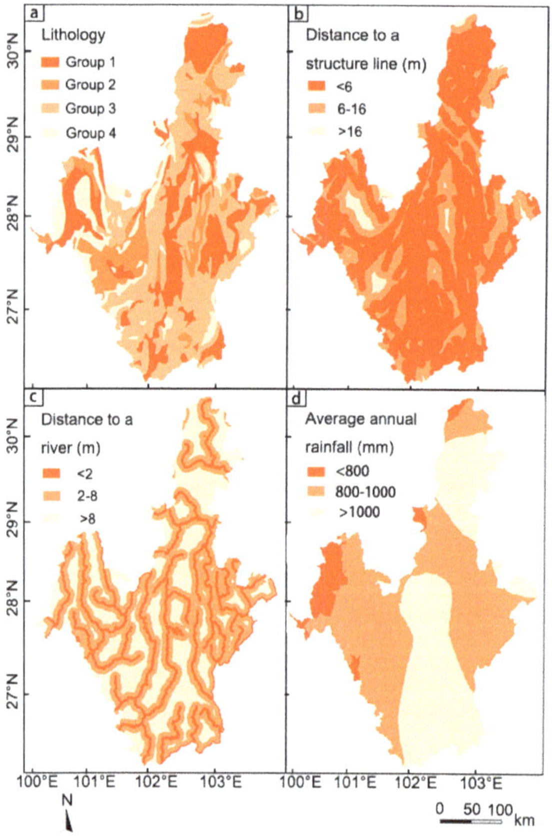

Figure 5. Zoning map of (**a**) lithology, (**b**) distance to a structure line, (**c**) distance to a river, (**d**) and average annual rainfall.

3. Methodology

3.1. Flowchart

The research methodologies applied in the present study are as shown in Figure 6. The flowchart consists of four major steps, as follows:

(a) Data preparation. The multitemporal DEM data, influencing factors, and historical landslide events are prepared;
(b) Data preprocessing. The multitemporal DEM data is compared to eliminate the noise based on coverage probabilities for confidence intervals;
(c) Susceptibility assessment of landslides. The assessment results for 1992, 2000, and 2009 are obtained using a LRM;
(d) Comparison and analysis. The assessment results for 1992, 2000, and 2009 are compared to obtain the impacts of the elevation difference and the historical landslide events on the susceptibility assessment.

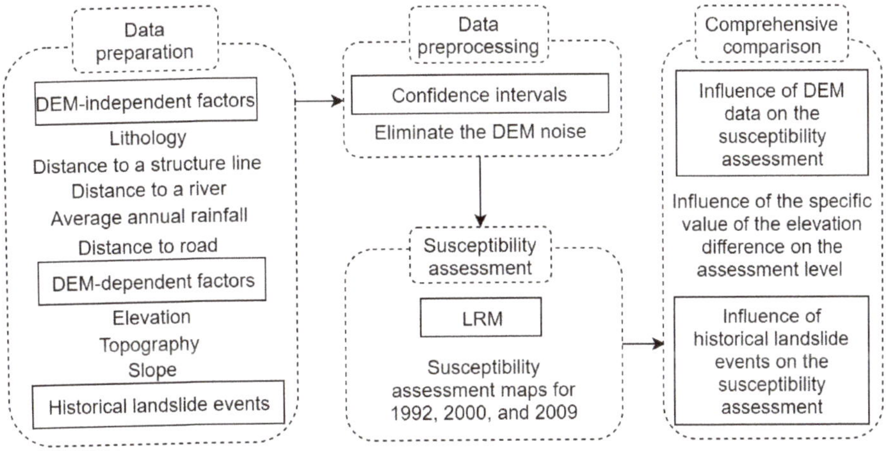

Figure 6. Methodology of research applied in this study. LRM—logistic regression model.

3.2. Confidence Interval

In the present study, because of the noise in the DEM data reducing the accuracy of data analysis significantly, not all DEM data are reliable. Therefore, the noise needs to be eliminated to determine the reliability of the DEM data. The confidence level of the DEM data is assessed to improve the accuracy of data analysis.

The confidence interval is a commonly-used interval estimation method used for sample statistics, which shows the confidence probability of the measured parameter value [42,44]. The elevation difference is considered as the sample data in the present study. The average value of elevation difference is μ, and the standard deviation is σ. The confidence probability is obtained as follows:

$$\Pr(c_1 \leq \mu < c_2) = 1 - \alpha \quad (1)$$

where α is the significance level and the interval (c_1, c_2) is the confidence interval. Therefore, the confidence interval of the average value is $(\mu - \sigma Z_{\alpha/2}, \mu + \sigma Z_{\alpha/2})$, where = $Z_{\alpha/2}$ is the corresponding standard score. In general, the confidence probability in the literature [42,45,46] is 90% or 95%, and $Z_{\alpha/2}$ is 1.645 or 1.96, respectively.

3.3. LRM

The LRM is a statistical model used to predict the probability of a categorical occurrence using one or more independent variables [47]. The purpose of LRM is to obtain the relationship between a dependent variable and multiple independent variables that have been identified.

In the present study, the independent variables are the influencing indices of landslides, and the dependent variable is the probability of the landslide occurring. By transforming the universal formula of LRM, Y is obtained:

$$Y = C_0 + \sum_{i=1}^{n} C_i I_i \tag{2}$$

where C_0 is the LRM constant coefficient, C_i is the LRM coefficient, I_i is the landslide index, $Y = \ln(P/(1 - P))$, and P is the probability of the landslide occurring. The LRM coefficient C_i and the LRM formula can be obtained using Statistical Product and Service Solutions (SPSS) software.

3.4. Ratio of Number of Landslides to Area (RNA)

RNA denotes the ratio of the number of historical landslide events to the area of the susceptibility levels at which historical landslide events are located. The RNA of the level i is obtained with the following equation:

$$RNA_i = n_i / A_i \tag{3}$$

where RNA_i is the number of landslides per area at a particular level i; n_i is the number of historical landslide events at level i; and A_i is the area of level i in the assessment.

4. Results

4.1. Elimination of Noise

As mentioned above, the DEM has a significant effect on susceptibility assessment through three indices, namely elevation, topography, and slope. The changes in the DEM from 1992 to 2000 and from 2000 to 2009 are obtained to explore the impact of the temporal variation of DEM data on susceptibility assessment. Meanwhile, the average values and standard deviation of the elevation differences are obtained from the topography changes revealed by the DEM data. The confidence probability is considered as 95% in this study. The corresponding standard scores and confidence intervals are, therefore, calculated (Table 3). After the noise is eliminated, the optimized map of topography changes from 1992 to 2000 and from 2000 to 2009 are obtained (Figure 7).

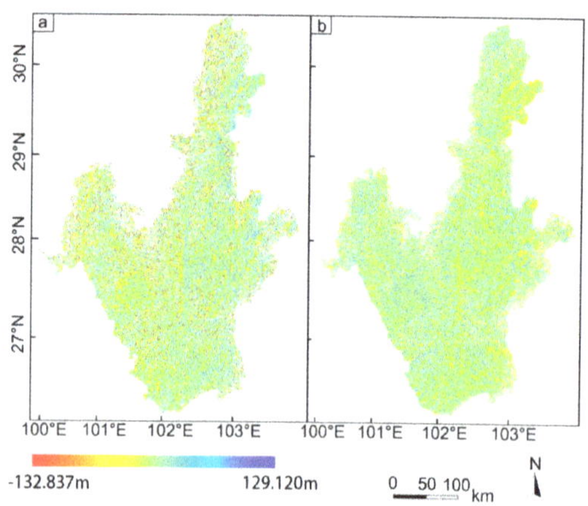

Figure 7. Topography changes revealed by the DEM data: (**a**) from 1992 to 2000; (**b**) from 2000 to 2009.

Table 3. Confidence interval of elevation differences for the periods ranging from 1992 to 2000 and from 2000 to 2009.

Data	Average Value (m)	Standard Deviation (m)	Standard Score	Confidence Interval (m)
From 1992 to 2000	−1.859	66.826	1.960	(−132.837, 129.120)
From 2000 to 2009	0.965	59.204	1.960	(−115.075, 117.005)

4.2. Susceptibility Assessments of Landslides

The susceptibility assessment maps of the study area in 1992, 2000, and 2009 are obtained in the GIS environment using LRM (Figure 8 and Table 4). The susceptibility is classified into four levels based on the natural breaks classification (NBC) method, which is based on natural groupings inherent in data. The NBC identifies groups of similar values and maximizes the differences between classes. The features are divided into classes whose boundaries are set based on relatively big differences in the data values. Here, level I, level II, level III, and level IV denote low, moderate, high, and very high susceptibility, respectively. The area percentages of various susceptibility levels in the study are shown in Table 4.

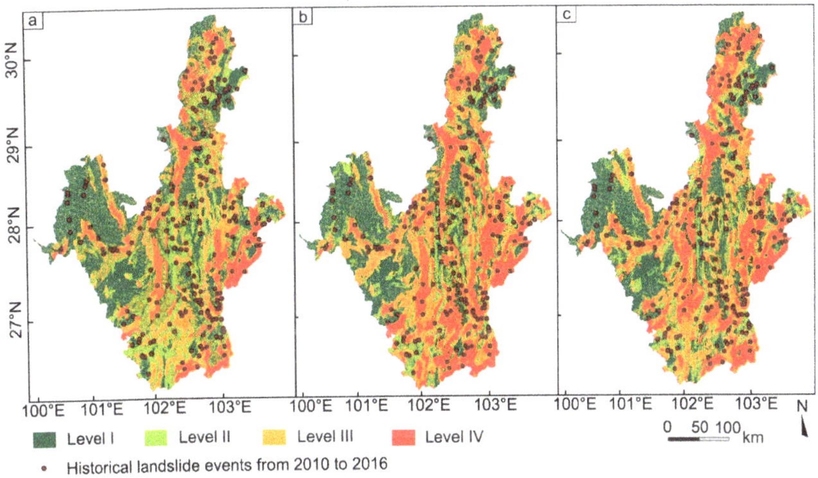

Figure 8. Susceptibility assessment maps for (**a**) 1992, (**b**) 2000, and (**c**) 2009.

Table 4. Susceptibility assessment of landslides using LRM scores.

Susceptibility Level	1992		2000		2009	
	Area (10^3 km^2)	Percentage (%)	Area (10^3 km^2)	Percentage (%)	Area (10^3 km^2)	Percentage (%)
Level I	26.445	34.23	22.342	28.91	23.378	30.21
Level II	26.127	33.82	17.969	23.25	15.781	20.39
Level III	13.573	17.57	17.128	22.16	20.747	26.81
Level IV	11.116	14.39	19.842	25.68	17.473	22.58

As shown in Figure 8 and Table 4, the results of the susceptibility assessment vary from year to year. The areas with low and moderate susceptibility (levels I and II, respectively) in 1992 are the largest, while the area with very high susceptibility (level IV) is the largest in 2000. The rapid decrease in the level IV area from 2000 to 2009 is noticeable. One possible explanation for the decrease is that the slopes where landslides occurred became stable in the short term, owing to the excessive landslides that occurred from 2000 to 2009; thus, the susceptibility level of the study area decreased after the hazards.

The susceptibility maps are obtained using different DEM-dependent factors, while the DEM-independent factors are constant in the assessment. Comparing these results illustrates the significant influence of the DEM data from different years on the susceptibility assessment.

4.3. Assessment Levels of Historical Landslide Events

Figure 8 and Table 5 show the susceptibility assessment levels at which historical landslide events occurred from 2010 to 2016.

Table 5. Historical landslide event levels.

Susceptibility Level	1992		2000		2009	
	Number	RNA * ($10^3/km^2$)	Number	RNA ($10^3/km^2$)	Number	RNA ($10^3/km^2$)
Level I	33	1.25	34	1.52	26	1.11
Level II	79	3.02	44	2.45	38	2.41
Level III	53	3.90	60	3.50	59	2.84
Level IV	53	4.77	80	4.03	95	5.44

* RNA denotes the ratio of number of landslides to area.

It can be clearly seen in Table 5 that the combined number of historical landslide events designated as levels III and IV is 106, 140, and 154, respectively, in 1992, 2000, and 2009. The RNA for level IV in 2009 is $5.44 \times 10^{-3}/km^2$, which is the largest among the RNA values over the three years. The most ideal assessment of landslides would be most landslides occurring in level IV areas. There is a significant positive correlation between the RNA and the accuracy of the assessment. Therefore, the results of the susceptibility assessment using the DEM for 2009 are more accurate than the results using the DEM for 1992 and 2000.

It is apparent from Table 5 that the small change in elevation over time has a great influence on the susceptibility assessment of landslides. However, in previous research, the survey data are not explained or the data are significantly different from the assessment time.

4.4. Differences between the Assessment Levels

Comparisons are performed using the susceptibility assessments of landslides in the study area from different periods. Figure 9 and Table 6 compare the differences between the assessment results for 1992 and 2000 and between those for 2000 and 2009, showing that the area where the level difference is zero is the largest area. Meanwhile, the area with increased susceptibility is much larger than the area with decreased susceptibility in the period ranging from 1992 to 2000, which is contrary to the results for the period ranging from 2000 to 2009.

Table 6. Differences in the assessment levels.

Level Difference	1992 and 2000		2000 and 2009	
	Area (10^3 km^2)	Percentage (%)	Area (10^3 km^2)	Percentage (%)
−3	0.041	0.053	0.516	0.667
−2	0.633	0.819	1.232	1.592
−1	5.433	7.030	13.308	17.193
0	41.923	54.244	48.466	62.614
1	26.954	34.876	12.799	16.535
2	1.928	2.495	0.893	1.154
3	0.373	0.483	0.190	0.245

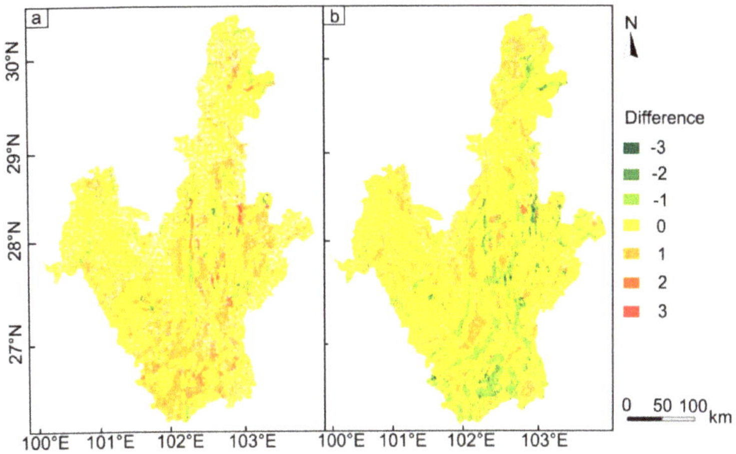

Figure 9. Maps of the differences between the assessment levels: (**a**) difference between 1992 and 2000; (**b**) difference between 2000 and 2009.

5. Discussion

5.1. Comprehensive Comparison of Assessment Results for 1992, 2000, and 2009

We compare the areas and RNA values of various assessment levels for the years 1992, 2000, and 2009 (Figure 10a,b), and the level differences from 1992 to 2000 and from 2000 to 2009 (Figure 10c). Figure 10 is quite revealing in several ways. First, the RNA for level IV in 2009 is larger than that in 1992 or 2000. Because of the positive correlation between the RNA and the accuracy of the assessment, the accuracy of the assessment is further exemplified in studies using the DEM for 2009. As Figure 10c shows, there is a significant difference between the two results. The area with increased susceptibility in the period ranging from 1992 to 2000 is larger than that in in the period ranging from 2000 to 2009. The observed increase in the area with increased susceptibility could be attributed to the frequent landslides from 2000 to 2009 (e.g., the various landslides caused by the 2008 earthquake).

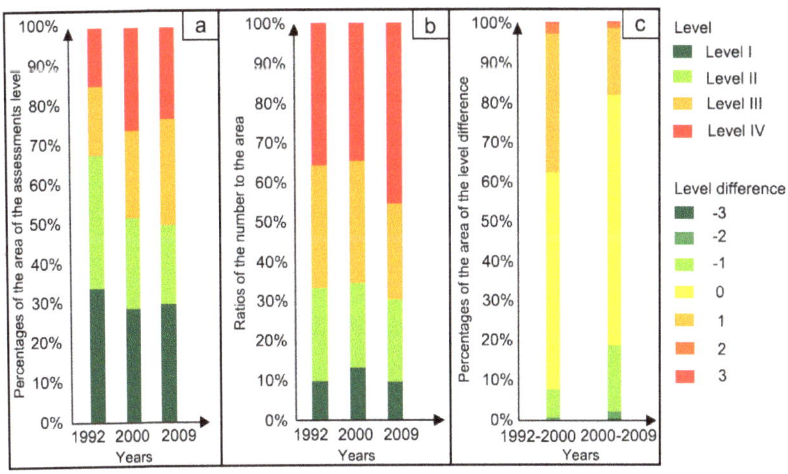

Figure 10. Comprehensive comparison of the percentages of (**a**) the assessment levels, (**b**) the RNA values of various assessment levels, and (**c**) the level differences.

5.2. Impact of Elevation Difference on the Susceptibility Assessment

In addition to the above analysis, we analyze the impact of elevation differences on the susceptibility assessment and the relationships between the elevation differences and the differences in assessment levels. Therefore, the areas with varying level and elevation differences are presented (Table 7).

Table 7. Numbers of historical landslide events at various assessment levels from 1992 to 2000 and from 2000 to 2009.

Area (km^2)			Level Difference						
			−3	−2	−1	0	1	2	3
Elevation differences (m)	From 1992 to 2000	−140 to −100	1.30	9.26	104.72	774.13	392.15	17.64	2.48
		−100 to −60	3.26	37.75	357.01	2921.89	1763.27	101.98	15.73
		−60 to −20	7.41	123.31	938.88	8591.50	5593.55	380.83	74.53
		−20 to 20	23.08	351.42	2576.43	17,433.58	11,260.52	882.75	170.13
		20–60	4.74	85.47	1016.90	8718.41	5753.17	417.77	85.52
		60–100	1.19	20.40	352.12	2867.55	1774.61	105.39	21.12
		100–1400	0.38	5.49	92.77	667.04	444.31	22.78	4.15
	From 2000 to 2009	−140 to −100	3.36	12.28	147.25	470.17	149.40	14.39	1.74
		−100 to −60	20.63	66.89	834.75	3121.54	955.57	58.83	13.3
		−60 to −20	90.56	246.75	2709.11	10,030.34	2757.22	155.02	37.61
		−20 to 20	299.11	629.70	5964.38	20,745.68	4914.82	370.04	74.94
		20–60	88.4	228.38	2748.58	9998.40	2771.69	202.98	40.93
		60–100	12.2	43.28	790.68	3456.15	1050.63	79.12	18.63
		100–1400	1.8	4.89	113.25	644.07	199.38	13.01	2.90

Table 7 demonstrates that the decrease in elevation results in a decrease in the susceptibility assessment level. There is a positive correlation between the elevation difference and the difference in the assessment level. Regardless of the level difference, the elevation difference is much larger in the area ranging from −20 to 20 m than in other areas.

5.3. Impact of Historical Landslide Events on Susceptibility Assessment

Most of the factors influencing susceptibility assessments have been explored in several studies [48–50]. However, much of the historical research overlooks the impact of historical landslide events on the assessment [51]. In contrast, because the slopes where landslides occur become stable in the short term following these events, the susceptibility of the area will consequently decrease. Therefore, historical landslide events have a significant effect on susceptibility assessments.

The numbers and percentages of historical landslide events at various assessment levels from 2000 to 2009 and the RNA values of the various assessment levels are investigated, allowing the change in the assessment levels of the historical landslide events to be obtained (Table 8 and Figure 11).

In Figure 11, the circles represent the percentages of historical landslide events or the RNA values at various assessment levels. It can be seen from Figure 11 that over one-third of the historical landslide events (38.71%) occur at a decreasing assessment level. A total of 54.24% of the historical landslide events are within the same assessment level. If the assessment level area is larger, then there may be more historical landslide events within that assessment level. To rule out the possible influence of area size, the RNA is considered to be the most important factor. The RNA at decreased levels ($81.31 \times 10^{-3}/\mathrm{km}^2$) is more than fifteen times greater than the RNA at increased levels ($5.25 \times 10^{-3}/\mathrm{km}^2$). Previous studies [52,53] do not take into account the impact of historical landslide events; therefore, the above results can establish the importance of historical landslide events for susceptibility assessments of landslides.

Table 8. Numbers and percentages of historical landslide events at various assessment levels and RNA values of various assessment levels.

Values			2000			
			Level I	**Level II**	**Level III**	**Level IV**
2009	Number	Level I	68	25	4	8
		Level II	2	35	28	20
		Level III	0	13	23	47
		Level IV	0	0	9	59
	Percentage	Level I	19.94%	7.33%	1.17%	2.35%
		Level II	0.59%	10.26%	8.21%	5.87%
		Level III	0	3.81%	6.74%	13.78%
		Level IV	0	0	2.64%	17.3%
	RNA (10^{-3}/km^2)	Level I	3.73	6.36	5.80	15.50
		Level II	0.59	3.93	9.50	36.83
		Level III	0.00	2.69	2.57	7.31
		Level IV	0.00	0.00	1.97	4.77

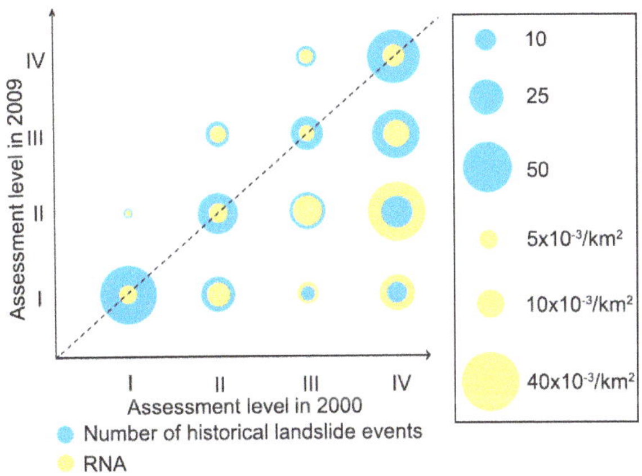

Figure 11. Changes in the assessment levels of historical landslide events from 2000 to 2009.

5.4. Advantages and Limitations

The elevation, topography, and slope—depending on the DEM—are usually the main factors affecting susceptibility [23]. The existing studies assume that the influencing factors for susceptibility are constant over time [15,54]. However, the DEM data always change significantly over time owing to landscape reshaping, such as landslides and engineering construction [30,31,55]. A susceptibility assessment is, therefore, closely related to the data source [56], and as such the susceptibility maps of the study area are not consistent.

Despite this, the existing studies have analyzed the susceptibility of landslides using a single time-related data source without considering whether the date of the data source is consistent with the study date [57,58]. Therefore, it is necessary to understand the temporal evolution of influencing factors (e.g., topography) and the influencing mechanisms of the factors on the susceptibility assessment [59,60]. The changes in influencing factors are significant in earthquake- and landslide-prone areas [61], such as the study area. Therefore, more research regarding the impacts of factors on the susceptibility of landslides needs to be undertaken. It is better not to evaluate susceptibility simply using constant factors.

The present study performs a preliminarily analysis of the impact of multitemporal DEM on susceptibility and shows that DEM variation has an impact on the susceptibility assessment using DEM-dependent factors. The analysis undertaken here can extend the knowledge regarding the impacts of multitemporal data on susceptibility and can aid preliminary studies on the mechanism causing the influence of factor variation.

DEM variation closely relates to local geotechnical properties, which change the susceptibility assessment of landslides. However, the reasons for DEM variation are debated. The mechanism explaining the consequent influences of geotechnical properties on DEM variation remains a scientific challenge. In addition, considering the large spatial extent and period (about 77,000 km^2, 17 years) in the present study, changes between the three susceptibility maps could not be completely explained by DEM variation. There are also data limitations and difficulties in analyzing the interactions of various factors.

6. Conclusion

In the present paper, DEM data for 1992, 2000, and 2009 are obtained to evaluate the susceptibility of landslides. Based on the DEM-dependent factors and the same DEM-independent factors, the susceptibility results are evaluated using LRM.

From the assessment, we find that the results are different by using DEM data from different times. The rapid decrease in the area of level IV from 2000 to 2009 is noticeable. The RNA for level IV in 2009 is larger than in 1992 and 2000. The area of increased susceptibility is much larger than that of decreased susceptibility based on the results of the assessment level differences from 1992 to 2000, which differ from the results from 2000 to 2009.

From the assessment results, we conclude that the DEM data have an impact on the susceptibility assessment in the study area. It is also worth noting that the influences of the specific values of the elevation differences on the assessment results are obtained. Meanwhile, the influence of historical landslide events on the susceptibility assessment is obtained by analyzing the hazard data and the differences between the assessment results. The most obvious finding to emerge from the analysis is that the assessment level of the area with historical landslide events decreases. The present study goes some way towards enhancing our understanding of the impacts and the mechanism of multitemporal factors on susceptibility.

Author Contributions: W.W., Z.H., and G.C. designed the study. J.L. and Y.L. performed the in situ investigation. Z.H. and J.L. wrote the manuscript. All authors have read and agreed to the published version of the manuscript.

Funding: This study was financially supported by the National Key R&D Program of China (Grant No. 2018YFC1505401), the National Natural Science Foundation of China (Grant No. 51478483, W.D. Wang; and Grant No. 41702310, Z. Han), and the Natural Science Foundation of Hunan (Grant No. 2018JJ3638) These financial supports are gratefully acknowledged. Thanks to the Scholarship of Central South University for Undergraduate Students Studying Abroad.

Conflicts of Interest: The authors declare no conflict of interest.

References

1. Wang, W.D.; Li, J.Y.; Qu, X.; Han, Z.; Liu, P. Prediction on landslide displacement using a new combinationmodel: A case study of Qinglong landslide in China. *Nat. Hazards* **2019**, *96*, 1121–1139. [CrossRef]
2. Han, Z.; Chen, G.Q.; Li, Y.G.; Zhang, H.; He, Y. Elementary analysis on the bed-sediment entrainment by debris flow and its application using the TopFlowDF model. *Geomat. Nat. Hazards Risk* **2016**, *7*, 764–785. [CrossRef]
3. Hong, H.; Pourghasemi, H.R.; Pourtaghi, Z.S. Landslide susceptibility assessment in Lianhua County (China): A comparison between a random forest data mining technique and bivariate and multivariate statistical models. *Geomorphology* **2016**, *259*, 105–118. [CrossRef]
4. Patriche, C.V.; Pirnau, R.; Grozavu, A.; Rosca, B. A Comparative Analysis of Binary Logistic Regression and Analytical Hierarchy Process for Landslide Susceptibility Assessment in the Dobrov River Basin, Romania. *Pedosphere* **2016**, *26*, 335–350. [CrossRef]
5. Cao, J.; Zhang, Z.; Wang, C.; Liu, J.; Zhang, L. Susceptibility assessment of landslides triggered by earthquakes in the Western Sichuan Plateau. *Catena* **2019**, *175*, 63–76. [CrossRef]

6. Huang, Y.; Zhao, L. Review on landslide susceptibility mapping using support vector machines. *Catena* **2018**, *165*, 520–529. [CrossRef]
7. Pham, B.T.; Pradhan, B.; Tien Bui, D.; Prakash, I.; Dholakia, M.B. A comparative study of different machine learning methods for landslide susceptibility assessment: A case study of Uttarakhand area (India). *Environ. Model. Softw.* **2016**, *84*, 240–250. [CrossRef]
8. Han, Z.; Chen, G.; Li, Y.; Tang, C.; Xu, L.; He, Y.; Huang, X.; Wang, W. Numerical simulation of debris-flow behavior based on the SPH method incorporating the Herschel-Bulkley-Papanastasiou rheology model. *Eng. Geol.* **2019**, *255*, 26–36. [CrossRef]
9. Han, Z.; Chen, G.Q.; Li, Y.G.; Wang, W.; Zhang, H. Exploring the velocity distribution of debris flows: An iterationalgorithm based approach for complex cross-sections. *Geomorphology* **2015**, *241*, 72–82. [CrossRef]
10. Li, F.; Wang, W.; Xu, J.; Yi, J.; Wang, Q. Comparative study on vulnerability assessment for urban buried gas pipeline network based on SVM and ANN methods. *Process Saf. Environ. Prot.* **2019**, *122*, 23–32. [CrossRef]
11. Pradhan, B.; Lee, S. Delineation of landslide hazard areas on Penang Island, Malaysia, by using frequency ratio, logistic regression, and artificial neural network models. *Environ. Earth Sci.* **2010**, *60*, 1037–1054. [CrossRef]
12. Chiri, H.; Abascal, A.J.; Castanedo, S.; Medina, R. Mid-long term oil spill forecast based on logistic regression modelling of met-ocean forcings. *Mar. Pollut. Bull.* **2019**, *146*, 962–976. [CrossRef] [PubMed]
13. Zhou, F.; Zhang, Q.; Sornette, D.; Jiang, L. Cascading logistic regression onto gradient boosted decision trees for forecasting and trading stock indices. *Appl. Soft Comput.* **2019**, *84*, 105747. [CrossRef]
14. Tsangaratos, P.; Ilia, I. Comparison of a logistic regression and Naïve Bayes classifier in landslide susceptibility assessments: The influence of models complexity and training dataset size. *Catena* **2016**, *145*, 164–179. [CrossRef]
15. Lin, G.-F.; Chang, M.-J.; Huang, Y.-C.; Ho, J.-Y. Assessment of susceptibility to rainfall-induced landslides using improved self-organizing linear output map, support vector machine, and logistic regression. *Eng. Geol.* **2017**, *224*, 62–74. [CrossRef]
16. Jacobs, L.; Kervyn, M.; Reichenbach, P.; Rossi, M.; Marchesini, I.; Alvioli, M.; Dewitte, O. Regional susceptibility assessments with heterogeneous landslide information: Slope unit- vs. pixel-based approach. *Geomorphology* **2020**, *356*, 107084. [CrossRef]
17. Bui, D.T.; Tsangaratos, P.; Nguyen, V.-T.; Liem, N.V.; Trinh, P.T. Comparing the prediction performance of a Deep Learning Neural Network model with conventional machine learning models in landslide susceptibility assessment. *Catena* **2020**, *188*, 104426. [CrossRef]
18. Li, Y.; Liu, X.; Han, Z.; Dou, J. Spatial Proximity-Based Geographically Weighted Regression Model for Landslide Susceptibility Assessment: A Case Study of Qingchuan Area, China. *Appl. Sci.* **2020**, *10*, 1107. [CrossRef]
19. Mokadem, N.; Boughariou, E.; Mudarra, M.; Ben Brahim, F.; Andreo, B.; Hamed, Y.; Bouri, S. Mapping potential zones for groundwater recharge and its evaluation in arid environments using a GIS approach: Case study of North Gafsa Basin (Central Tunisia). *J. Afr. Earth Sci.* **2018**, *141*, 107–117. [CrossRef]
20. Pham, B.T.; Tien Bui, D.; Pham, H.V.; Le, H.Q.; Prakash, I.; Dholakia, M.B. Landslide Hazard Assessment Using Random SubSpace Fuzzy Rules Based Classifier Ensemble and Probability Analysis of Rainfall Data: A Case Study at Mu Cang Chai District, Yen Bai Province (Viet Nam). *J. Indian Soc. Remote Sens.* **2017**, *45*, 673–683. [CrossRef]
21. Pourghasemi, H.R.; Teimoori Yansari, Z.; Panagos, P.; Pradhan, B. Analysis and evaluation of landslide susceptibility: A review on articles published during 2005–2016 (periods of 2005–2012 and 2013–2016). *Arab. J. Geosci.* **2018**, *11*, 193. [CrossRef]
22. Liu, K.; Song, C.; Ke, L.; Jiang, L.; Ma, R. Automatic watershed delineation in the Tibetan endorheic basin: A lake-oriented approach based on digital elevation models. *Geomorphology* **2020**, *358*, 107127. [CrossRef]
23. Deng, F.; Rodgers, M.; Xie, S.; Dixon, T.H.; Charbonnier, S.; Gallant, E.A.; López Vélez, C.M.; Ordoñez, M.; Malservisi, R.; Voss, N.K.; et al. High-resolution DEM generation from spaceborne and terrestrial remote sensing data for improved volcano hazard assessment—A case study at Nevado del Ruiz, Colombia. *Remote Sens. Environ.* **2019**, *233*, 111348. [CrossRef]
24. Lee, C.-F.; Huang, W.-K.; Chang, Y.-L.; Chi, S.-Y.; Liao, W.-C. Regional landslide susceptibility assessment using multi-stage remote sensing data along the coastal range highway in northeastern Taiwan. *Geomorphology* **2018**, *300*, 113–127. [CrossRef]

25. Schlögel, R.; Marchesini, I.; Alvioli, M.; Reichenbach, P.; Rossi, M.; Malet, J.P. Optimizing landslide susceptibility zonation: Effects of DEM spatial resolution and slope unit delineation on logistic regression models. *Geomorphology* **2018**, *301*, 10–20. [CrossRef]
26. Juliev, M.; Mergili, M.; Mondal, I.; Nurtaev, B.; Pulatov, A.; Hübl, J. Comparative analysis of statistical methods for landslide susceptibility mapping in the Bostanlik District, Uzbekistan. *Sci. Total Environ.* **2019**, *653*, 801–814. [CrossRef]
27. Hong, H.; Miao, Y.; Liu, J.; Zhu, A.X. Exploring the effects of the design and quantity of absence data on the performance of random forest-based landslide susceptibility mapping. *Catena* **2019**, *176*, 45–64. [CrossRef]
28. He, Q.; Shahabi, H.; Shirzadi, A.; Li, S.; Chen, W.; Wang, N.; Chai, H.; Bian, H.; Ma, J.; Chen, Y.; et al. Landslide spatial modelling using novel bivariate statistical based Naïve Bayes, RBF Classifier, and RBF Network machine learning algorithms. *Sci. Total Environ.* **2019**, *663*, 1–15. [CrossRef]
29. Khan, H.; Shafique, M.; Khan, M.A.; Bacha, M.A.; Shah, S.U.; Calligaris, C. Landslide susceptibility assessment using Frequency Ratio, a case study of northern Pakistan. *Egypt. J. Remote Sens. Space Sci.* **2019**, *22*, 11–24. [CrossRef]
30. Cucchiaro, S.; Cavalli, M.; Vericat, D.; Crema, S.; Llena, M.; Beinat, A.; Marchi, L.; Cazorzi, F. Geomorphic effectiveness of check dams in a debris-flow catchment using multi-temporal topographic surveys. *Catena* **2019**, *174*, 73–83. [CrossRef]
31. Pineux, N.; Lisein, J.; Swerts, G.; Bielders, C.L.; Lejeune, P.; Colinet, G.; Degré, A. Can DEM time series produced by UAV be used to quantify diffuse erosion in an agricultural watershed? *Geomorphology* **2017**, *280*, 122–136. [CrossRef]
32. Dou, J.; Yunus, A.P.; Bui, D.T.; Merghad, A.; Sahana, M.; Zhu, Z.F.; Chen, C.W.; Han, Z.; Pham, B.T. Improved landslide assessment using support vector machine with bagging, boosting, and stacking ensemble machine learning framework in a mountainous watershed, Japan. *Landslides* **2020**, *17*, 641–658. [CrossRef]
33. Gu, X.D.; Liu, X.H. *Rock Formations in Sichuan*; China University of Geosciences Press: Wuhan, China, 1997.
34. Xu, W.C. Research on Hydrogeological and Environmental Geological Survey Strategy in Sichuan Province. *Earth* **2015**, *10*, 229.
35. CGEIS. China Geological Environment Information Site. Bulletin of National Geological Hazards 2004–2016. Available online: http://www.cigem.gov.cn (accessed on 22 February 2017).
36. Trigila, A.; Iadanza, C.; Esposito, C.; Scarascia-Mugnozza, G. Comparison of Logistic Regression and Random Forests techniques for shallow landslide susceptibility assessment in Giampilieri (NE Sicily, Italy). *Geomorphology* **2015**, *249*, 119–136. [CrossRef]
37. Li, L.; Liu, R.; Pirasteh, S.; Chen, X.; He, L.; Li, J. A novel genetic algorithm for optimization of conditioning factors in shallow translational landslides and susceptibility mapping. *Arab. J. Geosci.* **2017**, *10*, 209. [CrossRef]
38. Reichenbach, P.; Rossi, M.; Malamud, B.D.; Mihir, M.; Guzzetti, F. A review of statistically-based landslide susceptibility models. *Earth-Sci. Rev.* **2018**, *180*, 60–91. [CrossRef]
39. Dai, F.C.; Lee, C.F.; Li, J.; Xu, Z.W. Assessment of landslide susceptibility on the natural terrain of Lantau Island, Hong Kong. *Environ. Geol.* **2001**, *40*, 381–391. [CrossRef]
40. Wang, Y.; Fang, Z.; Hong, H. Comparison of convolutional neural networks for landslide susceptibility mapping in Yanshan County, China. *Sci. Total Environ.* **2019**, *666*, 975–993. [CrossRef]
41. Zhu, A.X.; Miao, Y.; Yang, L.; Bai, S.; Liu, J.; Hong, H. Comparison of the presence-only method and presence-absence method in landslide susceptibility mapping. *Catena* **2018**, *171*, 222–233. [CrossRef]
42. Freire, A.P.C.F.; Elkins, M.R.; Ramos, E.M.C.; Moseley, A.M. Use of 95% confidence intervals in the reporting of between-group differences in randomized controlled trials: Analysis of a representative sample of 200 physical therapy trials. *Braz. J. Phys. Ther.* **2019**, *23*, 302–310. [CrossRef]
43. Fu, Q.X. Landslide Spatiotemporal Susceptibility Analysis of Chengdu-Yaan' Section in Sichuan-Tibet Railway. Master's Thesis, Central South University, Changsha, China, 2017.
44. Wang, Y.; Wang, J.; Balakrishnan, S.; Singh, A. Rate optimal estimation and confidence intervals for high-dimensional regression with missing covariates. *J. Multivar. Anal.* **2019**, *174*, 104526. [CrossRef]
45. Hespanhol, L.; Vallio, C.S.; Costa, L.M.; Saragiotto, B.T. Understanding and interpreting confidence and credible intervals around effect estimates. *Braz. J. Phys. Ther.* **2019**, *23*, 290–301. [CrossRef] [PubMed]

46. Noce, L.; Gwaza, L.; Mangas-Sanjuan, V.; Garcia-Arieta, A. Comparison of free software platforms for the calculation of the 90% confidence interval of f2 similarity factor by bootstrap analysis. *Eur. J. Pharm. Sci.* **2020**, *146*, 105259. [CrossRef] [PubMed]
47. Wang, W.D.; Li, J.; Han, Z. Comprehensive assessment of geological hazard safety along railway engineering using a novel method: A case study of the Sichuan-Tibet railway, China. *Geomat. Nat. Hazards Risk* **2020**, *11*, 1–21. [CrossRef]
48. Liu, M.; He, Y.; Wang, J.; Lee, H.P.; Liang, Y. Hybrid intelligent algorithm and its application in geological hazard risk assessment. *Neurocomputing* **2015**, *149*, 847–853. [CrossRef]
49. Nedumpallile Vasu, N.; Lee, S.-R.; Pradhan, A.M.S.; Kim, Y.-T.; Kang, S.-H.; Lee, D.-H. A new approach to temporal modelling for landslide hazard assessment using an extreme rainfall induced-landslide index. *Eng. Geol.* **2016**, *215*, 36–49. [CrossRef]
50. Stancanelli, L.M.; Peres, D.J.; Cancelliere, A.; Foti, E. A combined triggering-propagation modeling approach for the assessment of rainfall induced debris flow susceptibility. *J. Hydrol.* **2017**, *550*, 130–143. [CrossRef]
51. Sandric, I.; Ionita, C.; Chitu, Z.; Dardala, M.; Irimia, R.; Furtuna, F.T. Using CUDA to accelerate uncertainty propagation modelling for landslide susceptibility assessment. *Environ. Model. Softw.* **2019**, *115*, 176–186. [CrossRef]
52. Gayen, A.; Pourghasemi, H.R.; Saha, S.; Keesstra, S.; Bai, S. Gully erosion susceptibility assessment and management of hazard-prone areas in India using different machine learning algorithms. *Sci. Total Environ.* **2019**, *668*, 124–138. [CrossRef]
53. Park, H.J.; Jang, J.Y.; Lee, J.H. Assessment of rainfall-induced landslide susceptibility at the regional scale using a physically based model and fuzzy-based Monte Carlo simulation. *Landslides* **2019**, *16*, 695–713. [CrossRef]
54. Segoni, S.; Tofani, V.; Rosi, A.; Catani, F.; Casagli, N. Combination of Rainfall Thresholds and Susceptibility Maps for Dynamic Landslide HazardAssessment at Regional Scale. *Front. Earth Sci.* **2018**, *6*, 1–11. [CrossRef]
55. Xia, X.L.; Liang, Q.H. A new depth-averaged model for flow-like landslides over complex terrains with curvatures and steepslopes. *Eng. Geol.* **2018**, *234*, 174–191. [CrossRef]
56. Li, L.; Nearing, M.A.; Nichols, M.H.; Polyakov, V.O.; Guertin, D.P.; Cavanaugh, M.L. The effects of DEM interpolation on quantifying soil surface roughness using terrestrial LiDAR. *Soil Tillage Res.* **2020**, *198*, 104520. [CrossRef]
57. Shahri, A.A.; Spross, J.; Johansson, F.; Larsson, S. Landslide susceptibility hazard map in southwest Sweden using artificial neural network. *Catena* **2019**, *183*, 1–14. [CrossRef]
58. Lin, C.H.; Lin, M.L.; Peng, H.R.; Lin, H.H. Framework for susceptibility analysis of layered rock slopes considering the dimensions of themapping units and geological data resolution at various map scales. *Eng. Geol.* **2018**, *246*, 310–325. [CrossRef]
59. Scheltinga, R.T.; Coco, G.; Kleinhans, M.G.; Friedrich, H. Observations of dune interactions from DEMs using through-water Structure from Motion. *Geomorphology* **2020**, *359*, 107126. [CrossRef]
60. Tang, X.; Li, J.; Liu, M.; Liu, W.; Hong, H. Flood susceptibility assessment based on a novel random Naïve Bayes method: A comparison between different factor discretization methods. *Catena* **2020**, *190*, 104536. [CrossRef]
61. Nian, T.K.; Guo, X.S.; Zheng, D.F.; Xiu, Z.X.; Jiang, Z.B. Susceptibility assessment of regional submarine landslides triggered by seismic actions. *Appl. Ocean Res.* **2019**, *93*, 101964. [CrossRef]

applied sciences

MDPI

Article

Spatial Proximity-Based Geographically Weighted Regression Model for Landslide Susceptibility Assessment: A Case Study of Qingchuan Area, China

Yange Li [1], Xintong Liu [1], Zheng Han [1,*] and Jie Dou [2]

1 School of Civil Engineering, Central South University, Changsha 410075, China; liyange@csu.edu.cn (Y.L.); 1202140116@csu.edu.cn (X.L.)

2 Department of Civil and Environmental Engineering, Nagaoka University of Technology, Nagaoka 940-2188, Japan; douj888@vos.nagaokaut.ac.jp

* Correspondence: zheng_han@csu.edu.cn; Tel.: +86-1887-4163-071

Received: 27 December 2019; Accepted: 4 February 2020; Published: 7 February 2020

Featured Application: Landslide susceptibility assessment and other geological disaster assessment issues.

Abstract: Landslides pose a serious threat to the safety of human life and property in mountainous regions. Susceptibility assessment for landslides is critical in landslide management strategy. Recent studies indicate that the traditional assessment models in many previous studies commonly assume a fixed relationship between influencing factors and landslide occurrence within an area, resulting in an inadequate evaluation for the local landslides susceptibility. To address this issue, in this paper we propose a spatial proximity-based geographically weighted regression (S-GWR) model considering spatial non-stationarity of landslide data for assessing the landslide susceptibility. Spatial proximity is the basic input condition for the proposed S-GWR model. The challenge lies in defining the spatial proximity expression that shows the geographical features of landslides and therefore affects the model ability of S-GWR. Our solution chooses the slope unit as spatial adjacency, rather than the grid unit in DTM. The multicollinearity between landslide influencing factors is then eliminated through variance inflation factor (VIF) method and principal component analysis (PCA). The proposed model is subsequently validated by using data in Qingchuan County, southwestern China. Spatial non-stationary is identified for landslide data. A comparison with grid unit and four traditional evaluation models is conducted. Validation results using the area under the ROC (receiver operating characteristic) curve and success rate curve indicate that the spatial proximity-based GWR model with slope unit has the highest predictive accuracy (0.859 and 0.850 respectively).

Keywords: landslide susceptibility assessment; geographically weighted regression; spatial non-stationary; spatial proximity; slope unit

---

1. Introduction

Landslides are catastrophic natural hazards frequently posing risks to the major societal, economic, and environmental on an international scale [1]. According to the report from EM-DAT [2], 21,412 landslides occurred worldwide between 1900 and 2014, resulting in 38,521,499 fatalities, with 7,229,487,068 people affected and total direct economic losses exceeding $2.7 trillion.

Landslide susceptibility assessment has long been recognized as a useful tool for landslide hazard management through land use planning and better decision making in landslide prone areas [3]. It is generally based on heuristic, statistical, or deterministic models [4–8]. Heuristic models are subjective and much susceptible to the expectation of the results [9,10]. Deterministic models have been reported with higher accuracy, but are limited by the difficulty of obtaining detailed landslide database [11,12]. Statistical models are the most widely used models due to their simplicity and high efficiency [13,14]. Many remarkable studies on the above aspects have been made, laying a solid foundation for landslide susceptibility mapping. However, in general, most of the previous studies consider the relationship between triggering factors and landslide occurrence as a fixed effect within an area, whereas different degrees of parameter influence may occur, such that, with the change of location, the effect of parameters can be consequently changed. The uncertainties due to this varied relationship remain a scientific challenge.

The second law of geography suggests that there exists variability over space of a given relationship between variables widely in spatial data [15], which is the so-called spatial non-stationarity. In view of that landslide susceptibility assessment is heavily based on spatial data, the relationship between influencing factors and landslide susceptibility may also have the characteristics of spatial non-stationarity. Previous studies, e.g., [16], have also suggested that the effective parameters in the occurrence of a natural disaster phenomenon do not have the same importance in different parts of an area. The existence of spatial non-stationarity indicates that average relationships fitted to the whole study area of traditional models might be inappropriate since they do not accurately fit local conditions [17]. This spatial non-stationarity characteristics in the data pose difficulties in landslide susceptibility assessment based on the traditional models.

Geographically weighted regression (GWR), the most popular local regression format, shows great capability in dealing with spatial non-stationary relationships [18]. It allows the relationship between dependent and independent variables to vary over space, as well as that regression coefficients in the model are calculated for each spatial zone [19]. This method has been applied in various fields of study such as social economics, geography, and meteorology [20–22]. However, previous studies applying GWR model for the assessment of geological hazard susceptibility have not yet been reported.

One difficulty limiting the application of GWR model in landslide susceptibility assessment is spatial proximity. It is the basic input condition and core problem for GWR model, and the issue regarding an adequate expression for the spatial proximity at different locations directly affects the modeling ability of GWR model [23,24]. Spatial proximity is the distance relationship between two units in space, and the closer the distance, the greater the impact. The key to determining the spatial proximity is segmenting the study area into map units to effectively express the spatial adjacency relationship between landslide data. The relationship should satisfy the requirements of GWR for good internal homogeneity and between-units heterogeneity. The commonly used segmentation methods in previous studies relating to GWR model can be categorized into two major kinds, i.e., administrative units [25] and grid units [26]. Administrative unit is mostly used for social and economic issues, and its segmentation does not accord with the neighborhood characteristics of landslide data. As such, administrative unit is rarely used in geological hazard assessment. Grid unit is a popular mapping unit for susceptibility assessment since it is easily accessible, but it is not associated with geological environments. Slope unit is a relative new mapping unit for evaluating landslide susceptibility, which is generated according to hydrology theory and is the watershed area defined by drainage lines (valley lines) and water divide lines (ridge lines) [27]. It is the basic topographical unit of geological hazard occurrence. Slope unit has higher internal homogeneity and between-unit heterogeneity than grid unit. It is closely related to geological environment conditions. In this sense, slope unit provides an alternative solution for spatial proximity expression of the GWR model for landslide susceptibility assessment.

Two other key issues of GWR model are the multicollinearity elimination and the kernel function establishment. Previous studies, e.g., [28], indicated that GWR is highly susceptible to the effects of multicollinearity between explanatory variables, and collinearity among pairs of explanatory variables or multicollinearity among more than two variables often lead to problems such as parameter estimate instability and unintuitive parameter signs. These problems remain significant owing to the complicated conditions of landslide posing a high possibility of correlations between explanatory variables. Kernel function is based on the distances between observations and calibration units to place emphasis on observations that are closer in space [28]. The selection of kernel function type and the determination of its bandwidth are crucial to the spatial proximity modeling of GWR.

In this study, we attempt to propose a spatial proximity based on geographically weighted regression (S-GWR) model for landslide susceptibility assessment. The presented model resolves the spatial non-stationarity of landslide susceptibility assessment with GWR model. Firstly, we generate slope units to establish spatial adjacency. Then, variance inflation factor (VIF) method [29] and principal component analysis (PCA) method [30] were adopted to eliminate multi-collinearity, and kernel function was determined according to the characteristics of landslide data. Finally, we chose Qingchuan County, Sichuan Province, China, as the study area to validate the applicability of the model, and further compared the established model with the grid-unit GWR model and other evaluation models.

2. Study Area

The study area is the Qingchuan County in the transitional region between the Sichuan Basin and the Western Sichuan Plateau. This area has long been recognized as one of the most landslide-prone areas of China [31]. It locates between 32°12′~32°56′ N in latitude and 104°36′~105°38′ E in longitude, covering a total area of 3217 km^2. The minimum elevation of the Qingchuan County is approximately 500 m and the maximum is 3820 m, characterized by northwestern part with higher elevation than the southeastern. Slope gradient reaches a maximum of about 80°, with a mean value of 38°.

The tectonics and geological settings in the area are complex. Because of the neotectonics, soft-lithology and hard-lithology usually appears alternately. There are about eight types of lithological outcrops throughout the study region (as shown in Figure 1), including the sedimentary rock (limestones, sandstone, and conglomerate) from Cambrian to Jurassic age, magmatic (granite), metamorphic rock (shales, schists, gneiss) from Cambrian to Jurassic age and Quaternary loess unconsolidated sedimentary. Two main active faults cross the area: the Pingwu–Qingchuan fault located in the north and crossing the whole territory, and the Yingxiu–Beichuan fracture which belongs to the Longmenshan fault belt, is a thrust fault 60°–70° NW dipping. Bailong river, Qingzhu river and Qiaozhuang river are distributed in the area. The discharges of the three rivers are measured approximately 525, 30, and 40 m^3s^{-1}, respectively, serving as the main channel for atmospheric precipitation and groundwater drainage.

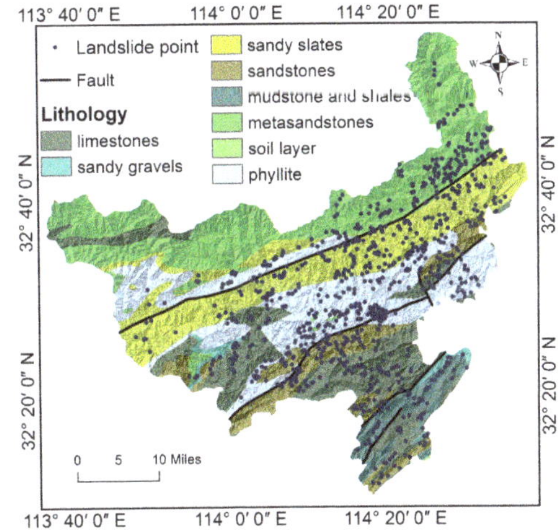

Figure 1. Lithological map of Qingchuan area and locations of landslide points.

3. Methodology

3.1. Flowchart of Research

The methodologies used in this study are as shown in Figure 2. The flowchart consists of three major steps.

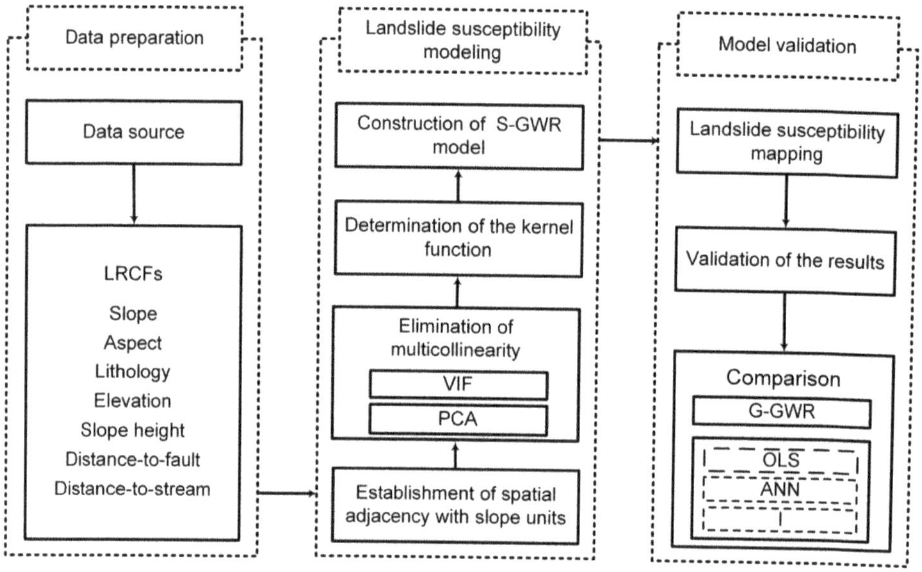

Figure 2. Methodology of research applied in this study.

The first step is dataset preparation, including the preparation of basic data and the determination and extraction of landslide-related causal factors (LRCFs);

The second step is the establishment of the landslide susceptibility assessment model, which is the most critical step. Firstly, slope units are generated to establish spatial adjacency of landslide data. After obtaining the data in the slope units, the multicollinearity between data is eliminated by variance inflation factor method (VIF) and principal component analysis (PCA). Then, the spatial kernel function of GWR model is determined, and finally the GWR model based on slope unit (S-GWR) is established for landslide susceptibility evaluation.

The last step is model validation, including applying the S-GWR model to the actual area and generating the landslide susceptibility. The results are then used for validation and compared with the grid-unit GWR model (G-GWR) and other evaluation models including ordinary least squares regression (OLS), artificial neural network (ANN), and information models (I).

3.2. Dataset Preparation

The basic data utilized in the study include the digital elevation model (DEM) of the study area with a spatial resolution of 10 m; a geological map with the scale of 1:100,000; aerial photographs taken by Ministry of Land and Resources. After field identification and interpretation of the color aerial photographs, 973 landslide points are identified, and the distribution is as shown in Figure 1.

Owing to that a variety of factors make a contribution to the landslide occurrence, an adequate selection of the effective parameters is essential to identify areas prone to landslide. We choose these factors in accordance with the well-illustrated criterion in many previous studies (as listed in Table 1). In detail, firstly, the actual conditions of research area are considered. The study area has a complex topography, so topographic factors, such as slope and slope height, provide terrain conditions for landslides. The alternating appearance of soft and hard lithology provides a material basis for the development of landslides. Besides, there are two large faults across the study area, and the landslide frequency may increase as the distance to a major or minor fault decreased [26]. There are also many river systems in the study area. In the dense-river-areas, with the increase of catchment area and reduction of erosion basis, the shearing force of river flow is enhanced, which gives rise to the steep gorges in the downstream and provides premise conditions for the growth of geological disasters.

Table 1. Criteria for a rational selection of landslide factors on the basis of literature review.

Criteria	Former Studies Using the Same Criterion for Landslide Susceptibility Assessment
Elevation	Kawabata et al. (2009) [32], Li et al. (2012) [33], Sabokbar et al. (2014) [26], Pradhan. (2013) [34], HawasKhan et al. (2019) [35]
Slope	Kawabata et al. (2009) [32], Li et al. (2012) [33], Sabokbar et al. (2014) [26], Pradhan. (2013) [34], HawasKhan et al. (2019) [35], Suh et al. (2019) [36], Zhuang et al. (2012) [37]
Slope height	Li et al. (2012) [33], Suh et al. (2019) [36], Zhuang et al. (2012) [37]
Aspect	Kawabata et al. (2009) [32], Li et al. (2012) [33], Sabokbar et al. (2014) [26], HawasKhan et al. (2019) [35], Zhuang et al. (2012) [37]
Distance-to-stream	Li et al. (2012) [33], Sabokbar et al. (2014) [26], Pradhan. (2013) [34], HawasKhan et al. (2019) [35]
Distance-to-fault	Li et al. (2012) [33], Sabokbar et al. (2014) [26], HawasKhan et al. (2019) [35], Suh et al. (2019) [36]
Lithology	Kawabata et al. (2009) [32], Li et al. (2012) [33], Sabokbar et al. (2014) [26], Pradhan. (2013) [34], HawasKhan et al. (2019) [35], Suh et al. (2019) [36]

Based on the analysis of the research area, the factors commonly used in previous studies on landslide susceptibility are further referenced (Table 1), and 7 variables (elevation, slope, slope height, aspect, distance-to-stream, distance-to-fault, and lithology) are finally identified for modeling. All these factors are processed within ArcGIS. Slope, elevation, slope height, and aspect are calculated using DEM. All data are directly obtained by ArcGIS without data transformation. Lithology map is digitized from the existing geological map. Lithology map is digitized from the existing geological map. Distance-to-fault is obtained by calculating the distance to the nearest fault, and distance-to-stream is the distance to the nearest stream.

3.3. Establishment of Spatial Proximity

Spatial proximity is the distance relationship between two units in space. GWR model establishes a model for each unit according to the spatial proximity, and the areas within a unit are considered homogeneous. Therefore, it is essential to determine the spatial adjacency relationship between units, i.e., the boundary of map units. Administrative boundaries [25] and grid boundaries [26] were commonly used as the spatial proximity expressions of GWR in previous studies, but they are inconsistent with the neighborhood characteristics of landslides. These boundaries could not perform well to express the heterogeneity between units and the homogeneity within units, affecting the modeling ability of GWR model.

Therefore, we incorporate the methodology of slope units to express spatial proximity. Slope unit is the basic unit of geological disasters. It divides the terrain into mapping units with similar hydrological and geomorphological conditions, and is shaped by similar processes occurring in the natural landscape under the same geo-environmental conditions [37]. A slope unit division map is formed by the GIS-based hydrologic analysis tool [38]. Firstly, reverse DEM is generated by subtracting the elevation value from the highest elevation value in each unit. Secondly, fill the DEM and reverse DEM, and the flow direction can be obtained by these filled DEMs. Then, by setting the minimum number of cells that flow to the calculating point, the watershed can be calculated. Eventually, by combining the watershed by DEM and the watershed by reverse DEM, slope units can be obtained.

3.4. Elimination of Multicollinearity

Collinearity among pairs of explanatory variables or multicollinearity among more than two variables in regression analysis is known to cause problems such as parameter estimate instability, unintuitive parameter signs, high coefficient of determination (R^2) diagnostics despite few or no significant parameters, and others [28]. Due to the huge and complex landslide data, the complicated relationship between geological and topographical factors should be considered. Therefore, the multicollinearity elimination in landslide susceptibility assessment is very important. Correlations in GWR parameters, both within a set of local parameter estimates for all locations (global multicollinearity) and among different parameter estimates at each location (local multicollinearity), are the symptom of multicollinearity among explanatory variables [28].

Global multicollinearity of the entire area can be easily distinguished by the variance inflation factor (VIF) method [29], which measures the effect of collinearity on the estimated variance of a regression coefficient. Local multicollinearity is the linear dependencies in the design matrix of local regression model. Principal component analysis (PCA) [30] is adopted to eliminate local multicollinearity since the diagnosis of local multicollinearity is very complicated. Through data transformation and processing, the influencing factors of landslide susceptibility can be grouped into less integrated factors, which not only maintains the main information of original factors, but also weakens the correlation among them. First obtained initial factor loading matrix. There are two main principles for selecting the principal component: (1) the principal component eigenvalue is greater than 1; (2) the cumulative contribution rate of principal components reaches 80%. Then, the factor rotation is performed so that the obtained factors have clear professional interpretation significance. Quartimax method [39], a common factor rotation method, was used for factor rotation.

3.5. GWR Modeling

GWR model extends the ordinary least squares regression (OLS) [40] by weighting the spatial dependence [41]. Based on the established spatial proximity, the coefficients of the model are estimated for each unit, and the value and symbol of the coefficients vary at different units [16]. This model is in the form of Equation (1)

$$y_i = \beta_0(\mu_i, v_i) + \sum_{j=1}^{p} \beta_{ij}(\mu_i, v_i)x_{ij} + \varepsilon_i \tag{1}$$

where (μ_i, v_i) represents the coordinates of an ith unit in space and p is the number of independent factors. x_{ij} is the jth independent variable of the ith unit. β_0 (μ_i, v_i) is the intercept parameter in position i and ε_i is the random error. β_{ij} (μ_i, v_i) is the local regression coefficient for the jth explanatory variable in position i, which varies with the change of spatial position and is a very important parameter for the embodiment of spatial non-stationary.

The establishment of the weight kernel function is an important step of GWR, which is used to determine the scope and degree of spatial dependence. The establishment process includes the selection of the type of kernel function and the determination of its bandwidth, and previous studies have found that the latter has a greater impact on the result of GWR than the former [42]. This paper used a common kernel function, Gauss kernel function [43], as the type of kernel function. Its function form is

$$w_{ij} = exp\left(-\left(d_{ij}/b\right)^2\right) \tag{2}$$

where w_{ij} is the weight for unit j in the neighborhood of unit i, and d_{ij} is the distance between the center point of the unit i and j as the measurement of spatial proximity degree. b is the bandwidth of the Guass kernel function. Many approaches are available for determining the bandwidth, including cross-validation (CV) method [42], Akaike information criterion (AIC) method [44]. Compared with CV method, AIC method is easier to avoid over-fitting problems, and the selected optimal model is often more effective. Therefore, the AIC method was adopted, and the bandwidth corresponding to the weight function with the minimum AIC value is the optimal bandwidth. Equation (3) is used for the calculation of AIC,

$$AIC = 2n\ln(\hat{\sigma}) + n\ln(2\pi) + n\left[\frac{n + tr(S)}{n - 2 - tr(S)}\right] \quad (3)$$

where $tr(S)$ is the function of b, and $\hat{\sigma}$ is the maximum likelihood estimation for GWR model.

3.6. Validation Processes

The GEZM validation process is important; without this, the study lacks scientific credibility. In this research, we first compare the prediction results with actual units free of or containing landslides [45]. Sample points with 30% of the number of units, accounting for totally 291 landslide points, are randomly selected. The values of the predicted result belonging to the low-susceptibility and very-low-susceptibility area are regarded as the stable points and the others are as the unstable points. Then, calculate the missing rate, which is the ratio of actual unstable slopes classified as stable slopes.

Although the missing rate is one indicator that evaluates the model results, the results are subject to the threshold value, which limits the accuracy of results [46]. Therefore, we introduced ROC curve [47] and success rate curve [34] for further evaluation. The two methods are independent of the specific decision threshold and are able to further verify the accuracy of the result. ROC curve method is a measure of the ability to discriminate landslides from non-landslide locations. Success rate curve, different from the ROC curve, only considers the prediction of the landslide samples. The areas under the curve (AUC) for ROC curve and success rate curve (0.5 to 1.0) are used to assess the accuracy of the models.

4. Results

4.1. Landslide Susceptibility Map Using the Proposed Model

Based on the hydrologic analysis tool in ArcGIS environment, we segment the topography of the study area into 55,899 slope units in total. The schematic diagram of slope units is shown in Figure 3.

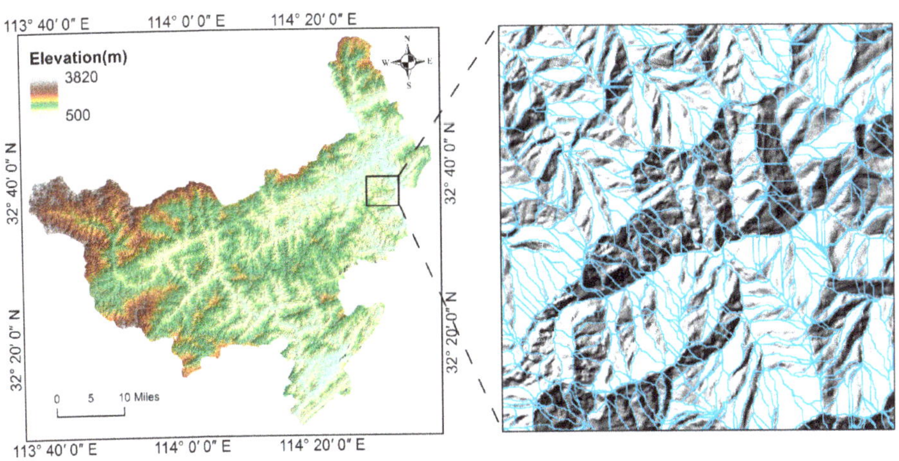

Figure 3. Schematic diagram of spatial proximity expression in slope units.

Firstly, in order to test the global multicollinearity, the variance inflation factor (VIF) value of each influencing factor is calculated (results shown in Table 2). A greater VIF value indicates a greater multicollinearity of the influencing factors. A value greater than 10 denotes that multicolinearity problems may exist. In this case, elevation factor for instant, the maximum VIF value is 1.88, which is apparently lower than 10. Table 2 indicates that all of the influencing factors used in the proposed S-GWR model pass the global multicollinearity test.

Table 2. Test results of variance inflation factor method of S-GWR.

Influencing Factors	VIF
Slope	1.61
Aspect	1.06
Elevation	1.88
Slope height	1.52
Lithology	1.35
Distance-to-fault	1.87
Distance-to-stream	1.12

Then, the local multicollinearity problem is subsequently processed with PCA method. According to the principles of selecting principal components, four principal components are selected in our study, and the components explained 81.10% of the total variances. After the factor rotation, four new principal components were obtained. As is shown in Table 3, the accumulative loadings of four principal components decrease systematically. The first component can be interpreted as geological factor, as it has high positive loadings of distance-to-fault and lithology. The second component has high positive loadings of slope and slope height, indicating that this component represents slope shape factor. In the same way, the third and the fourth component represent hydrographic factor and aspect factor, respectively.

Table 3. Rotated component loadings of the indicators on selected principal components.

Principal Components	PC1	PC2	PC3	PC4
Slope	0.128	0.858	0.047	0.133
Aspect	0.038	0.132	0.055	0.981
Elevation	0.675	0.374	0.374	−0.007
Slope height	0.106	0.861	−0.152	−0.010
Lithology	0.829	−0.030	−0.195	0.108
Distance-to-fault	0.860	0.122	0.110	−0.066
Distance-to-stream	0.070	−0.106	0.939	0.055
Accumulative loadings	2.706	2.211	1.178	1.194

The S-GWR model is constructed using the Gaussian kernel function and the AIC method. The neighbors of the model estimated by the AIC method is 1000. According to the prediction results, landslide susceptibility of Qingchuan County is mapped. The landslide susceptibility map is classified into five classes based on the natural breaks method [48], including very low, low, moderate, high, and very-high (shown in Figure 4). The landslide susceptibility map shows that the high-susceptibility area and very-high-susceptibility area are highly consistent with the actual landslide points, and very-low- to very-high-susceptibility classes occupy 46.37%, 25.67%, 17.62%, 8.69%, and 1.65% of study area.

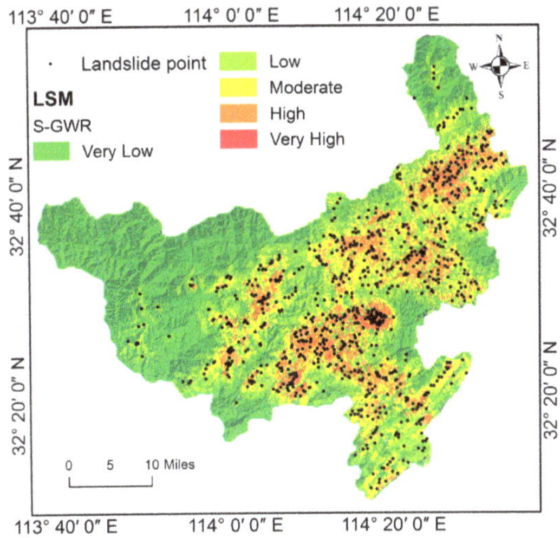

Figure 4. Landslide susceptibility mapping (LSM) by the proposed S-GWR model.

4.2. Validation of the S-GWR Models

4.2.1. Validation of the Spatial Non-Stationary

In order to validate the spatial non-stationary of the S-GWR model, we explore the spatial distribution of the regression coefficients between four principal components and landslide susceptibility. Results are shown in Figure 5. The coefficients estimated by S-GWR vary greatly with districts in space. This result implies a noticeable spatial non-stationarity of the relationship between four principal components and landslide susceptibility. The regression coefficients of each component have positive and negative values, which indicates that the relevance between each influencing factor and landslide susceptibility has varying direction and strength in space.

Spatial distribution of the regression coefficient of geological factors is approximately from northeast to southwest, similar to the distribution of actual lithology features and fault zones. The coefficient values of geological factors tend to decrease with the increase of the distance to fault zone and the large absolute values are mainly distributed in the metasandstones, phyllite, and sandy slates area. The distribution area with large absolute values of the coefficients of slope shape factor consist of the actual landslide points. In most of the eastern part of the study area, slope shape factor is positively correlated with landslide susceptibility, but the opposite effect exists in some central regions. The majority of the coefficient of hydrographic factor is negative, indicating that there is a significant negative correlation between the distance-to-stream and landslide susceptibility in most regions. This correlation is quite obvious in the central area, where most of the landslides occurred, such as Chaoba township, Courtyard Hui township, and the northeast area near the Bailong river. The aspect coefficient distribution map shows that the absolute value in the eastern region is larger than that in the western region. Compared with the first three principal components, the difference between the maximum and minimum values of aspect factor coefficients is the smallest.

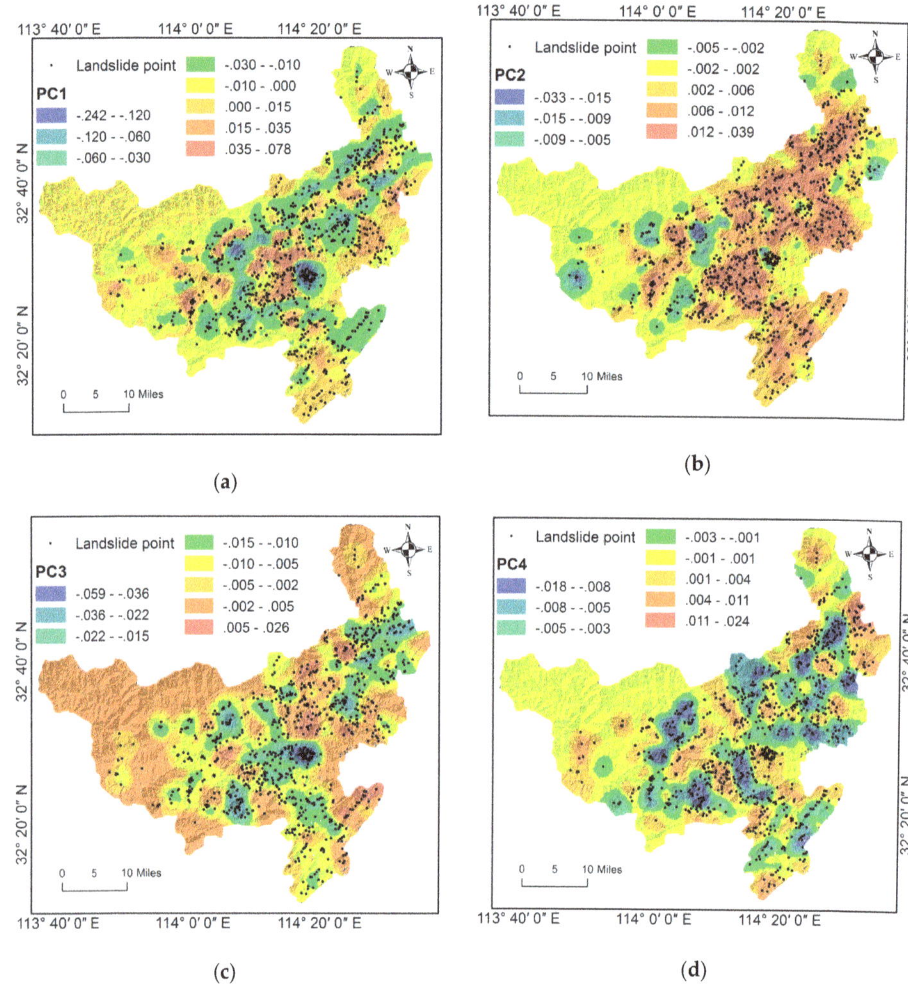

Figure 5. Spatial distribution of the principal component coefficient function estimate. (**a**) Regression coefficient of geological factor (PC1); (**b**) Regression coefficient of slope shape factor (PC2); (**c**) Regression coefficient of hydrographic factor (PC3); (**d**) Regression coefficient of aspect factor (PC4)).

4.2.2. Validation of the Accuracy

In order to validate the accuracy and effectiveness of the proposed S-GWR model, we compare the prediction results with actual units free of or containing landslides. Table 4 shows the prediction result of the proposed S-GWR model. Among the 291 actual landslide points, 36 points are classified as stable points with a missing rate of 12.37%. ROC curve and success rate curve are used for further evaluation. Figure 6 shows that the AUC value of ROC curve is 0.859, and the AUC value of success rate curve is 0.850.

Table 4. Sample evaluation contingency table of the proposed S-GWR model.

		Predicted Groups (Model)	
		Stable	Unstable
Actual Groups (inventory)	Stable	11,881 (72.43%)	4523 (27.57%)
	Unstable	36 (12.37%)	255 (87.63%)

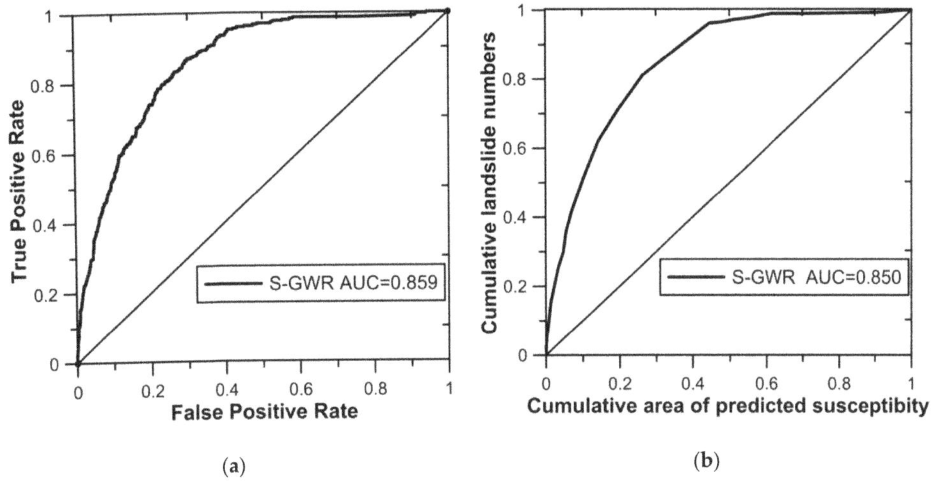

Figure 6. Validation of S-GWR model. (**a**) ROC curve; (**b**) success rate curve.

4.3. Comparison with Other Models

4.3.1. Comparison with Grid-Unit-Based GWR (G-GWR)

In order to verify the applicability of slope units in expressing spatial proximity of the GWR, the commonly used grid unit is adopted as a comparison. For the validity of contrast and simplicity of calculation, grid size is selected as 200 × 200 m, approximately equal to the slope unit, generating 80,264 grid units.

The following procedure is the same as the process of the proposed S-GWR. The variance inflation factor (VIF) values of each influencing factor are calculated, with the maximum test result of 2.14 (as shown in Table 5). It indicates that the original G-GWR passes the global multicollinearity test. Based on the principles of selecting principal components, G-GWR also obtained four components, including geological factor, slope shape factor, hydrographic factor, and aspect factor. The establishment of kernel function of G-GWR is the same as S-GWR.

Table 5. Test results of variance inflation factor method of G-GWR.

Influencing Factors	VIF
Slope	1.02
Aspect	1.83
Elevation	2.05
Slope height	2.00
Lithology	2.14
Distance-to-fault	1.10
Distance-to-stream	1.36

The established G-GWR model is then applied to the landslide susceptibility assessment in Qingchuan County. The sample evaluation contingency table of the G-GWR model is shown in

Table 6. It shows that 61 points of the 291 actual landslide points are classified as stable points, with a missing rate of 20.96%, which is higher than that of the S-GWR model with 12.37%. The ROC curves and success rate curves of the two models are shown in Figure 7. The AUC values of ROC curve and success rate curve of G-GWR are 0.837 and 0.827 respectively, both lower than those of S-GWR. The comparative verification of the S-GWR model and the G-GWR model shows that the S-GWR model has a better performance in classification precision, thus can better identify susceptible areas to landslide occurrence. This comparison indicates that slope unit is more suitable for GWR model in the expression of spatial relationship than traditional grid unit, in particular for the purpose of landslide susceptibility assessment.

Table 6. Sample evaluation contingency table of G-GWR model.

		Predicted Groups (Model)	
		Stable	Unstable
Actual groups (inventory)	Stable	13,449 (78.20%)	3750 (21.80%)
	Unstable	61 (20.96%)	230 (79.04%)

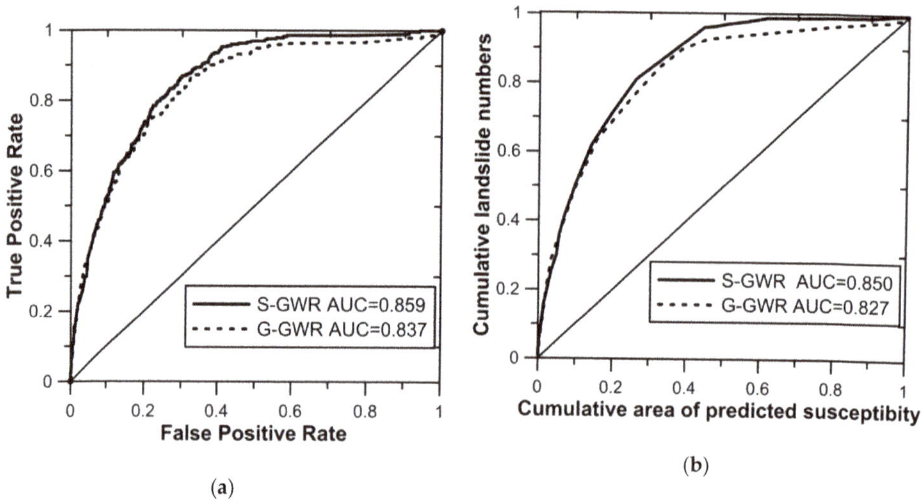

Figure 7. Validation of S-GWR and G-GWR models. (**a**) ROC curve; (**b**) success rate curve.

4.3.2. Comparison with OLS, ANN, and I Models

In order to further verify the applicability and accuracy of the landslide susceptibility assessment using the GWR model, we choose the ordinary least squares (OLS) model [40], artificial neural network (ANN) model [32], and information (I) model [49] as compared models, which are commonly used for the susceptibility assessment. The slope units and grid units are adopted as the sampling units of the three models respectively.

The ROC curve and success rate curve of eight models are shown in Figures 8 and 9. It can be seen from the figures that for the same unit type, the AUC values of ROC curve and success rate curve of GWR models are both higher than those of ANN, I, and OLS models. It shows that the GWR model performs better, and further implies the importance of spatial non-stationarity in landslide data sets. As to the results by the same model via different unit types, the AUC values of ROC curve and success rate curve of the models with slope unit are all higher than those of the models with grid unit. It indicates that adopting slope unit as mapping unit can express the landslide characteristics better than the traditional grid unit, highlighting that slope unit is more suitable for landslide assessment.

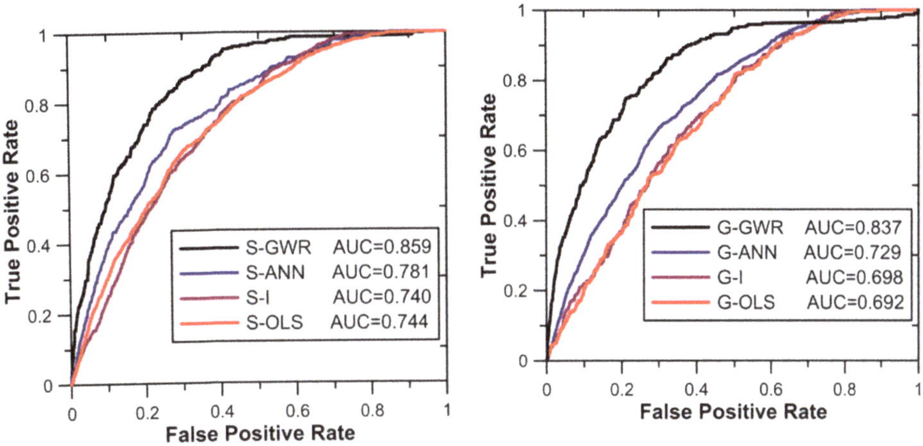

(**a**) (**b**)

Figure 8. ROC curves and area under the curves (AUC) for the susceptibility maps produced by GWR, ANN, I, OLS. (**a**) Slope unit; (**b**) grid unit.

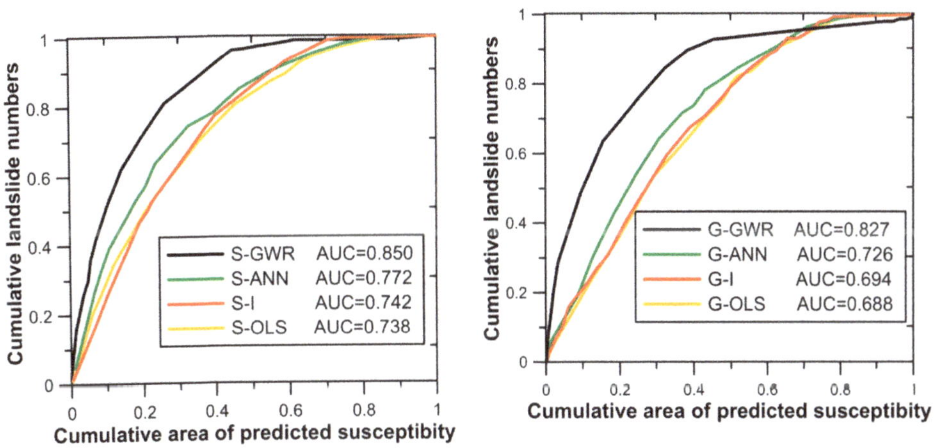

(**a**) (**b**)

Figure 9. Success rate curves and area under the curves (AUC) for the susceptibility maps produced by GWR, ANN, I, OLS. (**a**) Slope unit; (**b**) grid unit.

5. Conclusions

In this paper, a spatial proximity-based geographically weighted regression (S-GWR) model is proposed for assessing the landslide susceptibility. The presented model solves the issues of the spatial non-stationarity between landslide factors and its occurrence that usually neglected in previous landslide susceptibility assessment studies. In order to express spatial proximity properly, slope units are adopted. The multicollinearity between the data is eliminated through VIF method and principal component analysis.

The Qingchuan area in China after the 2008 Ms. 8.0 Wenchuan earthquake is selected as a case study to illustrate the effect and validity of the proposed S-GWR model. The result shows that the four influencing factors (including geological factor, slope shape factor, hydrographic factor, and aspect factor) in the S-GWR model all have noticeable spatial non-stationary effects on the landslide susceptibility. By quantitatively testing, the missing rate of S-GWR is 12.37%, which is much lower than that of G-GWR of 20.96%, and the AUC values of ROC curve and success rate curve of S-GWR (0.859 and 0.850, respectively) are both higher than those of G-GWR (0.837 and 0.827, respectively). Besides, the AUC values of the ROC curve and success rate curve of the GWR models are higher than those of ANN, I, and OLS models, and the accuracy of each model using slope unit is higher than those of the grid unit with similar cell size.

Our study verifies the importance of considering the spatial non-stationary and the applicability of the GWR model in the landslide susceptibility assessment. It also suggests that slope unit can better express the spatial relationship between landslide data and make the evaluation results more accurate.

Author Contributions: Conceptualization and methodology, Y.L. and Z.H.; Writing—validation, X.L.; Review and editing, J.D.; All authors have read and agreed to the published version of the manuscript.

Funding: This study was financially supported by the National Key R&D Program of China (grant no. 2018YFD1100401); the National Natural Science Foundation of China (grant no. 41702310); the Natural Science Foundation of Hunan (grant no. 2018JJ3638); and the Innovation Driven Program of Central South University (grant no. 2019CX011). These financial supports are gratefully acknowledged.

Acknowledgments: We would like to thank W.W. in Central South University, China, for his constructive suggestions. We also extend our gratitude to editor-in-chief and three reviewers for their insightful comments during the review stage while we submitted the manuscript.

Conflicts of Interest: The authors declare no conflict of interest.

References

1. Han, Z.; Su, B.; Li, Y.G.; Ma, Y.F.; Wang, W.D.; Chen, G.Q. Comprehensive analysis of landslide stability and related countermeasures: A case study of the Lanmuxi landslide in China. *Sci. Rep.* **2019**, *9*, 12407. [CrossRef] [PubMed]
2. Guha-Sapir, D.; Below, R.; Hoyois, P. *EM-DAT: International Disaster Database*; Catholic University of Louvain: Brussels, Belgium, 2014; Volume 27.
3. Pham, B.T.; Pradhan, B.; Bui, D.T.; Prakash, I.; Dholakia, M.B. A comparative study of different machine learning methods for landslide susceptibility assessment: A case study of Uttarakhand area (India). *Environ. Model. Softw.* **2016**, *84*, 240–250. [CrossRef]
4. Pourghasemi, H.R.; Yansari, Z.T.; Panagos, P.; Pradhan, B. Analysis and evaluation of landslide susceptibility: A review on articles published during 2005–2016 (periods of 2005–2012 and 2013–2016). *Arab. J. Geosci.* **2018**, *11*, 193–205. [CrossRef]
5. Jeong, J.S.; Ramírez-Gómez, Á. Optimizing the location of a biomass plant with a fuzzy-DEcision-MAking Trial and Evaluation Laboratory (F-DEMATEL) and multi-criteria spatial decision assessment for renewable energy management and long-term sustainability. *J. Clean. Prod.* **2018**, *182*, 509–520. [CrossRef]
6. Mileu, N.; Queirós, M. Integrating risk assessment into spatial planning: Riskote decision support system. *ISPRS Int. J. Geo Inf.* **2018**, *7*, 184. [CrossRef]
7. Abudeif, A.M.; Moneim, A.A.; Farrag, A.F. Multicriteria decision analysis based on analytic hierarchy process in GIS environment for siting nuclear power plant in Egypt. *Ann. Nucl. Energy* **2015**, *75*, 682–692. [CrossRef]
8. Dou, J.; Yunus, A.P.; Bui, D.T.; Merghadi, A.; Sahana, M.; Zhu, Z.; Chen, C.W.; Han, Z.; Pham, B.T. Improved landslide assessment using support vector machine with bagging, boosting, and stacking ensemble machine learning framework in a mountainous watershed, Japan. *Landslides* **2019**, 1–18. [CrossRef]
9. Zhu, A.; Pei, T.; Qiao, J.; Chen, Y.; Zhou, C.; Cai, G. A landslide susceptibility mapping approach using expert knowledge and fuzzy logic under GIS. *Prog. Geogr.* **2006**, *25*, 1–12. (In Chinese) [CrossRef]
10. Abella, E.A.C.; Van Westen, C.J. Qualitative landslide susceptibility assessment by multicriteria analysis: A case study from San Antonio del Sur, Guantánamo, Cuba. *Geomorphology* **2008**, *94*, 453–466. [CrossRef]
11. Wu, W.; Sidle, R.C. A distributed slope stability model for steep forested basins. *Water Resour. Res.* **1995**, *31*, 2097–2110. [CrossRef]
12. Montgomery, D.R.; Dietrich, W.E. A physically based model for the topographic control on shallow landsliding. *Water Resour. Res.* **1994**, *30*, 1153–1171. [CrossRef]

13. Hjort, J.; Luoto, M. Statistical methods for geomorphic distribution modeling. *Treatise Geomorphol.* **2013**, *2*, 59–73. [CrossRef]
14. Baeza, C.; Corominas, J. Assessment of shallow landslide susceptibility by means of multivariate statistical techniques. *Earth Surf. Proc. Land* **2001**, *26*, 1251–1263. [CrossRef]
15. O'Sullivan, D. Geographically weighted regression: The analysis of spatially varying relationships. *Geogr. Anal.* **2003**, *35*, 272–275. [CrossRef]
16. Arabameri, A.; Pradhan, B.; Rezaei, K. Gully erosion zonation mapping using integrated geographically weighted regression with certainty factor and random forest models in GIS. *J. Environ. Manag.* **2019**, *232*, 928–942. [CrossRef]
17. Leyk, S.; Norlund, P.U.; Nuckols, J.R. Robust assessment of spatial non-stationarity in model associations related to pediatric mortality due to diarrheal disease in Brazil. *Spat. Spatio Temporal Epidemiol.* **2012**, *3*, 95–105. [CrossRef]
18. Wheeler, D.C.; Páez, A. Geographically weighted regression. *Int. Encycl. Hum. Geogr.* **2009**, *47*, 407–414. [CrossRef]
19. Düzgün, H.S.; Kemeç, S. *Spatial and Geographically Weighted Regression*; Springer: Boston, MA, USA, 2008. [CrossRef]
20. Brunsdon, C.; Fotheringham, A.S.; Charlton, M.E. Geographically weighted regression: A method for exploring spatial nonstationarity. *Geogr. Anal.* **1996**, *28*, 281–298. [CrossRef]
21. Páez, A. Exploring contextual variations in land use and transport analysis using a probit model with geographical weights. *J. Transp. Geogr.* **2006**, *14*, 167–176. [CrossRef]
22. Tu, J. Spatially varying relationships between land use and water quality across an urbanization gradient explored by geographically weighted regression. *Appl. Geogr.* **2011**, *31*, 376–392. [CrossRef]
23. Openshaw, S. *The Modifiable Areal Unit Problem*; Geo Books: Norwich, UK, 1984. [CrossRef]
24. Wu, S. The Theory and Method of Geographically and Temporally Neural Network Weighted Regression. Ph.D. Thesis, Zhejiang University, Hangzhou, China, 2018. (In Chinese).
25. Zhang, J.; Zhao, R. Study on the influence factors of housing price in the urban area of Bohai Ring Megalopolis based on geographically weighted regression. *Territ. Nat. Resour. Study* **2019**, 87–93. (In Chinese) [CrossRef]
26. Sabokbar, H.F.; Roodposhti, M.S.; Tazik, E. Landslide susceptibility mapping using geographically-weighted principal component analysis. *Geomorphology* **2014**, *226*, 15–24. [CrossRef]
27. Ba, Q.; Chen, Y.; Deng, S.; Yang, J.; Li, H. A comparison of slope units and grid cells as mapping units for landslide susceptibility assessment. *Earth Sci. Inform.* **2018**, *11*, 373–388. [CrossRef]
28. Fotheringham, A.S.; Oshan, T.M. Geographically weighted regression and multicollinearity: Dispelling the myth. *J. Geogr. Syst.* **2016**, *18*, 1–27. [CrossRef]
29. Zhang, F. The Discussion on Solutions of Multicollinearity in Multilinear Regression Models. Master's Thesis, South China University of Technology, Guangzhou, China, 2010. (In Chinese).
30. Li, G.; Li, R.; Lu, Y.; Zhao, Y.; Yu, B. Using principal component analysis and geographic weighted regression methods to analyze AOD data. *Bull. Surv. Mapp.* **2018**, *493*, 50–56. (In Chinese) [CrossRef]
31. Han, Z.; Li, Y.G.; Du, Y.; Wang, W.D.; Chen, G.Q. Noncontact detection of earthquake-induced landslides by an enhanced image binarization method incorporating with Monte-Carlo simulation. *Geomat. Nat. Hazards Risk* **2019**, *10*, 219–241. [CrossRef]
32. Kawabata, D.; Bandibas, J. Landslide susceptibility mapping using geological data, a DEM from ASTER images and an Artificial Neural Network (ANN). *Geomorphology* **2009**, *113*, 97–109. [CrossRef]
33. Li, Y.; Chen, G.; Tang, C.; Zheng, L. Rainfall and earthquake-induced landslide susceptibility assessment using GIS and Artificial Neural Network. *Nat. Hazards Earth Syst. Sci.* **2012**, *12*, 2719–2729. [CrossRef]
34. Pradhan, B. A comparative study on the predictive ability of the decision tree, support vector machine and neuro-fuzzy models in landslide susceptibility mapping using GIS. *Comput. Geosci.* **2012**, *51*, 350–365. [CrossRef]
35. Khan, H.; Shafique, M.; Khan, M.A.; Bacha, M.A.; Shah, S.U.; Calligaris, C. Landslide susceptibility assessment using Frequency Ratio, a case study of northern Pakistan. *Egypt. J. Remote Sens. Space Sci.* **2019**, *22*, 11–24. [CrossRef]
36. Suh, J.; Choi, Y.; Roh, T.D.; Lee, H.J.; Park, H.D. National-scale assessment of landslide susceptibility to rank the vulnerability to failure of rock-cut slopes along expressways in Korea. *Environ. Earth Sci.* **2011**, *63*, 619–632. [CrossRef]

37. Zhuang, J.; Peng, J.; Iqbal, J.; Liu, T.; Liu, N.; Li, Y.; Ma, P. Identification of landslide spatial distribution and susceptibility assessment in relation to topography in the Xi' an Region, Shaanxi Province, China. *Front. Earth Sci.* **2015**, *9*, 449–462. [CrossRef]
38. Kamp, U.; Growley, B.J.; Khattak, G.A.; Owen, L.A. GIS-based landslide susceptibility mapping for the 2005 kashmir earthquake region. *Geomorphology* **2008**, *101*, 631–642. [CrossRef]
39. Rossoni, L.; Engelbert, R.; Bellegard, N.L. Normal science and its tools: Reviewing the effects of factor analysis in management. *Rev. Adm.* **2016**, *51*, 198–211. [CrossRef]
40. Zheng, W.; Xu, W.; Tong, F.; Shi, A. 3D geological visualization and numerical modeling of complicated slope. *Chin. J. Rock Mech. Eng.* **2007**, *26*, 1633–1644. (In Chinese)
41. Su, S.; Gong, Y.; Tan, B.; Pi, J.; Weng, M.; Cai, Z. Area social deprivation and public health: Analyzing the spatial non-stationary associations using geographically weighed regression. *Soc. Indic. Res.* **2016**, *133*, 819–832. [CrossRef]
42. Yan, G.; Liang, S.; Zhao, H. Improvement and implementation of slope element partitioning method based on GIS. *Sci. Geogr. Sin.* **2017**, *37*, 1764–1770. (In Chinese) [CrossRef]
43. Qin, W. The Basic Theoretics and Application Research on Geographically Weighted Regression. Ph.D. Thesis, Tongji University, Shanghai, China, 2007.
44. Zhang, L.; Yang, Y.; Liang, X. The diagnostic approach of multicollinearity in geographically weighted regression model. *Geomat. Spat. Inf. Technol.* **2017**, *40*, 28–31. (In Chinese)
45. Guzzetti, F.; Reichenbach, P.; Ardizzone, F.; Cardinali, M.; Galli, M. Estimating the quality of landslide susceptibility models. *Geomorphology* **2006**, *81*, 166–184. [CrossRef]
46. Qiu, H. Study on the Regional Landslide Characteristic Analysis and Hazard Assessment: A Case Study of Ningqiang County. Ph.D. Thesis, Northwest University, Xian, China, 2012.
47. Goetz, J.N.; Guthrie, R.H.; Brenning, A. Integrating physical and empirical landslide susceptibility models using generalized additive models. *Geomorphology* **2011**, *129*, 376–386. [CrossRef]
48. Jenks, G. Optimal Data Classification for Choropleth Maps. Ph.D. Thesis, University of Kansas, Kansas, KS, USA, 1977.
49. Akbar, T.A.; Ha, S.R. Landslide hazard zoning along Himalayan Kaghan Valley of Pakistan-by integration of GPS, GIS, and remote sensing technology. *Landslides* **2011**, *8*, 527–540. [CrossRef]

MDPI
St. Alban-Anlage 66
4052 Basel
Switzerland
Tel. +41 61 683 77 34
Fax +41 61 302 89 18
www.mdpi.com

Applied Sciences Editorial Office
E-mail: applsci@mdpi.com
www.mdpi.com/journal/applsci

www.ingramcontent.com/pod-product-compliance
Lightning Source LLC
Chambersburg PA
CBHW061825210926
51850CB00006B/812
9783036537870